Microbiology

Third Edition

BIOS INSTANT NOTES

Series Editor:
B. D. Hames
School of Biochemistry and Molecular Biology, University of Leeds, Leeds, UK

Biology
Animal Biology, Second Edition
Biochemistry, Third Edition
Bioinformatics
Chemistry for Biologists, Second Edition
Developmental Biology
Ecology, Second Edition
Genetics, Third Edition
Human Physiology
Immunology, Second Edition
Mathematics & Statistics for Life Scientists
Medical Microbiology
Microbiology, Third Edition
Molecular Biology, Third Edition
Neuroscience, Second Edition
Plant Biology, Second Edition
Sport & Exercise Biomechanics
Sport & Exercise Physiology

Chemistry
Analytical Chemistry
Inorganic Chemistry, Second Edition
Medicinal Chemistry
Organic Chemistry, Second Edition
Physical Chemistry

Psychology
Sub-series Editor: Hugh Wagner, Dept of Psychology, University of Central Lancashire, Preston, UK

Cognitive Psychology
Physiological Psychology
Psychology
Sport & Exercise Psychology

Instant Notes

Microbiology

Third Edition

S. Baker
School of Biological and Molecular Sciences,
Oxford Brookes University, Oxford, UK

J. Nicklin
Department of Biology,
Birkbeck College, London, UK

N. Khan
Department of Biology,
Birkbeck College, London, UK

and

R. Killington
Department of Microbiology,
University of Leeds, Leeds, UK

Taylor & Francis
Taylor & Francis Group

Published by:

Taylor & Francis Group

In US: 270 Madison Avenue,
        New York, NY 10016

In UK: 2 Park Square, Milton Park
        Abingdon, Oxon OX14 4RN

First published 2007

ISBN 0 4153 9088 5

A catalog record for this book is available from the British Library.

Library of Congress Cataloging-in-Publication Data

Microbiology/J. Nicklin ... [et al.]. -- 3rd ed.
        p. ; cm.
    Rev. ed. of: Microbiology/J. Nicklin, K. Graeme-Cook and R. Killington. 2nd ed. 2002.
    Includes bibliographical references and index.
    ISBN 0-415-39088-5 (alk. paper)
  1.   Microbiology--Outlines, syllabi, etc.   I.   Nicklin, J. (Jane)   II.   Nicklin, J.
(Jane). Microbiology.
    [DNLM: 1.   Microbiology--Outlines.   QW 18.2 M6234 2006]

QR62.M524 2006
579--dc22

                                                                                2006025141

Editor: Elizabeth Owen
Editorial Assistant: Kirsty Lyons
Production Editor: Simon Hill
Typeset by: Keyword Group Ltd
Printed by: Cromwell Press Ltd

Printed on acid-free paper

10 9 8 7 6 5 4 3 2 1

Taylor & Francis Group, an informa business   Visit our web site at http://www.garlandscience.com

CONTENTS

ABBREVIATIONS

A	adenine
ABC	ATP-binding cassette
ACP	acyl carrier protein
ADP	adenosine 5′-diphosphate
Ala	alanine
AMP	adenosine 5′-monophosphate
A-site	amino-acyl site (ribosome)
ATP	adenosine 5′-triphosphate
ATPase	ATP synthase
BHK	baby hamster kidney
Bp	base pair
C	cytosine
C-phase	Chromosome replication phase (bacterial cell cycle)
cAMP	cyclic adenosine 5′-monophosphate
CAP	catabolite activator protein
CAT	chloramphenicol acetyl transferase
CFU	colony-forming unit
CMV	cytomegalovirus
CNS	central nervous system
CoA	coenzyme A
CPE	cytopathic effect
CRP	cAMP receptor protein
CTL	cytotoxic T lymphocyte
Da	Dalton
d-Ala	D-alanine
DAP	meso-diaminopimelic acid
D-Glu	D-glutamic acid
DHA	dihydroxyacetone
DNA	deoxyribonucleic acid
dNTP	deoxyribonucleoside triphosphate
DOM	dissolved organic matter
D-phase	division phase (bacterial cell cycle)
Ds	double-stranded
EF	elongation factor
EM	electron microscopy
ER	endoplasmic reticulum
FAD	flavin adenine dinucleotide (oxidized)
FADH2	flavin adenine dinucleotide (reduced)
FMN	flavin mononucleolides
G	guanine
G-phase	gap phase (bacterial cell cycle)
GTP	guanosine 5′-triphosphate
HA	hemagglutination
Hfr	high frequency recombination
HMP	hexose monophosphate pathway
HSV	herpes simplex virus
I	inosine
ICNV	International Committee on Nomenclature of Viruses
Ig	immunoglobulin
IHF	integration host factor
Inc group	incompatible group (of plasmids)
IS	insertion sequence
Kb	kilobase
KDO	2-keto-2-deoxyoctonate
KDPE	2-keto-2-deoxy-6-phosphogluconate
Lac	lactose
LBP	luciferin-binding protein
LPS	lipopolysaccharide
MAC	membrane-attack complex
MCP	methyl-accepting chemotaxis protein
MEM	minimal essential medium
MHC	major histocompatibility complex
m.o.i.	multiplicity of infection
mRNA	messenger ribonucleic acid
MTOC	microtubule organizing centre
NAD+	nicotinamide adenine dinucleotide (oxidized form)
NADH	nicotinamide adenine dinucleotide (reduced form)
NADP+	nicotinamide adenine dinucleotide phosphate (oxidized form)
NADPH	nicotinamide adenine dinucleotide phosphate (reduced form)
NAG	N-acetyl glucosamine
NAM	N-acetyl muramic acid
NB	nutrient broth
NTP	ribonucleoside triphosphate
O	operator
OD	optical density
Omp	outer membrane protein
P	promoter
PCBs	polychlorinated biphenyls
PCR	polymerase chain reaction
PEP	phosphoenol pyruvate
Pfu	plaque-forming unit
PHB	poly-b-hydroxybutyrate
Phe	phenylalanine
Pi	inorganic phosphate

PMF	proton motive force	S	Svedberg coefficient
PMN	polymorphonucleocyte	snRNA	small nuclear ribonucleic acid
PPi	inorganic pyrophosphate	SPB	spindle pole bodies
PPP	pentose phosphate pathway	ss	single-stranded
PS	photosystem	T	thymine
PSI and II	photosystems I and II	TCA	tricarboxylic acid
P-site	peptidyl site (ribosome)	TCID	tissue culture infective dose
R	resistance (plasmid)	tRNA	transfer RNA
r	rho factor	Trp	tryptophan
RBC	red blood cell	TSB	tryptone soya broth
redox	reduction-oxidation	U	uracil
RER	rough endoplasmic reticulum	UL,US	unique long, unique short
RNA	ribonucleic acid	UDP	uridine diphosphate
rRNA	ribosomal RNA	UDPG	uridine disphosphate glucose
rubisco	ribulose bisphosphate carboxylase	UV	ultraviolet light

PREFACE

The third edition of *Instant Notes in Microbiology* has once again seen a complete rewrite of many chapters. Those on general microbiology, bacteriology and the relevant molecular biology have been completely revised and now have a more molecular focus. The chapter on bacterial infections has been also been revised. The chapters on fungi and protists have been updated and a new section on parasitic protists has been added. The chapter dealing with the viruses retains its structure, but has been revised.

In this book, the authors have chosen to use the word Bacteria to mean those members of the Kingdom Bacteria, and not in its older usage to denote non-eukaryotic microbes. For this we have used "prokaryotes" to include both the Bacteria and the Archaea. While not all taxonomists will be happy with these definitions, the terms have been used consistently throughout the book and reflect current thinking within microbiology – though no doubt this will have changed by the publication of the fourth edition.

The authors would like to thank the Department of Earth Sciences, University of Oxford and ABgene Ltd., Epsom, Surrey for help in the preparation of this book. However, we would also like to thank our families for their continuing support, as well as to colleagues and the reviewers of the second edition, who all gave us valuable feedback.

A1 THE MICROBIAL WORLD

Key Note

Microorganisms are found in all three major kingdoms of life: the Bacteria, the Archaea and the Eukarya. The presence of a nucleus defines the eukaryotes, while both the Bacteria and Archaea can be defined as prokaryotes. Apart from the nucleus, there are many physiological and biochemical properties distinguishing the prokaryotes from the eukaryotes.

What are microbes?

Microbes are a diverse group of organisms that can be divided into the viruses, unicellular groups (Archaea, Eubacteria, protista, some fungi and some chlorophyta) and a small number of organisms with a simple multicellular structure (the larger fungi and chlorophyta). These larger microorganisms are characterized by having a filamentous, sheet like or parenchymous thallus that does not display true tissue differentiation. Most microbes cannot be seen without the aid of a microscope.

Microbiology

Microbiology is defined as the study of microorganisms. The discipline now includes their molecular biology and functional ecology as well as the traditional studies of structure and physiology. The discipline began in the late 17th century with Leeuwenhoek's discovery of bacteria using simple microscopy of mixed natural cultures. Through the 1850s and 60s, Louis Pasteur's simple experiments using sterilized beef broth finally refuted the long-held theory of spontaneous generation as an origin for microbes, and microbiology moved into mainstream science.

The early days were characterized by studying environments like soil and sediments, natural fermentations and infections, and it was not until Robert Koch developed techniques for pure culture in the late 19th century that the science moved to a reductionist phase, where microbes were isolated and characterized in the laboratory.

Through the 20th century microbiologists focused on the discovery and characterization of many different microorganisms, including a new kingdom of microorganisms the Archaea, new eubacterial pathogens including *Legionella* and MRSA (methicillin resistant *Staphylococcus aureus*), and the complex of pathogens associated with HIV (human immune deficiency virus) including the fungal pathogen Pneumocystis. The discovery of the unique communities found in extreme environments with their temperature tolerant DNA (deoxyribonucleic acid) polymerase enzymes has further opened up the new field of molecular biology.

The rapid advances of techniques in molecular biology have allowed microbiology to return to the natural environments. Techniques, such as DGGE (denaturing gradient gel electrophoresis) and SCCP (single stranded conformation polymorphism) DNA chips and *in situ* hybridization now give us the tools to study microbial community ecology at the molecular level. Microbiology has returned to its roots!

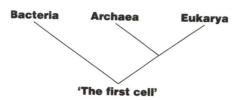

Fig. 1.   The three cell lineages evolved from a common ancestor.

Bacteria, Archaea and Eukaryotes

The microbial world has three main cell lineages within it, all of which are thought to have evolved from a single progenitor (*Fig. 1*). The lineages are formally known as domains and were established from the DNA sequence of genes common to all organisms (see Section B3). The three domains are the Bacteria (previously called the Eubacteria), the Archaea (previously called the Archaeabacteria) and the Eukarya. The defining property of the Eukarya compared to the Archaea and Bacteria is the presence of a nucleus. It is frequently convenient to group the annucleate lineages (the Bacteria and Archaea) together as the prokaryotes. The prokaryotes are, with a very few exceptions (see Section C6), all microorganisms, but the Eukarya include not only microbial fungi, chlorophyta and protists (see Section I) but also the macroorganisms such as higher plants and animals.

Prokaryotic cell structure is characterized by the absence of a nucleus, but it also lacks energy-generating organelles such as mitochondria and chloroplasts. Instead, prokaryotes generate energy by cytoplasmic substrate-level phosphorylation and oxidative phosphorylation across their cell membranes (see Section E3). Apart from these major differences, there are a multitude of distinctive biochemical and physiological properties, the most important of which are listed in *Table 1*. The differences that exist between the Bacteria and the Archaea (see Section C6) are discussed elsewhere in more detail.

Summary of subjects covered in this volume

Systematics
Bacterial systematics allows the microbiologist to name, classify, and identify Bacteria and Archaea in a rational way. The importance of molecular biology to microbiology is emphasized by the prominence of 16S rRNA (ribosomal ribonucleic acid) sequencing in microbial phylogeny.

General microbiology
Microbiology as a science has a long and varied history, but we are only just beginning to appreciate the full ecological, biochemical, and genetic diversity of microbes. Sophisticated methods have been developed to measure the growth of many of these microbes in the laboratory. Our understanding of microorganisms has improved so that we can now appreciate the fine structure of the prokaryotic cell, rather than just considering it as a bag of enzymes. The understanding of microbial cell division and movement has also led to important breakthroughs in eukaryotic biology. Although the microbiology of human disease is well studied, we are beginning to find that microbes play an essential global role in the biogeochemical cycling of the elements.

Microbial growth
The way in which most prokaryotic cultures divide in batch and continuous culture can be modelled mathematically to reveal the limitations imposed by

Table 1.   Some of the major differences between the prokarya and the eukarya

Prokaryotes	Eukaryotes
Organization of genetic material and replication	
DNA free in the cytoplasm	DNA contained in a membrane-bound structure
Chromosome – frequently haploid, single and circular	More than one chromosome – frequently diploid and linear
DNA complexed with histone-like proteins	DNA complexed with histones
May contain extrachromosomal DNA as part of the genome	Organelles may have separate chromosomes, but rarely find free DNA in the cytoplasm
Cell division by binary fission or budding	Cells divide by mitosis
Transfer of genetic information can occur via conjugation, transduction and transformation	Transfer of genetic information can only occur during sexual reproduction
Cellular organization	
Reinforcement of the cytoplasmic membrane with hopanoids (not Archaea)	Reinforcement of the cytoplasmic membrane with sterols
Cell wall made up of peptidoglycan and lipopolysaccharides or teichoic acids. Variety of cell wall constructions in Archaea	Often lack a rigid cell wall, but cell reinforced via a cytoskeleton of microtubules. Where cell wall is present it is made up of a thin layer of chitin or cellulose
Energy generation across the cell membrane	Energy generation across the membrane of mitochondria and chloroplasts
Internal membranes for specialised biochemical pathways (e.g. ammonia oxidation, photosynthesis)	Internal membranes in most cells (endoplasmic reticulum, Golgi apparatus etc.)
Flagella made up of a single protein (flagellin)	Multi-protein flagella made up of a 9 + 2 arrangement of fibrils
Ribosomes are small (70S)	Ribosomes are larger (80S)

laboratory conditions. From these models it can be shown that the design of any growth vessel should primarily optimize the oxygen requirements of the culture growing in it.

Molecular biology

Microbiology has always been intrinsic to advances in genetics, DNA metabolism and *in vitro* and *in vivo* genetic manipulation across the whole of biology. The principles of DNA replication, transcription to mRNA and translation to protein have all been characterised in *Escherichia coli* in the first instance. Coupled to our detailed knowledge of the control of transcription and the mechanisms of DNA transfer between cells, it is now possible to use bacteria as powerful tools in recombinant DNA technology.

Eukaryotic microbes

A number of distinct groups of eukaryotic microbes are considered using the classification based on the Tree of Life website. The general cell biology and cell division of eukaryotic microbes is then described, followed by a more detailed consideration in separate sections of the structure, physiology, and reproduction of the Fungi, and the photosynthetic and the non-photosynthetic protista. The beneficial and detrimental effects of each group of microbes on its environment are also examined, with a more extensive review of the taxonomy and pathogenicity of protistan parasites.

B1 PROKARYOTIC SYSTEMATICS

Key Notes

Classification and taxonomy	Classification is a method of organizing information. Microorganisms can be classified on their growth properties (e.g. chemolithotroph, denitrifier), but are formally classified using the Linnaean system. The full classification of a microorganism is its taxonomy. Microorganisms can be uniquely identified solely by the use of their genus and species names.
Identification of prokaryotes	Identification is the placing of new isolates into the taxonomic framework, normally to the level of genus and species. However, the definition of species is still less clear in the prokaryotes than it is in the higher eukaryotes.
Phylogeny of prokaryotes	The evolutionary relationship of a microorganism among and between taxa is its phylogeny. The reliance on DNA sequences to elucidate these relationships has led to the emergence of phylogenetics. The taxonomy of Bacteria and Archaea for the most part reflects their phylogeny.
Related topics	Identification of Bacteria (B2) Inference of phylogeny from rRNA gene sequence (B3) Prokaryotic diversity (C2) Bacterial niche diversity (C11)

Classification and taxonomy

With the advent of molecular methods, the distinction between identification, classification, and evolutionary relationships has become blurred. In bacteriology, a **classification** is simply a method of organizing information. This organization may have an underlying meaning, or no meaning at all. We could choose to classify bacteria on the color their colonies have when grown on agar plates, a classification that would give prominence to the few yellow and red colored bacteria, while the majority would be classified in a group of white to cream colony forms. In early microbiology, organisms were classified according to shape, with the bacillus shape (now also called a rod) forming the largest group, and cocci, filaments, and so on smaller ones. It is important to stress that classifications were arbitrary; however they do still have a use today. Microbiologists classify organisms according to their growth properties (anaerobe, chemolithotroph, methylotroph, etc., see Sections C2 and D1).

The primary means of classification in microbiology, in common with the rest of biology, is the Linnaean system. This is a hierarchical system, with major divisions sequentially separated down to the lowest level species (*Fig. 1*).

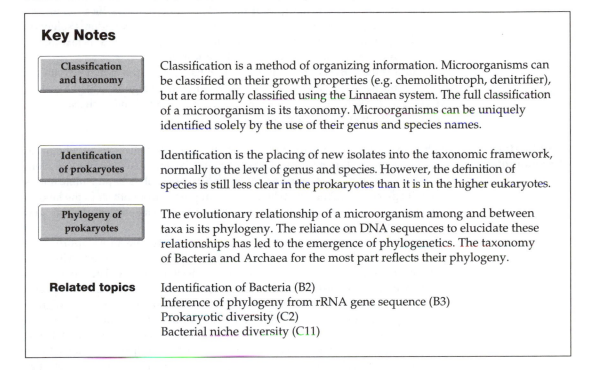

Domain
Kingdom
Phylum
Class
Order
Family
Genus
Species

Fig. 1.   The full Linnaean system of classification.

A full classification of *Escherichia coli* is Prokaryota (domain), Bacteria (kingdom), Proteobacteria (phylum), λ-proteobacteria (class), Enterobacteriales (order), Enterobacteriacaea (Family), *Escherichia* (Genus) and *coli* (Species). Each of these levels is described as a **taxon** (plural **taxa**). This system of classification allows biologists a unique identification across all the domains and kingdoms solely by using the appropriate genus and species. The full name of *E. coli* is its **taxonomy**, and the Linnaean system is a **taxonomic classification**.

In this text the correct domain and kingdom names are used to denote both the Archaea and Bacteria (prokaryotes) and the individual kingdoms. Slightly older systems give the kingdom names as Eubacteria (true bacteria) and Archaeabacteria, which in many ways is slightly more descriptive.

Identification of prokaryotes

With a taxonomy for prokaryotes in place, the microbiologist can now begin to place organisms within this framework, using the taxonomy to describe and identify the species. When first purified, an organism is described as an **isolate** and is generally given a number which helps to distinguish it from others in the laboratory. Isolates may be genetically identical, but have been taken from their natural environments at separate times or geographical locations.

Normally it is uncontroversial to place a new bacterium in an appropriate genus, so this isolate may become, for example, *Paracoccus* strain NCIMB 8944. The **species concept** is more difficult to apply to Bacteria and Archaea than in the animal kingdom because of the enormous genetic diversity, so the subspecies definition of **biovar** is often applied to organisms of the same species with slightly different properties (sometimes referred to as **strains**), for example the plague organism, *Yersinia pestis*, is divided into biovars, including Orientalis and Mediaevalis. The differences between strains of the same species are sometimes defined by subspecies, for example the plant pathogen *Erwinia carotovora* has subspecies including atroseptica and carotovora. In this case the subspecies defines members of the same species that cause different plant pathologies. In microbiological research, the problem of species definition means that it is always best to leave the original isolate code in place when writing about an organism, so that its history in various laboratories can be traced.

Although the definition of the individual Bacterial and Archaeal taxa down to genus level has gained consensus among microbiologists, there has yet to be agreement on what constitutes the fundamental unit of biological diversity, the species. Species definition in one genus (e.g. sharing less than a certain percentage of DNA homology) does not necessarily hold true in another, a problem particularly true in taxa dominated by human pathogens (see Secton B2). If we cannot properly and consistently define a species, then identification down to species level becomes problematic. The methods used for prokaryotic identification, that is the way in which we assign a classification to our new isolate, are explored further in Section B2.

Phylogeny of prokaryotes

Phylogeny is a description of the evolutionary relationships among and between taxa. In microbiology there is a heavy reliance on DNA sequence rather than morphology (as seen in the plants and animals) as data for phylogeny, and thus is normally referred to as **phylogenetics**. The classification of Bacteria and Archaea for the most part reflects their phylogeny as we currently understand it. There is, of course, always room for debate in any system, and a lively discourse continues over the exact phylogenetic and taxonomic position of many species and genera.

B2 IDENTIFICATION OF BACTERIA

Key Notes

Identification of Bacteria	Most identification in microbiology laboratories is deduced from the biochemistry of the new isolate, with sequencing data more frequently playing a part in the later stages of formalizing the classification of the isolate.
Identification from growth characteristics	The growth on media can be used to aid identification of isolates, in a selective or differential (diagnostic) manner. A numerical taxonomy can be built up by scoring the ability of the organism to grow on a range of sugars and its possession of key enzymes. The numerical taxonomy can then be used to identify the organism, either by consulting commercial libraries or by reference to *Bergey's Manual of Systematic Bacteriology*.
Other methods of identification	Microorganisms can be identified by the fatty acids that they produce under defined conditions (FAMEs – fatty acid methyl ester analysis), or by examination of their 16S rRNA sequence.
Identification of pathogens	Medical laboratories tend to rely more on biochemical classification from pure isolated culture rather than the molecular techniques used in non-clinical research. As PCR (polymerase chain reaction) becomes cheaper, more medical laboratories are beginning to use molecular techniques. The identification of pathogenic bacteria relies more on their disease-causing properties than their phylogeny, which leads to some anomalies in Bacterial classification.
Related topics	Prokaryotic systematics (B1) Inference of phylogeny from rRNA gene sequence (B3)

Identification of Bacteria

Although molecular methods (see Section B3) are becoming increasingly important in the identification and classification of bacteria, most identification work in clinical laboratories throughout the world is done using cheaper growth and biochemical methods. Furthermore, a full identification of any bacterium to publishable standard should include a polyphasic approach, i.e. including the characteristics of the strain determined molecularly, biochemically, and from growth studies. If bacteria can be grown from a clinical or environmental sample, the first step in identification or classification is growth studies, followed by an analysis of the enzymes that may be present in the strain, and lastly a molecular analysis of the genome.

Identification from growth characteristics

An unknown bacterial isolate may be subcultured on many different sorts of solid and liquid media to aid its identification. Broadly these media fall into two types: **selective media** allow the growth of one type of bacterium while inhibiting that of others; **differential** or **diagnostic media** usually contain some sort of visual indicator, a change in which is linked to a unique biochemical property of a group of microorganisms (see Section C3). If the strain can be purified, then a Gram stain and an examination of morphology might be performed. Once an overall picture of the organism's growth has been obtained, the ability to use sugars (and if acid is produced during their use) and an assessment of the possession of certain enzymes is performed in detail. These properties can be combined and scored against the known properties of other organisms to form a basic **numerical taxonomy**. Kits can be obtained commercially that semi-automate this process, but are limited to certain groups of Prokaryotes, particularly the enterics. With most environmentally isolated, non-pathogenic Bacteria and Archaea, guidance must be sought from the standard reference text for identification, *Bergey's Manual of Systematic Bacteriology*.

Other methods of identification

If an organism can be grown under the same conditions as many other reference strains, other methods of identification are also available to the microbiologist. The lipids of a pure culture can be extracted and esterified, then quantified by gas chromatography. This FAMEs then requires the GC (gas chromatography) trace to be compared by computer against other organisms grown in exactly the same way on the same media. This is a rapid and inexpensive procedure, but interpretation can be difficult and is unsuitable for microorganisms that grow under unusual conditions. The FAME profile of any organisms alters depending on the medium used for growth. An adaptation of this method is phospholipid-linked fatty acid analysis (PLFA), which is a more specific and sensitive technique.

As DNA sequencing becomes easier and cheaper, it has also become standard practice to complement the results from biochemical and physiological tests with the results of 16S rRNA gene sequencing (see Section B3). Although this practice blurs the distinction between identification and phylogeny further, the results can be obtained quickly and easily, but may be misleading if examined in isolation.

Identification of pathogens

Medical microbiology has, to some extent, fallen behind the rest of microbiology in its approach to the identification of disease-causing organisms. Due to the costs involved, routine identification of pathogens is still carried out by the classical bacteriological methods. This would normally entail the isolation or enrichment of bacteria from a clinical specimen using broth or agar, procurement of a pure culture from the primary culture and then identification of the bacterium by microscopy, growth characteristics and perhaps PCR. The reliance on the culture of pathogens means that some common pathogens are overlooked. For example, for many years the existence of *Campylobacter* (see Section H1) as the most frequent causative agent of food poisoning was not known because of the difficulty in cultivating this genus in the laboratory. As the cost of PCR continues to fall, medical microbiology will embrace such concepts as **viable but nonculturable**, (meaning they can be detected but cannot be cultured in the laboratory), which are currently accepted in disciplines such as environmental microbiology, with a concurrent increase in the reliability of diagnosis.

Pathogen identification and naming is driven by patient symptoms rather than the overall properties of the microorganism. For example, the enteric genera are all very closely related on a genetic level, but cause a variety of human diseases (*Table 1*). It is now becoming clearer from examination at the molecular level that the pathogenic members of the Gram positive genus *Bacillus* (*Table 2*) may even be the same species, but with different sets of plasmids.

Table 1.   Disease caused by species of the enteric bacteria

Genus	Disease
Escherichia	Enteropathogenic diarrhoea
Shigella	Shigellosis
Salmonella	Typhoid fever, gastroenteritis
Vibrio	Cholera, gastroenteritis
Klebsiella	Pneumonia
Yersinia	Plague

Table 2.   Disease caused by closely related species of Bacillus

Species	Disease
B. subtilis	Non-pathogenic
B. anthracis	Anthrax
B. cereus	Gastroenteritis
B. thuringiensis	Insect pathogen

B3 INFERENCE OF PHYLOGENY FROM rRNA GENE SEQUENCE

Key Notes

Bacterial phylogeny	The recent advent of molecular phylogenetic methods has made phylogeny more accessible to the laboratory microbiologist.
The molecular clock concept	The changes in DNA or protein sequence over long periods of time can be used to measure overall evolutionary change from a common ancestor. The evolutionary chronometer chosen should be universally distributed, functionally homologous, and possess sequence conservation. To be able to distinguish between rapid and slow periods of change, the molecule chosen should also have regions of conservation and hyper-variability.
Ribosomal RNA	Cytochrome *c* has been suggested as a suitable chronometer, but 16S rRNA sequence has gained widespread acceptance. The size is convenient for most sequencing protocols, but some doubts remain as to the validity of some phylogenies. About 185 000 Bacterial rRNA sequences have been entered in to the Ribosomal Database Project.
Acquisition of 16S sequence	16S rRNA genes from many different organisms are amplified using universal primer sets, though currently no primer set has been found capable of amplifying every single known species. In mixed populations, primer sets can sometimes generate false results due to the formation of chimeric PCR products from more than one template.
16S bioinformatics	Once a 16S rRNA PCR product has been amplified and sequenced, it must be placed in the context of its phylogenetic relationships with other sequences. It is first aligned against similar sequences, and then clipped so that all the data in the alignment are the same length. Phylogenetic trees can then be constructed by neighbour joining or maximum parsimony methods. These methods can still generate many different trees from the same data set, so bootstrapping is used to assign confidence levels to the existence of each branch of the tree.
Related topics	Prokaryotic systematics (B1) Identification of Bacteria (B2) Manipulation of Cellular DNA and RNA (F16)

Bacterial phylogeny

The relatedness of Bacteria to one another is discussed elsewhere in this book (see Section C6) in the context of diversity. Most of the major phyla (Gram positive, *Cyanobacteriacae*, and so on) have been deduced from classical taxonomic methods and have been in place for several decades. The more recent advent of **molecular phylogenetic methods** has made phylogeny more accessible to the laboratory microbiologist.

The molecular clock concept

A microbiologist should be able to place any organism in the context of its relationship to other organisms and its evolution from a common ancestor. To be able to do this an evolutionary clock must be identified, which reflects small changes in the organism over time. Cellular macromolecules, such as proteins and nucleotides, have the potential to act as **evolutionary chronometers**, but to be ideal they must meet the following criteria:

- **Universally distributed** – i.e. present in all known (and presumably yet to be discovered) organisms.
- **Functionally homologous** – i.e. the molecule must perform the same action in all organisms. Molecules with different functions could be expected to become too diverse to show any sequence similarity.
- **Possess sequence conservation** – i.e. an ideal chronometer should have regions of sequence that are highly conserved and thus expected to change only very slowly over long periods of time coupled with other regions with moderate- or **hyper-variability** to illuminate more recent changes.

Many macromolecular chronometers have been proposed, including cytochrome *c* (see Section E3), ATPase (see Section E3), RecA (see Section F13) and 16S/18S rRNA. Most of the protein chronometers have failed to satisfy the universality requirement, particularly when examining the extremophilic Bacteria and Archaea that lack a conventional electron transport chain. To date, the most widely used chronometer is 16S rRNA of Bacteria and Archaea along with its 18S rRNA equivalent in Eukarya.

rRNA

The small, medium, and large rRNA molecules are all ideal chronometers. They have been found to perform the same function in all known organisms and have regions of conservation as well as hypervariability. The interaction of RNA with ribosomal proteins means that rRNA buried deep within the protein structure is less likely to change, as any change must also be reflected in the protein sequence. Changes in either ribosomal protein or rRNA in these regions that lead to an unstable ribosome are lethal to the organism and do not persist in subsequent generations. However, a considerable part of the rRNA molecules do not have any direct interaction with the ribosomal proteins, and so can accumulate mutations much more easily, i.e. are hypervariable.

Of the three rRNA molecules available in prokaryotes (5S, 16S, and 23S, see 'RNA and the genetic code', Section F), 16S rRNA provides the ideal balance between information content (5S too short) and ease of sequencing (23S too long). The attractiveness of the molecule is shown by the number of entries in the **Ribosomal Database Project**, standing at 184 990 Bacterial sequences for release 9 in October 2005. There are few drawbacks to the use of the molecule, primarily that many Bacteria have more than one copy of the 16S rRNA gene on their genome, frequently with a different sequence. This can cause confusion and dispute, depending on which sequence is used, and emphasizes the ultimate goal of a polyphasic approach.

Acquisition of 16S rRNA gene sequence

DNA fragments containing all or part of the 16s rRNA gene are generally obtained by PCR. The primers are designed to anneal to the conserved regions within the gene and sometimes this enables the use of one primer set to amplify 16S from many phylogentically diverse bacteia (**a universal primer set**). However, no one set of primers can amplify all the genes from all the Bacteria and all the Archaea, and many primer sets have been designed that are phylum- or group-specific.

The use of combinations of these sets means that most microorganisms can be amplified from any environment, pure culture or mixed culture.

Although acquisition of sequence by PCR is quick, there are limitations imposed by the technique itself. PCR can generate **chimeras**, PCR products that are composed of the 5′ end of one species' gene coupled to the 3′ end of another. Although many computer programs exist to eliminate these false sequences from the final results, it is sometimes difficult to detect them when dealing with rare or undiscovered organisms. The existence of the division *Korarchaeota* is still in doubt for precisely this reason.

16S rRNA gene bioinformatics

Once the PCR product has been amplified and sequenced (see Section F16), it must be placed in the context of its phylogenetic relationships with other sequences. A preliminary idea of the close relatives can be gained by the use of an **alignment** to one or more known sequences. A program such as BLAST (National Center for Biotechnology Information, U.S. National Library of Medicine) can do this, though it is extremely limited in detail and will only give information relating to one other sequence at a time. To gain a true idea of phylogeny, the 16S rRNA gene sequence should be compared to as many other sequences as possible simultaneously. Only 10 years ago this was impossible to achieve on anything but a supercomputer, however recent advances in computing power now mean that most personal computers can carry out some or all of this process. Many web-based programs also allow free access to the more powerful computers that may be needed.

To begin with, the newly acquired sequence must be aligned with all or some of the sequences obtained previously. As there is some variation in length of 16S rRNA genes, gaps must be inserted to achieve a perfect alignment, though this can be done by programs such as CLUSTAL (European Bioinformatics Institute (EBI)). The aligned sequences are then **clipped** so that the 5′ and 3′ ends are equivalent bases and the alignment sent to a program capable of generating **phylogenetic trees**.

An ideal representation of phylogeny would be multi-dimensional, but given the constraints of our 3-dimensional universe in general, and the scientific predilection for presentation in 2-dimensional form on paper in particular, the 'tree' is a good compromise. Two main algorithms are used: **neighbor joining** and **maximum parsimony**. Neighbor joining is an evolutionary distance method, based on a matrix of differences in the dataset. The resulting tree has branches of lengths proportional to evolutionary distance, statistically corrected for back mutation. Maximum parsimony is a more difficult concept to grasp, in that the resulting tree has branches whose length is proportional to the minimum amount of sequence change necessary to enable the creation of a new branch.

For both methods, it is possible to generate trees differing in details, such as the number of branches, from one dataset. A process known as **bootstrapping** is applied to get an idea of the sum of all the possible trees, and this gives a confidence value for the presence of each branch. In addition, neighbor joining and parsimonious trees generated from the same dataset can give quite different results, and to date neither method is considered to be more 'right' than the other. Thus any tree should be considered as the best possible result with the data available, and should not necessarily overrule any other information.

C1 DISCOVERY AND HISTORY

Key Notes

The history of bacteriology

Robert Hooke (1660) and his contemporary Antonie van Leeuwenhoek are considered to be the first microbiologists, but the theory of spontaneous generation did not allow for the existence of microorganisms to be placed in their true context. However, Pasteur's swan-necked flask experiments (1861) showed that food-spoilage organisms were microscopic and airborne. Cohn (1875) founded the science of bacteriology, and was followed by other late Victorian scientists, such as Koch, Beijerinck, and Winogradsky. Bacteria became the model system for biochemistry throughout the 20th and early 21st centuries, culminating in the sequencing of *Haemophilus influenzae* by Ventner and colleagues in 1995.

Major subgroups

The Bacteria were first subdivided by use of the Gram satin, but now are separated into many phyla. The Bacteria are a separate kingdom from the Archaea, and each of their phyla contains species with broad nutritional, physiological and biochemical properties.

Related topics

Prokaryotic systematics (B1)
Prokaryotic diversity (C2)
The major prokaryotic groups (C6)

The history of bacteriology

It was not until the development of the first microscopes (by the Janssen brothers around 1590) that microbes were observed as minute structures on surfaces. Robert Hooke began showing the fruiting structures of moulds around the courts of Europe and published the first survey of microbes (*Micrographia*) in 1660. The first person to observe prokaryotes microscopically was Antonie van Leeuwenhoek in 1676. He published his observations on these 'animicules' to the Royal Society of London. However, the theory of spontaneous generation stopped much further investigation, since it claimed that the intervention of divine power led to the spontaneous creation of moulds and other spoilage organisms (including mice) should food be left unattended. The belief that living organisms could arise from otherwise inert materials began around the time of Aristotle (384–322 AD) and eventually this theory was disproved (but not entirely discarded) by the experiments of Pasteur in 1861 using swan-necked flasks. These allowed the preservation of beef broth for long periods, with spoilage only initiated once an airborne, invisible contaminant was reintroduced into the broth.

Ferdinand Cohn is attributed with founding the science of bacteriology, proposing a morphological classification for bacteria and using the term *'Bacillus'* for the first time in 1875. This was soon followed by the seminal work by Koch, with his system for firmly establishing the link between bacteria and disease over the period between 1876–1884 (Koch's postulates, see Section H1). Martinus Beijerinck developed the technique of enrichment culture, establishing the first pure culture of *Rhizobium* in 1889, a year before Winogradsky demonstrated the link between oxygen and nitrification in bacteria. Beijerinck also went on to establish the science of virology, while working for a company producing yeast in the Netherlands, firmly establishing the strong link between microbiology and biotechnology. By the turn of the 20th century, the first journal had been published (the American Society for Microbiology's *Journal of Bacteriology*, still in publication today), and many of the commonly researched organisms named, though not necessarily classified as they are today. Bacteria became the model systems for biochemistry throughout the 20th and early 21st centuries, culminating in the sequence of the first free living organism (*Haemophilus influenzae*) being decoded by Ventner and colleagues in 1995.

Major subgroups
The work of the early bacteriologists (Cohn, Koch, Beijerinck) was focused on identification and classification of bacteria (see Section B1). However, this work was based around the morphology (gross shape) of the organism, and since most bacteria are rods (see Section C2) and form white or cream colonies on agar plates, the success of this approach is limited. The development of the Gram stain in 1884 by the Danish physician Hans Christian Gram, allowed the separation of the prokaryotes into two classes, which was later found to be based around cell wall structure. Gram positive bacteria retained the purple dye (crystal violet) while the Gram negative bacterial cell wall allowed the stain to be washed away. The Gram positive Bacteria have remained a valid taxonomic group (see Section C6), but the Gram negative Bacteria have been proved to be more phenotypically and genetically diverse, and for a while included members of the Archaea.

Although we can now separate the Bacteria from the Archaea, and further subdivide the Bacteria on the basis of phylogeny, many physiological properties are broadly spread throughout the kingdom. Bacteria with the capability of causing a human pathology are not restricted to any single subgroup, nor is the ability to denitrify, grow anaerobically or photosynthesize. The majority of Bacterial species have unique properties, which may be very different from members of the same genus, but superficially similar to those that could be considered to be phylogenetically very distinct.

C2 PROKARYOTIC DIVERSITY

Key Notes

Morphological diversity	One of the most common shapes for Bacteria and Archaea is the rod or bacillus. Shape is dictated to some extent by the method of cell division, most commonly binary fission, though cocci, vibroid, spiral, filamentous and even star-shaped prokaryotes have been isolated. The smallest bacteria are spheres only 0.3 μm in diameter, while the largest is 0.75 by 0.25 mm. Large Bacteria are composed mostly of gas vesicles and seem to have little more cytoplasmic content then *E. coli*.
Habitat diversity	Prokaryotes can grow at extremes of temperature, pH, oxygen, and radiation dosage, and can be named after their ability to grow under certain conditions (e.g. heat preferential bacteria are thermophiles). Prokaryotes can grow at 4°C and below, while the highest recorded growth temperature is 96°C (Bacteria) or 110°C (Archaea). Organisms growing at the limits of life are known as extremophiles.
Related topics	Prokaryotes and their environment (C11)

Morphological diversity

The model organism used in bacteriology, *Escherichia coli*, is a rod-shaped organism, about 3 μm in length and 1 μm in diameter. Many very different Bacteria have this rod-like morphology, sometimes called a bacillus. Bacilli can be found in all the taxonomic groups of the Bacteria, as well as in the Archaea. This means that, unlike the classification of higher organisms, shape is not a reliable characteristic when classifying prokaryotes (*Table 1*), even though it is one of the few visible differences between cells when using a light microscope. However, the shape of some microorganisms has had an influence on the naming of some prokaryotes. The Gram positive Bacterial genus *Bacillus* is, of course, made up of rod-shaped species, while another Gram positive genus, *Streptococcus*, is made up of species of spherical bacteria (or cocci) about 1 μm in diameter.

Bacteria and Archaea are considered to be independent single-celled micro-organisms dividing by a process of binary fission (see Section D] which means that both mother and daughter cells have the same size after division. This means that most bacteria are regular in shape (spherical, cylindrical, etc.) but a few exceptions, such as the star-shaped bacterium *Stella humosa*, exist. Actinomycetes, such as *Streptomyces coelicolor*, grow as multinucleate filaments, in which the concept of a discrete cell becomes more esoteric.

Many prokaryotes form smaller cells when stressed, and these are thought to be resting stages. The smallest vegetative prokaryotes are the marine ultramicro-bacteria, such as *Sphingopyxis alaskensis*, whose diameter is less than 0.3 μm and has an estimated cellular volume of less than 1 μm^3. The mycoplasmas and chlamydia are the smallest bacteria found colonizing humans – *Mycoplasma genitalium* is only 0.2 by 0.3 μm. Although these seem tiny compared to Eukaryotic

Table 1.   Morphology and classification of selected prokaryotes

Shape (morphology)	Singular	Plural	Examples	Classification
Spherical or ovoid	Coccus	Cocci	*Streptococcus pneumoniae* *Deinococcus radiodurans* *Neisseria menigitidis* *Desulfurococcus fermentans*	Bacteria – Gram positive – Lactobacillales Bacteria – Gram variable – Deinococcus/Thermus Bacteria – Gram negative – β proteobacteria Archaea – Crenarchaeota – Thermoprotei
Cylindrical	Rod	Rods	*Escherichia coli* *Bacillus subtilis* *Methanobacterium oryzae*	Bacteria – Gram negative – γ proteobacteria Bacteria – Gram positive – Bacilli Archaea – Euryarchaeota – Methanobacteria
3D comma	Vibrio	Vibrio	*Vibrio cholerae*	Bacteria – Gram negative – γ proteobacteria
Curved rod	Spirillum	Spirilla	*Spirillum pleomorphum*	Bacteria – Gram negative – β proteobacteria
Filamentous			*Streptomyces coelicolor*	Bacteria – Gram positive – Actinobacteria
Star			*Stella humosa*	Bacteria – Gram negative – α proteobacteria
Flat square			'Haloquadratum walsbyi'	Archaea – Euryarchaeota – Halobacteria

cells (2 – 200 μm) the theoretical size limit of life was calculated by the US National Academy of Sciences Space Study Board to be a spherical cell of 0.17 μm. This minimum size assumes the presence of a minimum genomic complement plus the ribosomes and other sub-cellular components require for existence.

The largest known prokaryote is the sulfur bacterium *Thiomargarita namibiensis*, which is 750 μm in length and has a diameter between 100 and 250 μm. It and Bacteria such as *Epulopiscium fishelsoni* (a surgeonfish symbiont, 600 μm long and 75 μm at its widest) can just be seen with the naked eye. Under the light microscope, organisms such as *Bacillus megaterium* (1.5 × 4 μm) and the cyanobacterium *Oscillatoria* (8 × 50 μm) seem unusually large. These large bacteria have little more cytoplasm than a single *Escherichia coli*, and most of the cellular volume is made up of vacuoles, since diffusion is the main method of movement of all prokaryotic substrates and metabolites. All the cytoplasm of a large cell must be close to a cellular membrane, so the vacuole structure allows the organism to grow to larger sizes.

Habitat diversity   Prokaryotes can grow at extremes of temperature, pH, oxygen, and radiation dosage. This has allowed them to colonize all parts of the earth that may provide the cells with substrates for growth. The way in which microorganisms grow has given us a means of describing them, according to their ability to use certain compounds for growth (-trophy) or their tolerance of physiochemical conditions (-phily). Thus, an organism that can use many organic compounds to grow can be described as a heterotroph, and if it can tolerate atmospheric oxygen concentrations it will be called an aerophile. The range of descriptions is given in *Table 2*.

Bacteria and Archaea have been found growing in snow, while the upper temperature limit of Bacterial life appears to be around 96°C (seen in *Aquifex aeolicus*). However, the Archaea hold the record for the highest growth temperature, with *Pyrodictium abysii* growing at above 110°C. Such high temperatures are not found on terrestrial earth, but are restricted to deep-sea hydrothermal vents. Broadly speaking, the Archaea have more representative species growing at very high or very low temperature and pH, and are commonly held to be predominantly **extremophiles**.

Table 2. Commonly used description of micro-organisms. The terms can be combined to describe more than one physiological property, e.g. the Archaea Acidianus breyerli *can be described as a hyperthermophilic chemolithoautotroph. The term obligate can be used to denote those organisms that are restricted to one mode of growth, e.g.* Methanococcus capsulatus *can only grow on methane so is described as an obligate methanotroph.*

Growth requirement	Description	Adjective	Example
Carbon from organic compounds	Heterotroph	Heterotrophic	*Escherichia coli*
Carbon from carbon dioxide	Autotroph	Autotrophic	*Paracoccus denitrificans*
Carbon from C1-compounds	Methylotroph	Methylotrophic	*Methylobacterium extorquens*
Energy from light	Phototroph	Phototrophic	*Rhodobacter sphaeroides*
Energy from inorganic compounds	Lithotroph	Lithotrophic	*Acidithiobacillus ferrooxidans*
Energy from organic compounds	Heterotroph	Heterotrophic	*Escherichia coli*
Molecular nitrogen	Diazotroph	Diazotrophic	*Rhizobium leguminosarum*
Atmospheric oxygen	Aerophile	Aerobic	*Pseudomonas aeruginosa*
Reduced levels of oxygen	Microaerophile	Microaerophilic	*Campylobacter jejuni*
Absence of oxygen	Anaerobe	Anaerobic	*Clostribium tetanii*
High salt	Halophile	Halophilic	*Halobacterium salinarium*
High alkalinity	Alkaliphile	Alkaliphilic	*Bacillus pseudofirmus*
High acidity	Acidophile	Acidophilic	*Lactobacillus acidophilus*
Low temperature (<0 to 12°C)	Psychrophile	Psychrophilic	*Psychromonas profunda*
'Normal' temperatures (10 to 47°C)	Mesophile	Mesophilic	*Escherichia coli*
Moderately high temperatures (40 to 68°C)	Thermophile	Thermophilic	*Bacillus (Geobacillus) stearothermophilus*
High temperatures (68 to 98°C)	Hyperthermophile	Hyperthermophilic	*Thermus aquaticus*
Very high temperatures (90 to above 110°C)	Extreme hyperthermophile	Hyperthermophilic	*Pyrodictium ocultum*

C3 CULTURE OF BACTERIA IN THE LABORATORY

Key Notes

Growth media	Bacteria can be grown on surfaces, or in aqueous suspension on solid, or in liquid media respectively. Solid agar media are normally held in Petri dishes, and inoculated by streaking or plating with a loop or spreader. Inoculants may be mixed (composed of many species) or pure (composed of only one species). The process of inoculation while maintaining culture purity is called aseptic technique.
	Defined media contain known amounts of simple chemical compounds, also known as minimal or synthetic media. Many organisms will not grow on such media without the addition of vitamins and trace elements, while auxotrophs require the addition of amino acids as well. Universal growth media are complex, containing compounds whose exact chemical formula is variable or uncharacterized. By changing the composition of a medium, they can be selective for a group of microorganisms, or even diagnostic for a genus or species.
Storage and revival of microorganisms	Bacteria and Archaea can be stored in a 50% glycerol solution for many years, and can also withstand freeze-drying in the lyophilization process. Short-term storage is accomplished by streaking out onto solid media. Cells are revived from storage by growth in a complex medium to reduce stress.
Sterilization	Pasteurization will kill many pathogens, but does not kill all Bacteria. Autoclaving is used to sterilize media and other apparatus used in microbiology, but the harsh temperatures mean that the equipment must be made of glass, steel, or polypropylene. Heat sensitive media or apparatus can be Tyndalized, UV (ultra-violet) treated, gamma irradiated or filter sterilized if liquid.
Related topics	Looking of microbes (C5)
	Measure of microbial growth (D1)
	Batch culture in the laboratory (D2)

Growth media   The microbiologist has many choices to make if wishing to grow microorganisms in the laboratory. The biomass introduced into the growth medium is known as the **inoculant**, and one of the first questions that need to be addressed is whether to **inoculate** onto **solid** or into **liquid** media. The medium will be chosen to reflect the origin of the inoculant, which might be a **mixed culture** of many different species of microorganism, or a **pure culture** of only one species. Solid media are normally held in circular sterile plastic containers with lids (**Petri dishes**), the solidity being provided by **agar**. The culture is **streaked** or **plated** onto the surface of the medium using a sterile wire or plastic **loop**, or a sterile glass **spreader** respectively.

Microorganisms have an enormous metabolic diversity (see Section C2 and C5), so require a medium made up of components to suit. Some organisms can synthesize all their cellular components from simple carbon and energy sources, such as glucose, and so will grow on a **minimal (synthetic/defined)** media. Such a medium will contain sources of nitrogen, phosphate, sulfur, calcium, magnesium, potassium, and iron as inorganic salts, and may be supplemented with **trace elements**, such as zinc, manganese, boron, cobalt, copper, nickel, chromium, and molybdenum. These elements are required in minute quantities by the organism, and are used as prosthetic groups in some enzymes (e.g. alcohol dehydrogenase contains zinc). A truly minimal medium will not contain any **vitamins**, but organisms will often grow more quickly if provided with riboflavin, thiamine, nicotinic acid, pyridoxine HCl, calcium pantothenate, biotin, folic acid, and vitamin B_{12}. Again, these are provided in minute quantities, enough to be used during enzyme synthesis but not in sufficient amounts to act as a carbon and/or energy source.

When growing some **auxotrophs** (see Section F10) or some bacteria with a requirement for amino acids, it may be necessary to supplement the minimal medium with some or all of the 20 possible acids. This is common in pathogens, where a protected lifestyle in an animal host has led to the loss of one or all the enzymes involved in amino acid synthesis pathways. Identifying amino acid requirements is often time-consuming and costly, in which case a **complex medium** can be used.

Complex media are defined in the sense that absolute quantities of buffering ions are added to a solution, as well as known amounts of plant, animal, or yeast extracts. The common complex medium Lauria Bertoni Broth (LB) contains 5 g L^{-1} NaCl, but carbon, energy, trace elements, and other growth factors are provided by 5 g L^{-1} 'yeast extract' and 10 g L^{-1} '**Tryptone**'. Tryptone is casein (milk solids) digested with pancreatic enzymes, so the exact composition, in terms of molar concentrations of amino acids, short peptides, and so on, is unknown and will vary between manufacturers. Similarly, **yeast extract**, a hydrolysate of baker's yeast (*Saccharomyces cerevisiae*) is of unknown composition. Pathogenic bacteria associated with bacteraemia are frequently grown on a complex medium containing whole or partially hydrolyzed blood, which provides not only the essential growth factors, but can give an indication of the presence of a haemolytic organisms by clear patches in the blood red medium.

Liquid media can be placed in a variety of containers, appropriate to the oxygen requirements of the organism to be grown. Facultative anaerobes and anaerobes can be grown in bottles, with gentle shaking to mix the culture, while the more commonly used aerophiles are generally grown in batch culture (see Section D2) in **Erlenmeyer flasks**. These flasks are adapted chemistry apparatus, conical flasks between 5 mL and 5 L in volume. They are filled to 10% of the total volume so that the liquid medium provides sufficient surface area for oxygen transfer to the culture (see Section D2). The inoculation of liquid and solid media and the transfer of cultures from one container to another without the ingress of contaminating organisms have become known as **aseptic technique**.

Media have been developed over the last 100 years in both composition and utility. **Selective** or **differential** media are of a composition that only allows the growth of one type or group of organisms. For example, a minimal medium containing methanol as a carbon and energy source will select for methylotrophs, those organisms able to use reduced C1 compounds. The medium can be enhanced further to be **diagnostic**. For example, the **Baird Parker** solid medium will allow the growth of only a handful of genera, including *Micrococcus* and *Staphylococcus*, but only *Staphylococcus aureus* will grow as gray-black shiny colonies with a narrow white entire margin surrounded by zone of clearing 2–5 mm. This 'egg yolk' colony form is used as a first indication of the presence of potential pathogens before more detailed tests are carried out.

Most microbiological media are adapted to the study of aerobes, but the true anaerobes, particularly those that are damaged by exposure to oxygen (some *Clostridia* and members of the Archaea), need special culture conditions. The total exclusion of oxygen is difficult, but a series of methods named after their inventor (the **Hungate** techniques) achieve this.

Storage and revival of microorganisms

The Bacteria and Archaea are remarkably resistant to extreme conditions and many can survive freezing or desiccation without ill effect, even in their vegetative states. This means that many, but by no means all, prokaryotes can be stored as pelleted biomass for decades. In the laboratory, long-term storage of biomass frozen at $-70°C$ in a 50% glycerol solution is often used, although this can sometimes cause cell death. Freezing in dimethyl sulfoxide (DMSO) can be alternative, but prokaryotes rarely require cryopreservation in liquid nitrogen as eukaryotic cells can do. Larger laboratories may purchase a freeze-drying apparatus to **lyophilise** cultures. This is by far the most efficient method of long term storage. Short term lab storage (for up to a week) is normally done by streaking the biomass out onto solid media held in a suitable container (Petri dish or 30 mL bottle). After storage, the cells are in a starved state and must be revived with a complex medium so as little stress as possible is placed on them.

Sterilization

Once a microbiological experiment has been completed, the live organisms should be safely destroyed. Similarly, before an experiment starts all living cells present should be inactivated so that only the inoculum desired is present. This poses problems for the microbiologist, particularly one working with pathogens, thermophiles, or sporulating bacteria. The process of Pasteurization ($71.7°C$ for 15 seconds) will kill many common human bacterial pathogens without affecting the medium a great deal and so is therefore used in food preparation. Moist heat in the form of steam or boiling will kill most vegetative cells as well as some viruses, but thermophiles and endospores will survive. **Autoclaving** contaminated equipment ($121°C$ for 15 minutes under 15 psi (photosystem 1) pressure) kills all cells as well as endospores, but is not suitable for use with polycarbonate (a common material for making plastic containers) and will cause many medium components to caramelize. Several methods (outlined below) have been used to deal with heat-sensitive components.

- **Tyndalization** – repetitive heating to 90–100°C for 10 minutes, followed by cooling for one to two days. Allows endospores to germinate in the medium, which are then killed by the heating.
- **Ultra violet radiation** – kills living cells but does not penetrate opaque containers or large volumes of solution well.
- **γ radiation** – kills living cells but causes brittleness in polycarbonate and polypropylene.
- **Filtration of media** – $0.22\,\mu m$ filters can be suitable for aqueous solutions of heat labile chemical constituents, but it is difficult to filter large (> 500 mL) quantities effectively while maintaining sterility.

C4 ENUMERATION OF MICROORGANISMS

Key Notes

Obtaining a pure culture

The pure culture of a single species of prokaryote is central to the interpretation of many microbiological experiments. This is normally done by streaking on an agar plate to obtain single colonies, or diluting with sterile media until it is possible to grow a culture from a single cell.

Counting prokaryotes

The number of bacteria per millilitre of sample is important for industrial, food, and medical standards, as well as in microbiological research. This can be expressed as a total count of all cells living and dead, or a viable count of those cells that can be expected to grow. Total counts can be estimated with a haemocytometer or via indirect methods such as quantitative PCR (QPCR), but viable counts present more of a problem.

Classically, viable counting is done by serial dilution, plating, and estimation of colony forming units (cfu) or the most probable number (MPN) method. These have been enhanced by biochemical and molecular techniques such as quantitative reverse-transcription PCR (QRT-PCR) and estimation of ATP (adenosine 5′-triphosphate). One of the most accurate ways of determining the ratio of living to dead cells is via the use of dyes and fluorescence-activated cell sorting (FACS).

Related topics

Manipulation of cellular DNA and RNA (F16)
Looking at microbes (C5)

Obtaining a pure culture

The concept of the pure culture is central to classical bacteriology, and is a central tenant of Koch's postulates (see Section H1). If an organism can grow on agar, it can be streaked to obtain single colonies, each of which should have arisen as a result of a single prokaryotic cell (*Fig. 1*). This works by means of the **dilution** effect of each round of streaking and sterilization. The first inoculum onto the plate might transport millions of bacteria to the agar but, each time the loop is dragged across the plate, these cells are removed further from their neighbours. Coupled with sterilization of the loop, fewer than ten cells are present in the last line of streaking.

Many prokaryotic cells will not grow on agar plates. If this is found to be the case then a similar process of dilution is carried out in liquid broth to obtain a pure culture. One millilitre of the primary liquid culture is taken with a sterile pipette and added to 9 mL of fresh sterile medium (a dilution of ten, written as 10^{-1}). The 10^{-1} dilution is mixed well, then 1 mL of that is removed and again added to 9 mL of fresh sterile medium (a dilution of 100, written as 10^{-2}). This process is repeated a further ten or 12 times, and then the dilutions are incubated. The lower dilutions will show no growth, while the others will be turbid. The lowest dilution that shows growth is likely to have arisen from inoculation with less than ten cells,

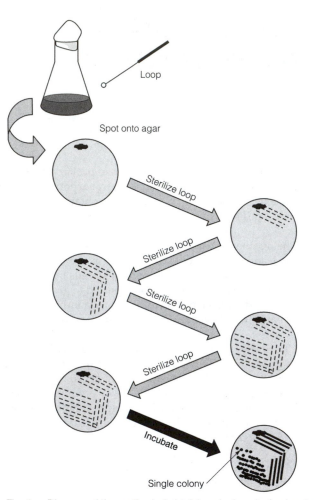

Fig. 1.   Diagram of the method of obtaining single colonies by streaking. Dashed lines show the path of the loop on the surface of the agar.

so repetition of this process will eventually lead to a pure culture. This approach can be further enhanced by making ten-fold replicates of each dilution. This should mean that the tube with growth in, where less than three or four of the other replicates are growing, must have arisen from a single cell. This concept can be extended into a means of actually estimating the numbers in the original culture (most probable number (mpn) method, see below).

Counting Prokaryotes

In microbiology, the number of cells per millilitre of sample is often important. We can be fairly sure of finding most species of prokaryote in a sample, provided that sample is large enough and was taken from a habitat that allows growth of that organism. It is likely that we might find *Salmonella* on eggs, but the important question we need to answer is 'Are the eggs safe to eat?' To be safe, there must be less than the intoxicating dose of *Salmonella* in an amount of egg that is likely to be consumed raw. The number of cells thought to cause *Salmonella* food poisoning is around 40, so we could reasonably expect safe eggs to carry less than, for example, 40 cells per dozen eggs. Such limits exist in industrial standards for most foods, so a good estimation of bacterial numbers is crucial to the food industry. In medicine, the presence of only one or two *Staphylococcus aureus* per 10 cm^2 of human skin could be considered quite normal, but 10^4 cells per mm^2 might reveal the underlying cause of a serious skin condition. In environmental microbiology the relative numbers of organisms per mL of a sample of river water might indicate the dominant species.

The numbers of prokaryotes in a sample can be expressed in two ways: the **total count** or the **viable count**. The former estimates the number of cells, alive or dead, the latter only those capable of growing under the conditions tested. Total counts are made by diluting the sample in a known amount of buffer and then counting the number of cells in each well of a **haemocytometer**. The haemocytometer is a specialized microscope slide and cover slip in which a grid of known size is displayed while viewing under the microscope. The count of cells per grid can then be multiplied up to reveal the number of cells per mL in the original sample. Flow cytometry is a method, similar in concept, in which the number of particles in a small sample is electronically counted by passing a laser shining across a capillary approximately one cell wide.

More recently these methods have been complemented by **quantitative PCR (QPCR)** (see Section F16). This method allows the counting of the number of copies of individual genes. For example, if the copy number per mL of the 16S rRNA gene is estimated, this can give an idea of the bacterial numbers. Although there are drawbacks to this method (we cannot be sure that all Bacteria have only one copy of this, or any other gene, per genome) by carrying out parallel experiments on the same sample using specific primer sets, we can simultaneously estimate the relative numbers of many different taxonomic groups, including the eukaryotes.

The most commonly used and informative method for enumeration of prokaryotes is the viable count. Whether cells are alive or dormant can be related to the presence of ATP. Thus, an estimate can be made of the overall activity of a sample by measuring ATP (chemically). Similarly, only live cells can transcribe DNA, so an estimation of viability can be made by measuring rRNA or mRNA (messenger RNA) concentrations, both by reverse transcriptase **QPCR (QRT-PCR)**. This presupposes that only active cells would accumulate these substances. Classically, viable counts are made by **serial dilution** (*Fig. 2*).

This agar plate-based method gives a result in **colony forming units** (cfu) mL^{-1}. This is not equivalent to the true viable count, as the numbers only reflect those species that are capable of forming visible colonies under the conditions of medium and incubation, chosen for the experiment. The number of cells in the original sample is estimated by back-calculating the number of dilutions made from the plate that has the highest number of easily discernible colonies. Normally this figure is less than 200 colonies per plate, but varies according to colony size. The equivalent method for prokaryotes not capable of growth on agar is the **most probable number (MPN)** technique, in which the pattern of growth in replicates of liquid cultures at various dilutions is used to deduce the number in the original sample.

One of the most accurate ways of counting any microscopic particle is by use of **FACS** (fluorescence-activated cell sorting). Fluorescent dyes can be obtained which differentially stain living and dead cells. The sample is introduced into a narrow capillary of only one cell width in diameter, and passes by a laser.

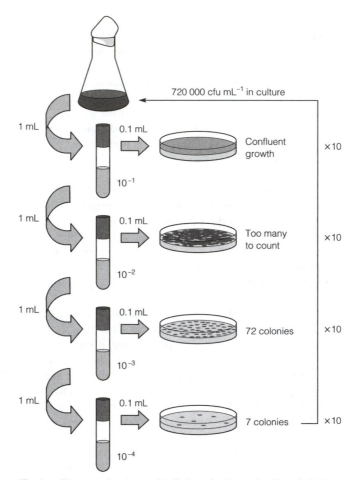

720 000 cfu mL^{-1} in culture

1 mL · 0.1 mL · Confluent growth · ×10 · 10^{-1}

1 mL · 0.1 mL · Too many to count · ×10 · 10^{-2}

1 mL · 0.1 mL · 72 colonies · ×10 · 10^{-3}

1 mL · 0.1 mL · 7 colonies · ×10 · 10^{-4}

Fig. 2.   Diagram showing serial dilutions for the estimation of viable count.

The laser excites the fluorescent dye and the excitation passes to a detector. The detector is linked to a gate just downstream of the laser, which will switch to move the cell into a receptacle. Non-fluorescent cells do not cause the gate to open and pass into a second vessel. The number of times the excitation gate opens can be counted and related to the flow rate past the laser to give an exact number of cells in the sample.

C5 LOOKING AT MICROBES

Key Notes

Light microscopy – old school!

The small size of microbes means that a microscope is needed to visualize almost all of them. For some of the smallest microbes the light microscope is at the limit of its resolution but contrast in light microscopy can be enhanced using stains or phase contrast, dark field, or fluorescence microscopy.

New developments in light microscopy

New microscopic techniques are greatly improving our ability to visualize microbes using light microscopy. Differential interference microscopy, atomic force microscopy, and confocal scanning microscopy are all techniques that create three dimensional images with improved depth of field over conventional light microscopy.

Transmission and scanning electron microscopy

Electron microscopy utilizes electrons rather than photons to image specimens, and has a much greater resolution capability than light microscopy. Electron microscopy was used extensively for the study of subcellular structures but required extensive pre-treatment of tissues (fixation and stabilization) to stabilize them for the electron beam. The technique has now improved to resolution at the molecular level, particularly when the chemical fixation techniques are replaced by cryo fixation and stabilization, which reduce tissue artefacts caused by traditional sample processing.

Related topics

Prokaryotes systematics (B1)
The Bacterial cell wall (C8)

Microscopy – old school!

Of the organisms classed as microbes, only the larger members of the fungi and chlorophyta are easily visible with the naked eye. A few members of the microbes are between 200 and 500 μm in size and can be seen using a hand lens, but by far the largest group of microbes have cell sizes of between 1 and 10 μm. This last group require substantial magnification before they can be viewed by the human eye and therefore different types of microscope have to be used to see these microbes.

The first microscopes were based on simple, single lenses, which provided sufficient magnification to see yeasts, protists, and larger Bacteria. The early microscopists Robert Hooke (1665) and Antonie van Leuwenhoek (1684) both observed and drew microbes using these simple single-lens microscopes. Present day light microscopes are compound microscopes, they have an objective lens (up to 100-fold magnification) and an eyepiece lens (usually 10-fold magnification) (*Fig. 1*), which together give a total magnification of 1000x. At maximum magnification an optical oil is used between the sample and the lens to optimize light collection and improve resolution (*Fig. 2*). The resolution limit of the light microscope is about 0.2 μm (i.e. you can differentiate two small black spots 0.2 μm apart), limited by the physics of glass lenses. This level of resolution and magnification allows us to see most of the

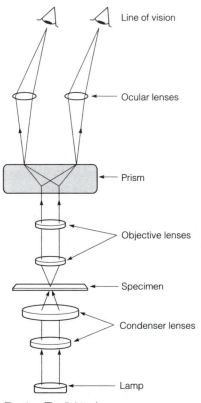

Line of vision

Ocular lenses

Prism

Objective lenses

Specimen

Condenser lenses

Lamp

Fig. 1. The light microscope.

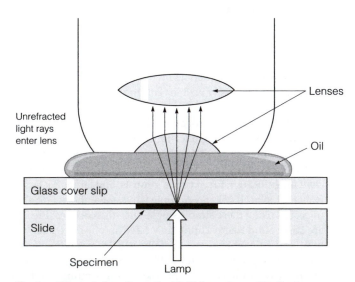

Lenses

Unrefracted
light rays
enter lens

Oil

Glass cover slip

Slide

Specimen   Lamp

Fig. 2.   Effect of oil on the path of light through an objective lens.

prokaryotes using a light microscope, particularly if stains are used to maximize contrast between the background and the specimen.

A number of stains can be used to visualize microbial cells for light microscopy, including crystal violet and safranin which are both used in the Gram stain (see Section C1). Bacterial spores can be stained using malachite green, and fungal structures can be stained using cotton blue in lactophenol. All stains should be treated with great care as, by their very nature, they are toxic to cells, including yours!

Other microscopic techniques can be used to maximize contrast between specimen and background including phase contrast, fluorescence, and dark field microscopy. Phase contrast microscopy takes advantage of the change in phase of light that occurs when light passes through a cell. This change in phase alters the refractive index of the sample relative to the background, and when this difference is amplified using a phase plate in the microscope the resulting image has enhanced contrast (*Fig. 3*). Dark field microscopy uses side illumination of the specimen. Only scattered light from the specimen is seen through the microscope (*Fig. 4*) and the object appears light on a dark background. Fluorescence microscopy takes advantage of the fact that some molecules will emit light (fluoresce) when irradiated with light of another wavelength. Some microbes contain naturally fluorescent compounds (chlorophylls and other pigments). Fluorescent dyes can be used to tag specific structures or processes. Light from the ultra-violet spectrum is commonly used to excite fluorescence and this requires a separate source of illumination.

New developments in microscopy

Recent improvements in light microscopy include differential interference microscopy, atomic force microscopy and confocal scanning microscopy. All these techniques create 3-dimensional images with improved depth of field over those of conventional light microscopy.

Differential interference contrast microscopy uses a polarized beam of light that is split into two and both light beams are passed through the specimen.

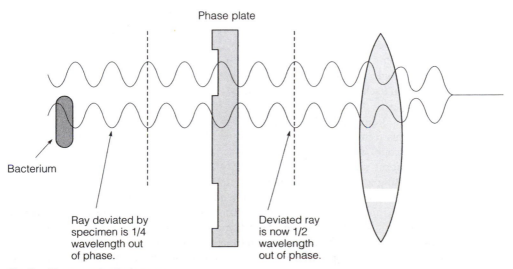

Phase plate

Bacterium

Ray deviated by specimen is 1/4 wavelength out of phase.

Deviated ray is now 1/2 wavelength out of phase.

Fig. 3.   Phase contrast microscopy.

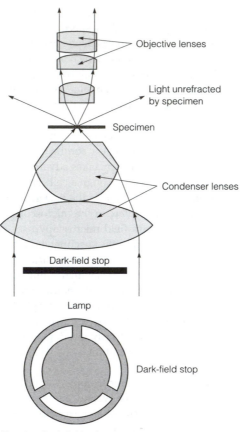

Fig. 4. Dark field microscopy.

The beams are then reunited in the objective lens. The image they produce is created from an interference effect caused by small changes in phase caused by passage through the specimen. Cell organelles viewed with DIC (differential interference contrast) microscopy have a 3-dimensional quality.

In atomic force microscopy, a living, hydrated specimen is scanned using a microscopic stylus, so small that it records minute repulsive forces that exist between itself and the specimen. The stylus records changes in topography as it scans across the specimen (*Fig. 5*). Data are then processed by computing to create detailed 3-dimensional images. No chemical fixatives or coatings need be used with this technique and therefore the artefacts seen in SEM (scanning electron microscope) (see below) are avoided.

Confocal scanning laser microscopy (CSLM) uses laser light source and computing to create 3-dimensional digital images of thick specimens. The precision of the laser beam, focused through a pinhole ensures that only a single plane of a specimen is illuminated at one time (*Fig. 6*). By adjusting focus, different layers of a specimen can be viewed and complex images can be created from digital data. Fluorescent staining and artificial colors linked to depth or density differences in the specimen can be used to enhance the image. CSLM is particularly useful in the study of microbial ecology of biofilms and soils.

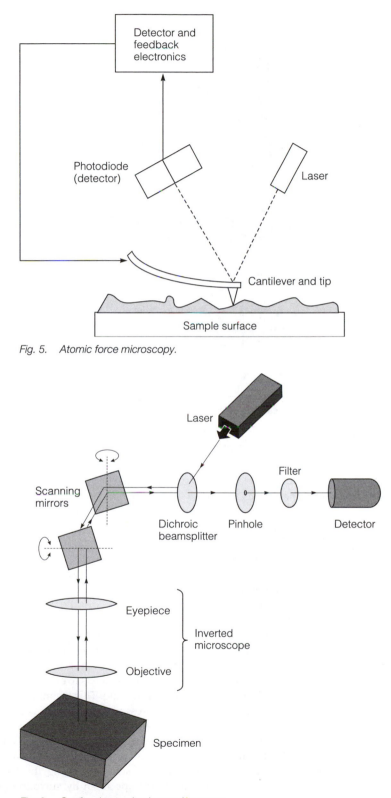

Fig. 5.   Atomic force microscopy.

Fig. 6.   Confocal scanning laser microscopy.

Column of transmission
electron microscope

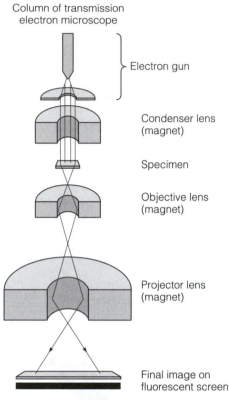

Fig. 7.   Transmission electron microscope.

Transmission and scanning electron microscopy

Transmission electron microscopy utilizes electrons rather than photons to image specimens. This technique was originally developed from X-ray crystallography and it took some years to develop the technique for biological specimens. In the electron microscope electron beams are focussed by a series of electromagnetic lenses (*Fig. 7*). The microscope is kept under high vacuum to ensure unimpeded travel of electrons to and through the specimen. The electrons pass through the specimen on to a phosphor screen for visualization (we cannot see electrons with our eyes). Biological specimens are inherently unstable under vacuum, and therefore have to be fixed (usually gluteraldehyde and paraformaldehyde, followed by osmium tetroxide) dehydrated, and stabilized in resins before they can be placed in a vacuum. The depth of specimen that electrons can pass though is very small, which means that the biological specimen has to be sliced into thin sections, usually with a diamond knife, to produce 1 micron sections. These sections are electron transparent and specimen contrast has to be enhanced using metal stains, for example lead citrate and uranyl acetate. This techniques was used extensively for the study of subcellular structures, and now has moved on to resolution at the molecular level, particularly when the chemical fixation techniques described above are replaced by cryo fixation and stabilization which reduce tissue artefacts caused by traditional sample processing. The electron microscope can also be used to study virus and organelle structure using the technique of negative staining. Contrast is brought to the specimen by surrounding it with phosphotungstic acid (see Section L1).

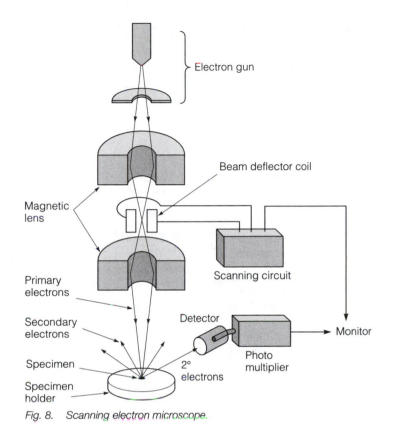

Fig. 8.   Scanning electron microscope.

External structures of microbes can be viewed at high magnification using the technique of scanning electron microscopy. The specimen is stabilized by either chemical or cryo fixation, covered by a layer of conductive material (carbon or gold/platinum) and the specimen is scanned with an electron beam. Electrons are either reflected or passed through the specimen and a digital image is built from the composite data, creating three dimensional structures (*Fig. 8*). This technique has been further developed for examination of specimens at low vacuum and ambient conditions.

C6 THE MAJOR PROKARYOTIC GROUPS

Key Notes

The Prokaryotes	The Prokaryotes can be divided into two kingdoms, the bacteria and the Archaea. This classification was first proposed because of the differences in 16S rRNA sequence.
The Proteobacteria	This phylum contains the largest number of known species, and is further subdivided into alpha (α), beta (β), gamma (γ), delta (δ), and epsilon (ϵ). The enteric group of the γ-proteobacteria include *Escherichia coli* and many well known pathogens. Other frequently encountered bacteria in this Phylum include *Pseudomonas* (γ-proteobacteria) and Campylobacter (ϵ-proteobacteria).
The Gram positive bacteria	The two main groups in this phylum are called the low GC and high GC Gram positives (or Actinomycetes), a phylogenetically valid classification based on the %GC content of their genomes. Low GC Gram positive organisms include *Bacillus, Clostridium*, and *Lactobacillu/Streptococcus*, while high GC organisms include *Mycoplasma* and *Streptomyces*.
Cyanobacteria	The Cyanobacteria are a uniformly phototrophic phylum, but differ fundamentally from apparently similar photosynthetic Proteobacteria, such as *Rhodococcus*. Examples include *Synechococcus, Anabaena*, and *Prochloron*. The Phylum includes many species exhibiting differentiation of vegetative cells to form gas vesicles or heterocysts.
Planctomycetes	The Planctomycetes, such as *Brocardia annmmoxidan*, are unusual in having budding rather than binary fission and genomic material bounded by a membrane. These are the only known prokaryotes to have a true nuclear organelle.
Spirochetes	Although there are many non-pathogenic representatives, this Phylum includes the pathogens *Treponema pallidum* (syphilis) and *Borrelia burgdorferi* (Lyme disease). They are highly motile and helical in shape, with an unusual form of motility using endoflagella.
Deinococcus/Thermus	The two representative genera of this Phylum both resist extreme conditions: *Deinococcus radiodurans* can resist high doses of radiation while Thermus aquaticus grows at temperatures up to 80°C.
Aquifex and hypertherm ophilic Phyla	Several Phyla near to the root of the 'tree of life' contain only a few species, most of which are thermophiles. This property coupled to the primitive nature of these bacteria supports the notion that life began at higher temperatures than we experience today.

Other Phyla	The great diversity of the bacteria is reflected in many other Phyla not discussed here. These include the numerically abundant members of the *Flavobacteria* as well as *Verrumicrobia*, *Cytophaga*, green sulfur bacteria, *Chloroflexus* and *Chlamydia*.
The Kingdom Archaea	Thermophiles, halophiles, and other extremeophilic Archaea are well known, but this Kingdom also includes many mesophiles. It is becoming apparent that there is a similar or even greater physiological and biochemical diversity in the Archaea compared to the Bacteria.
The Phylum Crenarchaeota	Most crenarchaeotes cultured in the laboratory are extremophiles capable of growth above 80°C. The best known examples are *Sulfolobus solfataricus* and *Pyrodictium abyssi*, the latter holding the current record for biological growth at high temperature (110°C). The morphology of *Pyrodictium* spp. is unusual, with disc shaped cells interconnected by hollow tubes of unknown function (cannulae).
The Phylum Euryarchaeota	This Phylum of the Archaea includes both mesophiles and extremophiles, most notable among which are the methanogens and *Pyrococcus furiosus*.
The 'Korarchaeota' and 'Nanoarchaeota'	Several Phyla in both the Archaea and the bacteria have been proposed on the basis of the existence of environmental 16S rRNA sequences. The 'Korarchaeota' have been identified in this way, but there is growing evidence that the signature sequences are artefactual. In contrast, the 'Nanoarchaeota' has such a small genome that evolutionary pressure may have resulted in an incorrect classification of these few isolated cultures.
Related topics	Prokaryotic systematics (B1) Identification of Bacteria (B2) Inference of phylogeny from rRNA sequence (B3) Composition of a typical prokaryotic cell (C7) Cell Division (C9) Bacterial Flagella and movement (C10) Prokaryotes and their environment (C11)

The prokaryotes The prokaryotes consist of many thousands of known species, to which some order has been applied with the advent of 16S rRNA sequencing (see Section F16). The resulting phylogenetic tree (*Fig. 1*) has been used as an indication of how life started. To view the two kingdoms in the prokaryotes as primitive is an oversimplification, as both the Bacteria and the Archaea are extremely well adapted to their environments. A brief survey of the Bacteria and Archaea cannot hope to encompass their diversity, and may even give a misguided view that the Bacteria are predominantly pathogenic and the Archaea are predominantly thermophilic. However, the best studied Bacteria are pathogenic, despite the numerous examples of beneficial species, and the best studied Archaea are thermophiles.

The proteobacteria The largest phylum in the Bacteria is the proteobacteria, which include most of the species commonly encountered in the microbiology laboratory. The proteobacteria are further subdivived into alpha (α), beta (β), gamma (γ), delta (δ)

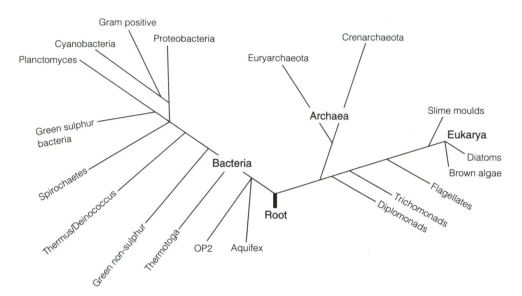

Fig. 1.   Phylogenetic tree showing the relationship of the Bacteria, Eukarya and Archaea.

and epsilon (ε) classes. A few well-known examples from each subdivision are shown in *Table 1*. Key members of the phylum are discussed further below.

Escherichia coli *and the enterics*
The rod-shaped facultatively anaerobic *Escherichia coli* (γ–proteobacteria) has become the model organism for microbiology and the standard vector for molecular biology. It is essential in human food digestion as it is the main producer of vitamin K from undigested food in the large intestine. Despite these obvious benefits, it is also the most common cause of urinary tract infections and can also cause pneumonia or even meningitis. In the mind of the general public, it is closely linked with outbreaks of diarrhea, normally caused by the strain 0157:H7

Table 1.   Some members of the proteobacteria

Sub division	Example genus/species	Notes
α-Proteobacteria	*Rhodobacter sphaeroides*	Photosynthetic organism
	Paracoccus denitrificans	Denitrifying bacterium
	Rickettsia prowezeki	Causes typhus
β-Proteobacteria	*Bordatella pertusis*	Whooping cough
	Burkholderia cepacia	Causes both onion bulb rot and is an opportunistic human pathogen
	Neisseria meningitides	Causes bacterial meningitis
γ-proteobacteria	*Acetobacter*	Used in the manufacture of vinegar
	Escherichia coli	The model bacterium
	Pseudomonas aeruginosa	Opportunistic pathogen
δ-proteobacteria	*Aeromonas*	Food spoilage organism
	Bdellovibrio	Preys on other Gram negative bacteria
	Myxococcus	Exhibits differentiation
ε-proteobacteria	*Campylobacter jejuni*	Most common food-borne pathogen
	Wolinella succinogenes	Possible symbiont of cattle

Table 2.   Pathogenic activities of the enteric bacteria

Species	Disease
Escherichia coli	Enterotoxin induced diarrhoea
Salmonella typhimurium	Typhoid fever
Shigella dystenteriae	Dysentery
Klebsiella pneumoniae	Pneumonia
Yersinia pestis	Plague
Proteus vulgaris	Urinary tract infections
Serratia marcescens	Mammary gland inflammation
Vibrio cholerae	Cholera

(Enteropathogenic *E. coli*). Most mammals have *Escherichia* species in their digestive systems but they are rarely dominant.

The genus *Escherichia* has a single species, and is grouped with other medically important bacteria known as the **enterics**. This is a phylogenetically distinct group (Enterobacteriaceae) which includes *Salmonella, Shigella, Proteus* and *Enterobacter*. The significance of these organisms in disease (*Table 2*) has become a barrier in rationalizing their taxonomy. *Escherichia* and *Salmonella* share over 50% genomic identity and *Shigella* over 70%, which would normally place them all as species in the same genus.

The **Pseudomonas** *group*

The straight rods of *Pseudomonas* (γ–proteobacteria) are often encountered in enrichment cultures, due to their nutritional versatility – some species can utilize more than 100 different organic compounds. The genus *Pseudomonas* also includes a significant number of plant pathogens. Despite their ubiquitous nature, they are still associated with some serious human diseases, mostly opportunistic infections of wounds and cystic fibrosis. The best studied in the group is *Pseudomonas aeruginosa*, characterized by, among other things, the ability to produce a fluorescent pigment on certain media. Collectively the group including *Pseudomonas, Ralstonia, Xanthomonas, Commamonas, Burkholderia* and *Brevundimonas* is known as the **pseudomonads**, though these genera do not share any distinct phylogeny within the proteobacteria.

Campylobacter

Despite the media fascination with *Salmonella* and *E. coli* food poisoning, the most common food-borne infections are caused by the *Campylobacter* species. Part of the reason why this genus is overlooked is because its members are obligately microaerophilic and difficult to grow in laboratories more used to dealing with exclusively aerobic organisms. The closely related organism *Helicobacter pylori* has been linked, either as the causative agent or as an opportunist, to gastric ulcers. The rod/spirillar Gram negative bacteria are classed as ε– proteobacteria.

The Gram positive Bacteria

The Gram positive Bacteria can be separated into two subgroups based on the percentage of guanine and cytosine in their genomes (see *Table 3*) compared to adenine and thymine. The **low GC** Gram positives have much less than 50% G + C, while the **high GC** Gram positives have far more than 50% G + C. The latter are sometimes referred to as the actinobacter.

Table 3.   Some members of the Gram positive Bacteria

Sub division	Example genus/species	Notes
Low GC	Staphylococcus aureus	Causes a variety of pus-forming diseases in humans
	Lactobacillus delbrueckii	Used in the production of yoghurt
	Streptococcus pyogenes	Implicated in 'Strep throat'
	Listeria monocytogenes	Causes listeriosis in humans
	Bacillus subtilis	The Gram positive equivalent of E. coli but is less frequently associated with disease
	Clostridium perfringens	Invasive organism causing gas gangrene
	Geobacillus sterothermophilus	Thermotolerant, growing from 30 to 70°C, has extremely heat-resistant endospores
	Mycoplasma genitalium	Causes non-gonoccocal urethritis
High GC	Corynebacterium diphtheriae	Pathogen, causing diphtheria
	Propionobacterium acnes	Members of this genus are prominent in the fermentation of many foods, including production of holes in Swiss cheese. This species may cause acne but is also resident on normal skin.
	Mycobacterium tuberculosis	Causative agent of tuberculosis
	Streptomyces coelicolor	Produces antibiotics against Gram negatives. Gives the soil its smell (produce of geosmins)
	Bifidibacterium lactis	Obligate anaerobe found in the intestine of breast-fed infant humans
	Rhodococcus rhodochrous	Metabolically diverse bacterium capable of metabolizing nitriles and cyanides

Bacillus

Bacillus subtilis (low GC) is one of the oldest named bacteria, first identified by Ferdinand Cohn in 1872. The genus is characterized by the formation of endospsores (see Section C9). The spores survive so efficiently that it is possible to recover *Bacillus subtilis* from most environments. Rod-shaped *Bacillus* is a physiologically diverse genus, recently rationalized by the removal of many thermotolerant strains *Geobacillus*. Closely related species include *Bacillus anthracis* the causative agent of anthrax and *Bacillus thuringiensis*, which produces commercially employed insect larvicides.

Clostridium

The members of this genus (low GC) are phylogentically distinct but morphologically similar to *Bacillus*. They are sometimes referred to as **clostridia**, and lack any respiratory chain. This renders them obligate anaerobes, generating ATP solely by substrate-level phosphorylation (see Section E). Although organisms such as *Clostridium tetani* (tetanus), *C. perfringens* (gas gangrene, food poisoning), *C. botulinum* (botulism) cause medical problems, *Clostridium acetobutylicum* can be used to biologically synthesize butanol, ethanol, isopropanol and other alcohols rather than using fossil fuel-based alternatives.

Lactobacillus/Streptococcus

These low GC organisms are significant in the human and animal food industry, being involved in the fermentation of cheese, yoghurt, sauerkraut, olives, and silage. Their metabolism tends to produce lactic acid so, along with the genera *Pediococcus, Micrococcus*, and others, they are often grouped together as the **lactic acid bacteria**. *Lactobacillus* species are rarely associated with human disease, but *Streptococcus* cause throat infections and dental caries.

Mycoplasma

The members of *Mycoplasma* (high GC) are remarkable in the Bacteria in that they do not have cell walls at any stage in their life cycle. They are considered to be among the smallest organisms capable of growth outside a host cell, and can have a genome of less than 600 000 bp. Although they are phylogentically true Gram positive organisms, they do not retain the crystal violet-iodine complex in the Gram stain (see Section C8) through lack of a cell wall and so appear to be Gram negative. The absence of a rigid wall also renders the cells **pleiomorphic**, where they can display different shapes under different physiological or environmental conditions, or even in the same culture, appearing as small cocci, swollen rods or branched filaments. Lastly the lack of cell wall renders them resistant to antibiotics interfering with cell wall synthesis, such as penicillin and vancomycin, so other antibiotics such as kanamycin must be used in combating the pathogenic species.

Streptomyces

The members of this low GC genus produce many antibiotics including streptomycin, spectinomycin, neomycin, tetracycline, erythromycin and chloramphenicol. Despite an enormous research effort to isolate and characterize the genes and proteins associated with antibiotic production, the significant role *Streptomyces* has in soil ecology has yet to be fully explored. The cells form filaments that can appear to be much like the fungal mycelium and, as the colony matures, aerial filaments with **sporophores** are formed. When released these **conidia** are similar in size but distinct from the endosprores of *Clostridia* and *Bacillus*.

Cyanobacteria

The Cyanobacteria are a uniformly phototrophic phylum, but differ fundamentally from apparently similar photosynthetic Proteobacteria such as *Rhodococcus*. They can be subdivided into five groups according to their cellular morphology (filamentous or unicellular, branching, non-branching etc.), and in this case the morphology has significance in their phylogeny. All produce complex membranes in which photosynthesis takes place. The cells are green-blue in color due to the presence of chlorophyll α and phycobilins on the cell membrane, giving the phylum its present name. The filamentous varieties can show cellular structural differentiation, with a few cells forming **gas vesicles** (to aid in floating) or **heterocysts** (sites of nitrogen fixation). The genera *Synechococcus*, *Anabaena*, and *Prochloron* are the best studied.

Planctomycetes

The plantomycetes lack a peptidoglycan cell wall, but instead rely on a well developed S-layer for rigidity (see Section C8). They are characterized by a stalk for attachment to surfaces, and divide by **budding** rather than binary fission. Although there is little doubt that members of the phylum, such as *Gemmata* are prokaryotes, their genomic material is bounded by a membrane. The Annamox organism *Brocardia annamoxidans* (used in waste-water treatment) also has a true organelle in the form of the anammoxosome, where the anaerobic oxidation of ammonia to molecular nitrogen is carried out.

Spirochetes

The spirochetes stain Gram negative, but form a separate phylum from *Escherichia coli* and close relatives. They are highly motile and helical in shape with an unusual form of motility using endoflagella (see Section C9). The most well known example of this phylum is *Treponema pallidum*, which causes syphilis; however the majority are free living or obligately symbiotic non-pathogenic organisms.

T. pallidum cannot be grown as a free living organism in the laboratory but must be grown with animal cell culture. The other notable pathogen from this phylum is *Borrelia burgdorferi*, which causes the tick-borne **Lyme disease**.

Deinococcus/Thermus

The phylum containing only the genera *Thermus* and *Deinococcus* includes industrially significant bacteria. *Thermus* species have the ability to grow at temperatures of up to 80°C, and *T. aquaticus* is the source of one of the most important proteins in molecular biology, *Taq* polymerase (see Section F16). The extremely radiation-resistant organism *Deinococcus radiodurans* can survive exposure to gamma rays better than *Bacillus* endospores, while still retaining a vegetative mode. The allied resistance to chemical mutagens (see Section F10) shows that this is achieved with a very efficient DNA repair system. This phylum contains no known human or animal pathogens.

Aquifex and hyperthermophilic phyla

Many hyperthermophilic species of Bacteria have been isolated that are phylogenetically close to the hypothetical root of the tree of life (*Fig. 1*). These isolates have been arranged into several distinct phyla containing only one or two genera such as *Aquifex, Thermotoga, Methanopyrus,* and *Pyrolobus*, all characterized by having optimal growth temperatures above 80°C. This property, coupled to the primitive nature of these Bacteria, supports the notion that life began at higher temperatures than we experience today.

Other phyla

The great diversity of the Bacteria is reflected in many other phyla not discussed here. These include the numerically abundant members of the Flavobacteria as well as Verrumicrobia, Cytophaga, Green sulfur bacteria, Chloroflexus and Chlamydia.

The kingdom Archaea

Until recently the Archaea were regarded as a solely extremophilic species, growing in environmental niches unsuitable for other forms of life, as extreme thermophiles, psycrophiles, halophiles or natronophiles. With the advent of rRNA sequencing, we are now seeing species within the Euryarachaeota in aboreal forest soils, in temperate lakes, indeed in most of the places we expect to see Bacteria. This should have been no great surprise since we have known about the methanogenic Archaea in our digestive systems for many decades. Despite having as great an ability to behave as chemoorganotrophs and chemolithotrophs as the Bacteria and a greater diversity in nucleic acid replication, there are currently no known human Archaeal pathogens.

As has been emphasized in other sections, the Archaea have many similarities physiologically to both the Bacteria and the Eukarya (see *Table 4*). However, our understanding of the ecology, physiology, and biochemistry of this kingdom does not begin to approach the wealth of biological knowledge we have for such organisms as *E. coli* and *Saccharomyces cerevisiae*.

The phylum Crenarchaeota

Those Crenarchaeota that have been cultured in the laboratory are extremophiles with growth optima in excess of 80°C, including *Desulfurococcus, Thermoproteus, Sulfolobus*, and *Pyrodictium*. Presumed psycrophilic Crenarchaeotes have also been identified via rRNA sequencing suspended in Antarctic waters. The majority of the bacteriology and enzymology of this division has focused on the biotechnological applications of Crenarchaeotal proteins, so our overall understanding of these organisms is relatively poor.

The best known example of the Crenarchaeota is *Sulfolobus solfataricus* (the first Sulfolobus species was *S. acidocaldarius*, discovered by Thomas Brock in 1970).

Table 4. Selected features of the Archaea, compared to the Eukarya and Bacteria

	Bacteria	Archaea	Eukarya
Cell wall lipids	Isoprenoids ester bonded to glycerol	Isoprenoids ether bonded to glycerol	Isoprenoids ether bonded to glycerol
Light mediated ATP synthesis	Photosynthetic	Bacterioruberins catalysed system (e.g. Bacteriorhodopsin)	Photosynthetic
Chromosome(s)	At least one circular, rarely linear	Circular	Linear
Nuclear membrane	Absent	Absent	Present
Histones	Absent	Present	Present
DNA replication	Unique	Similar to Eukarya	Similar to Archaea
MRNA	May be polycistronic	May be polycistronic	Monocistronic
RNA polymerase holoenzyme	$\alpha_2\beta\beta'\sigma$	Up to 13 subunits	More than 33 subunits
rRNA	5S, 16S, 23S	5S, 16S, 23S	5S, 18S, 23S
Transcription	Unique. Generally initiated by derepression	Simplified Eukarya-like mechanism	Generally initiated by activation
Translation initiation	Via Shine Delgarno sequence	Sometimes via Shine Delgarno sequence	Scanning mechanism

S. solfataricus was isolated from a hot spring emerging from the volcanic area around Naples in Italy. Morphologically the cells are nondescript, irregularly shaped cocci, but this archaeon is both an extreme thermophile with a growth optimum of around 80°C, and an acidophile, growing at pH 3. It has proved easy to work with in the laboratory, especially as few airborne contaminants survive in the culture medium. Work on this organism pioneered the concept that the RNA polymerase of Archaea has more in common with eukaryotes (e.g. the use of basal transcription factors) than Bacteria.

The Crenarchaeote *Pyrodictium abyssi* is another extremely thermophilic acidophile, and holds the current record for biological growth at high temperature. Its optimum growth temperature is between 97 and 105°C, but can still divide at 110°C. An obligate thermophile, it will not grow below 80°C, and can tolerate temperatures of up to 140°C. Like *Sulfolobus* it is an acidophile, growing at an optimum pH of 5.5, but is also a strict anaerobe with a requirement for sodium chloride (0.7–4.2% w/v). Such conditions cannot be found anywhere within the terrestrial aqueous ecosystems on earth, so it is unsurprising that this bug was first isolated from a **black smoker** in the deep oceans. Volcanic fissures under the sea release large amounts of sulfide, hydrogen, and metals at high temperature, giving a plume of black liquid across the sea bed. As *P. abyssi* uses hydrogen as an energy source and sulfur as an electron acceptor instead of oxygen, these are ideal conditions for growth. As well as being biochemically remarkable, the morphology of the Archaeon is unusual as well. The cells themselves are disc-shaped, and are interconnected with up to 100 of their neighbors by hollow tubes, known as **cannulae**. The function of these fine tubes hasn't been fully explored, but it is thought that they extend into the Archaeal equivalent of the periplasm, but not into the cytoplasm itself.

The phylum Euryarchaeota

This phylum also includes extremophiles, but also numbers mesophiles including some of the **methanogens** among its members. The biochemistry of methanogenesis is detailed in Section E3 and include such genera as *Methanococcus*, *Methanobacterium*, and *Methanosarcina*. The phylum is physiologically diverse, with representatives in the extreme thermophiles (e.g. *Picrophilus* and the cell wall free *Thermoplasma*), the halophiles (*Halococcus*, *Natronococcus*) and many oligotrophic marine Archaea.

Pyrococcus furiosus is an extremely thermophilic euryarchaote and was the source of the *Pfu* polymerase, used as an alternative to *Taq* polymerase in the PCR reaction. Although *Taq*, of bacterial origin, and *Pfu*, of Archaeal origin, can be substituted for one another in the *in vitro* technique of PCR, the two proteins illustrate a difference between the two kingdoms. *Pfu* is the main replicative DNA polymerase of *Pyrococcus furiosus*, but has little in common with the eubacterial equivalent, DNA polymerase III (see Section F5). *Taq* is a DNA polymerase I-type protein, with a function in DNA repair (see Section F11), but it is not yet known whether *Pfu* or other proteins perform the equivalent role in *P. furiosus*. *P. furiosus* is attractive for PCR as it has the quickest doubling time of the Archaea, replicating its genome in only 37 minutes. To do this it must be growing at between 70 and 103°C, and studies on this organism reveal that even at its maximum growth temperature, the DNA within the cell remains double stranded without breaks, something that cannot be achieved if pure DNA is boiled in water. It is also almost as resistant to radiation as *Deinococcus*.

The methanogens are a diverse group of the Euryarchaeota, including methanogenic thermophiles, acidophiles, halophiles, and mesophiles. They are linked by the ability to produce methane from C1 compounds or acetate (see Section E3) and an obligately anaerobic way of life. The model organism for the genus is *Methanocaldococcus jannaschii*, one of several methanogens to have had their genomes sequenced. Comparative genomics reveals that *M. jannaschii* has similar enzymes to the Bacteria in its central metabolic pathways (apart from the specialist ones involved in methane generation) but a greater similarity to the Eukaryotes in DNA replication, transcription, and translation. Notwithstanding, 50% of the genes of *M. jannaschii* have absolutely no counterpart in any organism so far sequenced. The organism was named after Holger Jannasch, a pioneer in the field of deep sea microbiology. The organism was isolated from a white smoker in the East Pacific, and consequently the growth requirements of *M. jannaschii* are strict: carbon dioxide; hydrogen; a few mineral salts; no oxygen; temperature of around 85°C and a significant amount of pressure. This barophile can withstand up to 200 atmospheres of pressure.

The 'Korarchaeota' and 'Nanoarchaeota' An indication of the growing interest in the Archaea is indicated by the controversy surrounding the existence of other major divisions in this kingdom. Environmental 16S rRNA gene sequences have been obtained that do not group with either of the established Archaeal kingdoms, but no organism has been isolated associated with any of these sequences. It has yet to be firmly established whether this group is artefactual, perhaps formed by other environmental 16S rRNA gene sequences ligating during PCR. The 'Nanoarchaeota' have a single putative representative, *Nanoarchaeum equitans* (an extreme thermophile isolated from an undersea hot water system off the coast of Iceland) which has the smallest completely sequenced genome of only 490 885 bp. Taxonomists are discussing whether it represents a new archaeal phylum or if the rapid evolution of a similar lineage of the Euryarchaeota (the phylum Thermococcales) might provide a more rational explanation.

C7 COMPOSITION OF A TYPICAL PROKARYOTIC CELL

Key Notes

The prokaryotic cell	The model Bacterial cell is that of *Escherichia coli*, a rod-shaped cell about 2 µm long and 1 µm wide. With a very few exceptions, there are no discernible subcellular components in prokaryotic cells. At the biochemical and physiological level, Bacteria have some similarity to the Archaea, yet the Archaea have similarities to both the Bacteria and the Eukaryotes.
The cytoplasmic membrane	The Bacterial cell membrane is made up of a lipid bilayer with proteins buried within it. The bilayer is impermeable to most molecules except gases such as oxygen, nitrogen and CO_2, while passage of larger molecule and ions is mediated by protein. The membrane does not confer rigidity to the cell but is reinforced with hopanoids. The Archaeal cell membrane has a similar structure but has a unique monolayer structure.
The cytoplasm	Subcellular structures in the aqueous cytoplasm include ribosomes and storage structures (inclusion bodies, carboxysomes, and magnetosomes). Activity specific structures also exist, for example thykaloid membranes or chlorosomes involved in photosynthesis. The examination of subcellular structure in prokaryotes is sometimes complicated by the process of electron microscopy and has been implicated in the formation of artefacts such as the mesosome.
Genomic material	The nucleoid is the tightly packed chromosomal material of prokaryotes, while plasmids are thought to float free in the cytoplasm. The chromosomal complement of a prokaryotic cell is always haploid and plasmid copy number is often more than one.
The periplasm	The periplasm of Gram negative Bacteria is of lower water content than the cytoplasm. It is packed with proteins and plays a major role in secretion, environmental sensing, and many other key pathways.
The cell wall	All prokaryotic cells have a strong cell wall (apart from *Mycoplasma* and *Thermoplasma*). Bacteria are covered in peptidoglycan. The Archaea have a far greater variety in the composition of the outer wall but has a Bacterial orthologue in pseuodmurein.
Endospores	*Clostridium* and *Bacillus* are able to form spores within the cytoplasm in response to cell starvation. Endospores have the capacity to survive heat, radiation, and chemical disinfectants
External features	All prokaryotes are covered in some form of slime capsule (or glycocalyx). In addition they may have filaments extending from the cell wall such as flagellae for movement, or much finer fimbriae for attachment.

Related topics	Prokaryotic diversity (C2)
	The major prokaryotic group (C6)
	The Bacterial cell wall (C8)
	Cell division (C9)
	Bacterial flagella and movement (C10)
	Genomes (F2)

The prokaryotic cell The model Bacterium is *Escherichia coli*, normally found in the colon of humans and other mammals. It has a limited capacity for independent existence, is motile and rod shaped, about 2 μm long and 1 μm wide. Its cell wall is Gram negative in structure (for a comparison of Gram negative and Gram positive Bacteria, see Section C8). Morphologically *E. coli* resembles many other Bacteria and some members of the Archaea, although the fine structure of the cell wall, biochemistry, and molecular biology differ considerably. The single circular chromosome (see Section F2) is attached to the cytoplasmic membrane (*Fig. 1*) and this DNA is free in the cytoplasm and not bound by any membrane. The word 'prokaryote' is derived from the Greek *pro* (before) and *karyote* (literally a nut's kernel, but biologically adapted to mean nucleus), so this prenucleate state defines the prokaryotes. There are generally no discernible subcellular components in Bacterial cells (compared to eukaryotic cells, Section C6), the exception being the presence of layered membrane bodies of unknown function in the methanotrophic and nitrifying Bacteria (Section C11). The main subcellular differences are that the following are **absent** from Bacterial and Archaeal cells:

- Nuclear membrane.
- Chloroplasts and mitochondria – energy generation takes place across the cell membrane rather than in any specialized bodies.
- Golgi apparatus or endoplasmic reticulum – free ribosomes in the cytoplasm translate mRNA.

Bacteria and Archaea come in a variety of shapes and sizes (see Section C2), and are classified according to their biochemical rather than morphololological differences (see Section B). The differences between the Bacteria and Archaea are broad, covering the machinery used for replication, transcription, translation, and

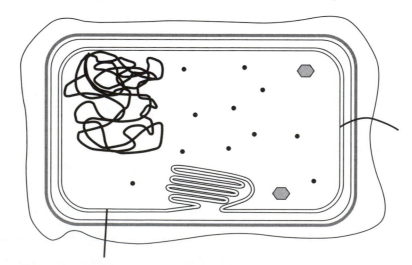

Fig. 1.   A generalized diagram of a Bacterial or Archaeal cell.

central metabolism (see Section F). Overall they are linked by a common cell structure and share many ecological niches, but in this respect are a product of convergent evolution.

The cytoplasmic membrane

In common with all biological membranes, the bacterial membrane between the cytoplasm and the cell wall is composed of a **lipid bilayer** (*Fig. 2*), made up of two tiers of phospholipids. These phospholipids are arranged so that their hydrophilic heads face into the cytoplasm or the outside world, while the hydrophobic tails are embedded within the membrane. Many proteins are partially buried within the membrane, or pass right through it. These proteins have many final functions but all mediate transport in one form or another – f1f0 ATPase regulates the transport of H^+, sugar transporters regulate the flow of carbohydrate, and so on. The lipid bilayer is impermeable to H^+, as without this property the organism would be unable to generate ATP (see Section E3). The current model of the cell membrane allows passage of:

- Gases (O_2, CO_2, N_2) via **passive diffusion** directly through the membrane.
- Water soluble (Na^+, K^+, except where these substitute for H^+ in extremophiles) ions via small pores in the membrane.
- Water itself up or down the **osmotic gradient**.
- Small molecules via **facilitated transport** through protein channels but following an osmotic, chemical or potential gradient.
- Molecules via **active transport**, normally at the expense of ATP.

The cytoplasmic membrane does not give rigidity or shape to the cell as this is provided by the tougher cell wall. Bacterial cell membranes lack the sterols eukaryotes have but instead stabilize membrane structure with **hopanoids**.

The Archaeal cell membrane performs the same roles as the Bacterial equivalent, but it is unique in biology with a predominantly monolayer phospholipid structure (*Fig. 2*).

The cytoplasm

Around 80% of the cytoplasm is water, the other fifth is protein, lipid, nucleic acid, carbohydrate, inorganic ions, and other low molecular weight compounds. The cytoplasm is the medium for enzyme activity and contains the ribosomes (see Section F8). Other subcellular structures that may be present are **inclusion bodies** (storage granules made of compounds such as the membrane-associated pol-β-hydroxybutyrate or cytoplasmic polyphosphate), **carboxysomes** (sites of CO_2 fixation) and **magnetosomes**. The magnetosome is found in magnetotactic bacteria such as *Magnetospirillum* and are thought to play some role in biomineralization.

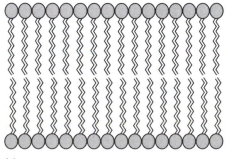

(a)

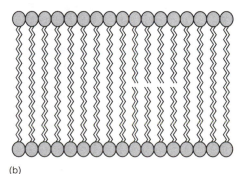

(b)

Fig. 2. Bacterial and Archaeal cell membranes. The bacterial membrane (a) is made up of a lipid bilayer, while the Archaeal membrane (b) has transmembrane phopsholipids as well as some regions of bilayer. The phospholipids of Archaea are also branched.

Although Bacteria and Archaea lack true organelles, some species have developed membranous configurations to aid some biochemical processes. Cyanobacteria, such as *Anabaena,* carry out oxygenic photosynthesis at **thykaloid membranes**, which are lined with **phycobilisomes**. The green sulfur bacteria, such as *Chlorobium*, perform anoxygenic photosynthesis in **chlorosomes**. Unlike the thykaloid membranes, these are separate from the cell membrane. Another method of anoxygenic photosynthesis uses spherical or lamellar systems attached to the cell membrane (e.g. purple Bacteria such as *Rhodopseudomonas*).

The major drawback in the study of the fine structure of prokaryotes is the method used to view such small objects. Electron microscopy (EM) has provided us with many of the details given here and in other texts, but the harsh methods used to fix and develop the samples is prone to the generation of artefacts. An example of this is the **mesosome**, a small vesicle seen in many EM preparations, often seen budding from the cell membrane. It is possible that these have one of the proposed roles in cell division, cell wall synthesis or chromosome resolution, but it seems more likely that mesosomes are created during the EM drying and fixation process.

Genomic material   The Bacterial chromosome(s) are often found in one area of the cell, normally close to the cytoplasmic membrane. It is tightly supercoiled and is associated with chromatin, proteins much like those found in eukaryotes but lacking basic histones. In contrast, it is thought that the chromatin of Archaea is very much like that of eukaryotic microorganisms. The tightly packed nuclear material is referred to as the **nucleoid** in prokaryotes, and is complemented by plasmids floating free in the cytoplasm. The size of some plasmids approaches that of chromosomes and the distinction between the two is sometimes esoteric. However, the chromosomal complement of a prokaryotic cell is always **haploid** and plasmid copy number is often more than one.

The periplasm   Gram negative Bacteria have a second phospholipid bilayer (the outer membrane) enclosing the cytoplasmic membrane. A **periplasm**, the space between the outer and cytoplasmic membranes, is thought to be aqueous but has much lower water content than the cytoplasm. An examination of the estimated total protein content suggests that the periplasm may even have a gel-like state. It is the site of several important reactions, taking a major role in the generation of ATP, environmental sensing, denitrification and many other metabolic pathways that interact directly with the electron transport chain (see Section E3). There has been continued debate about the existence of similar environments in Gram positive Bacteria, with recent suggestions that a gel-like stratum exists outside the Gram positive cell membrane, but is not bounded by a another membrane.

The cell wall   Prokaryotic cells are strong compared to eukaryotic single cells. The basis of their strength comes from the cell wall, which is present in all prokayotes apart from the Bacterium *Mycoplasma* and the Archaeal *Thermoplasma*. The Bacteria are covered in a semi-rigid compound called **peptidoglycan** (see Section C8) in most cases, with the exception of the *Chlamydia*. In these Bacteria the cell wall rigidity is provided by cysteine-rich membrane proteins interlinked with disulphide bonds. The Archaea have a far greater variety in the composition of the outer wall, but it has a Bacterial orthologue in **pseuodmurein**.

Endospores   A small number of Bacterial genera, most significantly *Clostridium* and *Bacillus*, are able to form spores within the cytoplasm in response to cell starvation.

These endospores have been shown to be able to survive desiccation for centuries, and their formation is discussed in detail in Section C9. The ability of endospores to survive heat, radiation and chemical disinfectants coupled to the pathogenic nature of some members of *Clostridium* and *Bacillus* has led to the efficient sterilization procedures (see Section C3) developed for bacteriology and medicine.

External features   All prokaryotes secrete a covering of polysaccharide or polypeptides sometimes known as the **slime capsule** (*Fig. 3*). The actual size and composition of this **glycocalyx** varies with the metabolic state of the organism. In many pathogens it is the first line of defense against host recognition. The slime capsule may also help the cell to resist drying out, and aid in cell attachment. The interaction of the glycocalyx with other bacteria from the same or different species helps to form biofilms such as plaque on teeth, and the bacterial filaments sometimes seen in aquatic environments. A more ordered structure is also present in some Bacteria and Archaea. A crystalline glycoprotein S-layer can form a mesh surrounding the cell, and again has been linked to attachment, biofilm formation and the repulsion of host defenses.

Bacterial motility is often provided by the rotation of **flagellae**, long filaments extending outside of the glycocalyx. These are highly complex structures, whose composition and role in chemotaxis is discussed further in Section C10. After suitable staining, these can often appear to be the dominant feature of cells such as those of *Salmonella* (*Fig. 3*). Prokaryotes are also coated with other finer hair-like appendages that are generally classed as **fimbriae**. There are currently seven different types known, identified on the basis of their protein composition, length, and diameter. They are often seen as broken filaments surrounding the cell in electron micrographs, suggesting they are brittle and dispensable. Some are pathogenicity factors, but it is thought that most function in attachment. Finer still are the fibrils which coat the surface of bacteria such as *Streptococcus*. Again the limitations of the electron microscope make it difficult to say whether these are a separate type of hairy coat or an artefact from the drying of the glycocalyx. The largest filamentous structures outside the cell are the sex pili generated by F^+ bacteria prior to transduction (see Section F12).

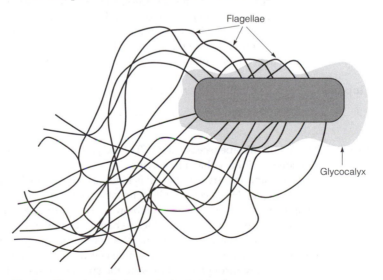

Fig. 3.   The external features of a bacterial cell.

C8 THE BACTERIAL CELL WALL

Key Notes

Structure of the cell wall

Gram positive Bacteria have a thick peptidoglycan layer outside the cell membrane to confer rigidity, whereas the Gram negatives have a thinner layer within the periplasm. The Gram negative outer membrane bounds the periplasm.

Peptidoglycan

Bacterial peptidoglycan is composed of a backbone of alternating sugars (N-acetylglucosamine and N-acetylmuramic acid) with a β1–4 linkage. This sugar polymer is crosslinked to similar molecules by an interbridge of amino acids and diaminopimelic acid. Lysozyme can break down peptidoglycan, which results in the internal osmotic pressure of the cell causing lysis. Teichoic and techuronic acids are dispersed throughout Gram positive peptidoglycan, and may have a function in ion binding. Not all Archaea have an equivalent to peptidoglycan, but those that do possess pseudopeptidoglycan.

The Gram stain

This staining procedure was invented by Christian Gram in around 1894, and has since been shown to have phylogenetic relevance. Cells stained with crystal violet and iodine are washed with alcohol, which decolorizes Gram negative cells. Counter staining with carbol fuschin gives a pale pink color to Gram negatives, while Gram positives retain the deeper purple of the crystal violet/iodine complex.

The Gram negative outer membrane

The outer Gram negative membrane is a lipid bilayer, but is modified to lipopolysaccharide (LPS). When purified from the rest of the cell this is sometimes called endotoxin due to its strong antigenic properties. The lipid part of LPS (Lipid A) anchors the structure to the outer membrane and is covalently linked to the core polysaccharide (R antigen or polysaccharide). The core is topped by the O antigen which is made of of 3 to 5 repeating sugars, depending on the species. The outer membrane is less selective than the cell membrane, with exit and entry points provided by porins.

Related topics

Identification of Bacteria (B2)

Structure of the cell wall

The cell membrane (see Section C7) of the prokaryotic cell alone is not sufficient to provide the rigidity that a free living organism requires. In Gram positive bacteria the shape and integrity of the cell is maintained by a thick single layer of **peptidoglycan** (*Fig. 1*). However, the Gram negative cell wall has evolved a greater level of complexity, with a second membrane outside the cell membrane to form a periplasm (see Section C7), in which a chemically similar but thinner peptidoglycan layer appears. The outer membrane is composed mainly of lipopolysaccharide. Both Gram negative and Gram positive cell walls have many proteins embedded in, penetrating through or attached to the surfaces of them, Some of these serve in

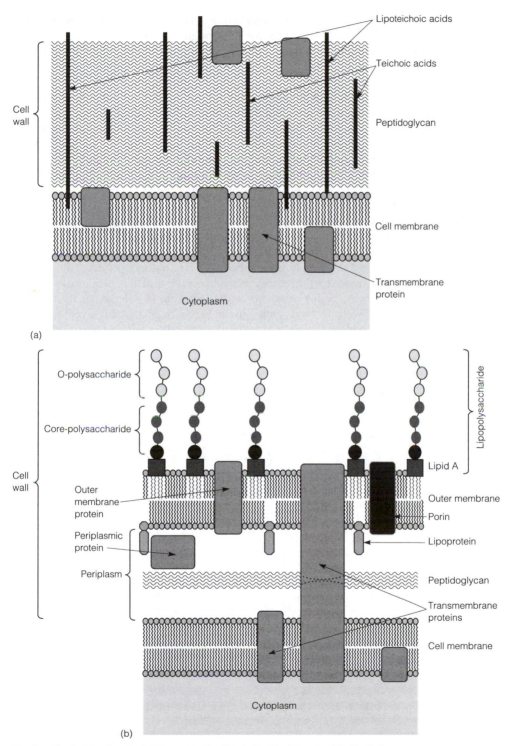

Fig. 1.   Bacterial cell walls. (a) Gram positive Bacteria; (b) Gram negative Bacteria.

the detection of environmental signals (see Section C10), growth substrates and secreted proteins to allow signals to pass through this thick outer layer.

The man differences between gram positive and Gram negative Bacterial cell walls are shown in Table 1.

Peptidoglycan

Peptidoglycan or **murein** is made up of two sugar derivatives, *N*-**acetylglucosamine** (NAG) and *N*-**acetylmuramic acid** (NAM) with a β1-4 linkage. Chains of alternating NAM and NAG are cross linked by amino acids such as L-alanine and *D*-glutamic acid as well as diaminopimelic acid (DAP). The way in which the cross links are formed differs in Gram negative and Gram positive Bacteria (*Fig. 2*). In most Gram negative Bacteria there is a direct **interbridge** between polypeptide side-chains emerging from adjacent NAG/NAM polymers, while in most Gram positive Bacteria the interbridge linking side-chains is a glycine pentapeptide. The structure of peptidoglycan is highly conserved among the bacteria as a whole, the only variations being slight changes in the interbridge. Peptidoglycan is resistant to many chemical challenges, but is easily broken down by **lysozyme**, which breaks the bonds between NAG and NAM. Without the constraining polymer, the osmotic potential is too much for the cell membrane to contain and the cell bursts. Lysozyme is present in many bodily secretions as the first form of defense against Bacterial invasion. Those cells without peptidoglycan (Archaea, *Mycoplasma*) avoid this lysis.

The Gram positive Bacteria have substances called **teichoic** and **teichuronic acids** interspersed with the peptidoglycan polymer. Teichoic is a broad term covering polymers containing glycerophosphate or ribitol phosphate, and the

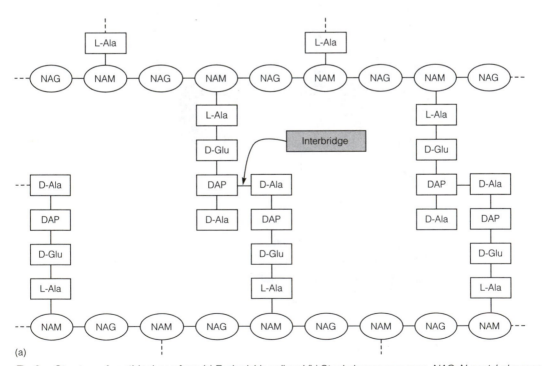

(a)

Fig. 2.   *Structure of peptidoglycan from (a) Escherichia coli and (b) Staphylococcus aureus. NAG: N-acetyl-glucosamine; NAM: N-acetylmuarmic acid; DAP: diaminopimelic acid. The remainder of the abbreviations refer to the D or L forms of the amino acids listed in Section E1.*

Continued

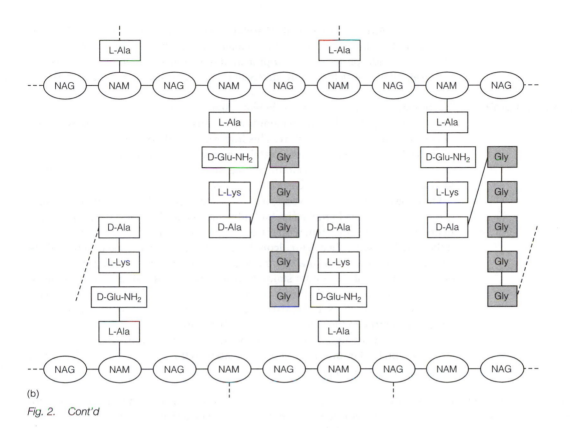

(b)

Fig. 2.   Cont'd

primary function appears to be to bind essential divalent cations such as Mg^{2+}, maintaining the local ionic environment of the cell. The effect of the teichoic acids overall is to give the cell a slightly negative charge.

Although our knowledge of the Archaeal cell wall is not as highly developed as that of Bacteria, we do know that some Archaea have a compound very similar to the murien, named **pseudopeptidoglycan**. This has a backbone containing NAG, but

Table 1.   Comparison of the cell walls of Gram positive and Gram negative Bacteria

Gram positive cell wall	Gram negative cell wall	Comments
Thick peptidoglycan layer	Thin peptidoglycan layer	Absent in
Polysaccharide	Lipopolysaccharide	
Teichoic acid	Absent	
Teichuronic acid	Absent	
Peptidoglycolipids	Absent	Found in *Corynebacterium, Mycobacterium, Nocardia*
Glycolipids	Absent	
Absent	Lipoprotein	
Absent	Phospholipid/phosphoprotein	
Absent	Porin	
Absent	Periplasm	

alternating NAM is replaced with *N*-acetyltalosaminuronic acid and is β1-3 linked instead of β1-4. However, many Archaea maintain rigidity with the use of a mixture of polysaccharide, protein and glycoprotein. It is extremely difficult to generalize on Archaeal cell wall structure, as the diversity is far higher than that of the Bacteria.

The Gram stain
The primary method of distinction between the two main groups of Bacteria (Gram positive and negative) is used in identification, classification, and taxonomy immediately after morphological study. This important method is based on the ability of the cell wall to retain or lose certain chemicals. The staining procedure was devised in 1884 by Christian Gram and has since proved to have first a biochemical and then a phylogenetic basis.

Once the cells have been fixed to a glass microscope slide (normally by heating gently over a Bunsen flame) a **crystal violet** stain is used to flood the preparation. In order to complex the stain with the cell wall, a solution of **iodine** is then added. If the slide is now washed with **alcohol**, the complexed crystal violet stain will be washed out of the thin-walled Gram negative cells, but not through the thicker Gram positive wall. A counter stain of **carbol fuschin** then stains all Gram negative cells a pale pink, while the Gram positive cells retain their deep violet color. These color differences can clearly be seen under a light microscope. Most bacteria react true to their phylogeny with this stain, with only a few species such as *Paracoccus* behaving abnormally. Some attempt has been made to classify the Archaea with the Gram stain, but unfortunately in this kingdom the variability in staining begins at a subgenus level.

The Gram negative outer membrane
In contrast to the Gram positive Bacteria, the model Bacterium *E. coli* does not present a coat of peptidoglycan to the outside world (*Fig. 1*). Instead the murein is suspended in the periplasmic space and there is a second outer membrane of **lipopolysaccharide** (LPS). Our detailed knowledge of the outer membrane structure of Gram negative Bacteria has been driven by the antigenic properties of LPS, so the details frequently presented about all Gram negatives are in reality more a representation of pathogenic enteric Bacteria such as *E. coli, Salmonella*, and so on. Thus for the general *Escherichia, Salmonella, Shigella, Pseudomonas, Neisseria*, and *Haemophilus*, the term LPS is interchangeable with '**endotoxin**'.

The outer membrane has some features in common with the cytoplasmic membrane in that it is a lipid bilayer and is considered to be a fluid mosaic. However, rather than being composed of phospholipid alone, there are many lipids with polysaccharide attached. This alters the chemical and physical properties of the membrane, in that it is much more porous to much larger molecules. This porosity is enhanced by the presence of many **porin** and transport proteins. The chemistry of this LPS membrane is complex and variable between species. In *Salmonella* the lipid part of the molecule (**lipid A**) is phosphorylated glucosamine linked to long carbon chains which, in contrast to membrane lipids, may be branched. Lipid A is covalently linked to the **core polysaccharide** (*Fig. 3*) but if separated from the rest of the molecule is frequently toxic to mammals, even if the bacterium of origin is non-pathogenic. Lipid A serves to anchor the LPS to the rest of the outer membrane.

The **core polysaccharide** (or R antigen or R polysaccharide) is the same in all members of a particular genus and is made up of heptose and hexose sugars plus ketodeoxyoctonate (KDO). Finally the **O-antigen** (or somatic antigen or O-polysaccharide) is made up of 3 to 5 repeating sugars. The composition of the repeat varies from species to species, and can be repeated from 1 to 40 times. This is the major antigenic determinant of pathogenic cells.

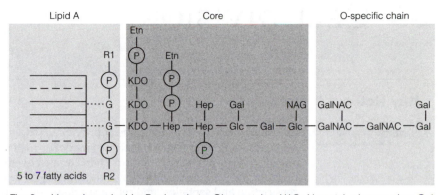

Fig. 3.　Lipopolysaccharide. P: phosphate; Glucosamine: NAG: N-acetyl- glucosamine; Gal: galactose; Glc; glucose; Hep: heptose; Etn: ethanolamine; R1 and R2: phoshoethanolamine or aminoarabinose.

C9 CELL DIVISION

Key Notes

Binary fission	The ability of many prokaryotes to approximately double in biomass, before dividing, means that it is easy to model growth mathematically. The four stages of binary fission are characterized by: growth and replication; FtsK ring construction and chromosome separation; FtsK ring constriction and septation; and cell separation. Cell division must be accompanied by the efficient dispersal of at least two complete genomes between mother and daughter cells, which is attained by replication and site specific recombination.
Sporulation	Gram positive Bacteria, such as *Bacillus subtilis*, differentiate to form resting cells within the cytoplasm known as endospores. At the beginning of spore development, DNA moves to the central axis of the cell and becomes tightly coiled, which is quickly followed by DNA separation and protoplast formation. The mother cell engulfs the protoplast to make a forespore, and coat synthesis begins. A primordial cortex forms while the mother cell cytoplasm becomes increasingly dehydrated. Cortex and spore coat synthesis is completed, and the mature spore becomes increasingly resistant to heat, chemicals, and radiation. Finally the spore is liberated, as what remains of the mother cell undergoes autolysis. When the spore reaches a suitable environment, the process is reversed with activation of the spore followed by germination only a few minutes later.
Related topics	Prokaryotic diversity (C2) The major prokaryotic groups (C6) Measurement of microbial growth (D1)

Binary fission

Most Bacteria and Archaea cells accumulate biomass until a critical volume is reached when they divide into two identical daughter cells. This division occurs in a single plane of symmetry and to some extent dictates the simplicity in cell shape (see Section C2) and the prevalence of cell numbers of $2n$ when cells appear in chains or clusters. The link of **binary fission** to cell growth calculations is discussed in Section D2. The process of cell division is outlined in *Fig. 1*, and can be divided into four steps: growth and replication; FtsK ring construction and chromosome separation; FtsK ring constriction and septation; and lastly cell separation. The process is poorly understood and involves many proteins and an extremely efficient control system. What is immediately apparent is the central role played by **FtsK**, a structural and functional analogue of the eukaryotic tubulins. FtsK was named after the filamentous temperature sensitive mutants that were first recognized in *E. coli*. At some temperatures they were unable to septate and became long, multinucleate filaments, a phenomenon seen in some wild type lactobacilli and in *Streptomyces*.

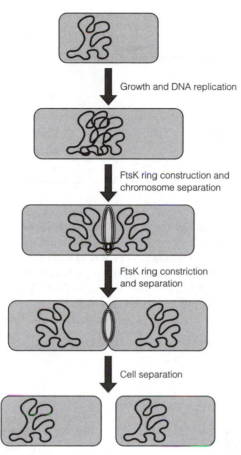

Fig. 1.   Bacterial cell division.

During the process of chromosomal replication, FtsK monomers begin to move to a point midway between the two ends of the cell. It is unclear how this happens, but inhibitors (the proteins MinC, D and E) have been identified at the poles preventing the initiation of unwanted sepatation. The FtsK proteins form a ring and it is certain that at this point other proteins are recruited to the locality to deal with the **site specific recombination** events (see Section F13) that must take place to allow the separation of the two replicated chromosomes. The formation of a septum begins with constriction of the FtsK ring across the width of the cell, and the septum forms completely and the daughter cells separate (*Fig. 1*). There are undoubtedly many other mechanisms governing the act of cell division, since the entire contents of the cell, including ribosomes, proteins, lipids, cell membrane, cell wall, and any extra-chromosomal material must also be divided between the two cells, as both daughter cells are immediately and fully functional. It is remarkable that this process is repeated every 40 minutes by *E. coli*. The universality of FtsK-like proteins across both the Bacteria and Archaea suggest that this is an evolutionarily conserved, fundamental mechanism for single-celled life.

Table 1.   Endospore and vegetative forms of Bacillus subtilis

Property	Vegetative cell	Spore
Heat, radiation and chemical resistance	Low	High
Metabolism	High	Low
mRNA content	High	Low
Lysozyme sensitivity	High	Zero
Calcium dipicolinate	Low	High
Water as % of volume	80	10–25
pH of soluble content	7	~ 5.7

Sporulation

The formation of endospores is a survival mechanism practiced by a Gram positive Bacterial genera (see Section C8), the best studied of which is *Bacillus subtilis*. It can be regarded as a specialized case of cell division, with some of the mechanisms being similar and performed by protein orthologues (for example, the sporulation protein SpoIIIE is very similar to FtsK). The spore and vegetative cell have very different properties, outlined in *Table 1*.

The process of spore formation can be divided into seven stages (*Fig. 2*):

1. *Beginning of spore development*. DNA moves to the central axis of the cell and becomes tightly coiled.

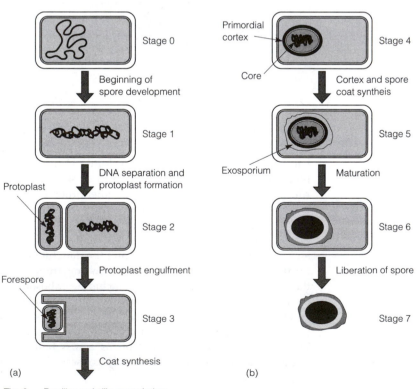

(a)   (b)

Fig. 2.   Bacillus subtilis sporulation.

2. *DNA separation and protoplast formation.* One complete chromosome becomes enclosed by the cell membrane and a **protoplast** is formed.
3. *Protoplast engulfment.* The mother cell engulfs the protoplast to make a **forespore**. The chromosome of what will become the spore is now surrounded by two membranes.
4. *Coat synthesis.* Wall material is deposited between the membranes and a **primordial cortex** forms around them. The material outside the cortex (the mother cell cytoplasm) becomes increasingly dehydrated.
5. *Cortex and spore coat synthesis.* Incorporation of calcium and dipicolinic acid.
6. *Maturation.* The spore becomes increasingly resistant to heat, chemicals and radiation. The cortical layers become distinct.
7. *Liberation of spore.* The mother cell undergoes autolysis.

More than 50 genes and their products are specifically involved in sporulation, all coordinated both spatially and temporally over about 6–7 hours. The reverse of the process, first **activation** and then **germination** takes only minutes.

C10 BACTERIAL FLAGELLA AND MOVEMENT

Key Notes

The flagellum	A single cell may have many flagellae spread all over the surface of the cell (peritrichous flagellae), the flagellae may be polar, found at one (monotrichous if single) or both ends (amphitrichous if only two in total). Lipotrichous flagellae grow in tufts from one position on the cell surface. All bacterial flagellae are made up of the protein flagellin.
The structure of the bacterial flagellum	The flagellar motor proteins are buried in the cytoplasmic membrane. The shaft of flagellin passes through the peptidoglycan via the P-ring and the outer membrane via the L-ring. Power for the motor is provided as part of the overall bioenergetics of the cell, with protons from the electron transport chain flowing through the motor. This flow is converted into rotation.
Movement with flagellae	Bacterial movement is either in the form of a run (movement in one direction) or a tumble (rotation in a fixed position). The location of the flagellae (peritrichous or polar) dictates which direction of rotation of the flagellae cause runs or tumbles.
Other types of movement	Bacteria are small enough to be influenced by Brownian motion and convection currents from heat sources. Other more active forms of motion exhibited by different types of prokaryotes include gliding, secretion of slime, and a ratchet system of cell membrane proteins. Spirochetes have a unique form of twisting motion, generated by the rotation of axial filaments or endoflagellae located between the cell membrane and the cell wall.
Related topics	The Bacterial cell wall (C8) Signal transduction and environment sensing (F9)

The flagellum

The most common means of locomotion in prokaryotes is via the rotation of flagellae. These act as propellers, allowing the cell to swim through liquid media. A single cell may have many flagellae spread all over its surface of the cell (**peritrichous flagellae**), the flagellae may be **polar**, found at one (**monotrichous** if single) or both ends (**amphitrichous** if only two in total). **Lipotrichous** flagellae grow in tufts from one position on the cell surface. Each flagellum is not straight but is helix shaped, with the length wavelength and amplitude of the helix varying from species to species. Although the presence or absence, the properties and position of the flagellae can be used as taxonomic characters, all bacterial flagellae are composed of the same protein: **flagellin**. However, much like cell wall composition, the situation in Archaea is very different, in that there are many different proteins in this kingdom that perform analogous functions.

The structure of the bacterial flagellum

The flagellin filament is only visible using the light microscope if it is complexed with another compound to make it thick enough. The filament is only 20 nm in diameter, but can be up to 20 μm in length. In Gram negative Bacteria, the **motor proteins** (providing rotation) are buried in the cytoplasmic membrane, with a drive shaft passing through the peptidoglycan in the periplasm via the **P ring** and the outer membrane via the **L ring**. The drive shaft is attached to the flagellin filament by means of a flexible hook (*Fig. 1*). Flagellae are biosynthesized from the MS ring upwards, with motor proteins, P-ring, L-ring and the hook sequentially added. The hook has a cap on top, and the flagellin filament elongates between these two substructures.

The flagellar motor is an adapted proton uniport (see Section E3). The cell must have an electron potential gradient across the membrane, and an ingress of protons through the motor cause the MS ring, and thus the flagellin, to rotate. This is an example of a simple **bioenergetic process** (see Section E3).

Movement with flagellae

Bacterial movement in solution can be divided into two main actions: the **run** is where the bacterium moves in a straight line towards an attractant. If the attractant

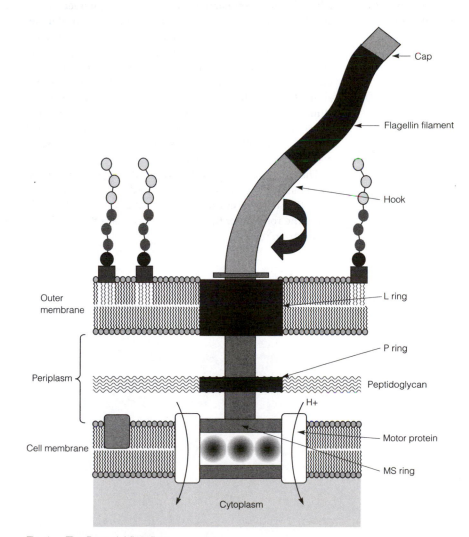

Fig. 1.   The Bacterial flagellum.

moves or disappears, then the cell **tumbles** (moves around randomly on the spot) before beginning another run. Random tumbling is the only form of steering that bacteria have at their disposal. The movement towards an attractant is discussed more in Section F9. Depending on how the filaments are attached to the cell, flagellae are used in three main ways to achieve locomotion.

- **Peritrichous**. Fagellae rotate anticlockwise and bundle together. Allows the bacterium to move forward (run). Clockwise rotation causes the flagellae to un-bundle, and pushes the bacterium in every direction simultaneously (tumble).
- **Polar**. If the Bacterium has a reversible motor for its flagellum, anticlockwise rotation causes a run. Clockwise rotation can lead to either a run in the reverse direction, or a tumble. If the flagellar motor can only rotate anticlockwise, then the cell can only reorientate itself by stopping and allowing Brownian motion to randomly knock it into a suitable position to move in another direction.

Other types of movement

All prokaryotes appear to move under the light microscope, as they are sufficiently small to be shaken by random bombardment of local molecules, in addition to the convection currents generated by the light source. Other active forms of movement do exist and, although flagellae are widespread throughout the Bacterial and Archaeal kingdoms, other forms of locomotion have evolved. Many Bacteria have a **gliding motility**, which allows flagellae-free bacteria to move rapidly across surfaces. In the cyanobacteria, this gliding is achieved by the secretion of slime, although the exact way in which this allows the Bacterium to move is poorly understood. *Flavobacterium johnsoniae* also glides, but appears to achieve this by using cell wall proteins to grab onto the surface and haul the Bacterium along a few nanometres at a time.

The spirochetes exploit their helical morphology to generate a unique form of movement, dependent on internalized flagellae. These corkscrew-like cells have **axial filaments** or **endoflagellae** located between the cell membrane and the cell wall. The rotation of these structures causes the whole cell to wriggle and rotate, and motion results.

C11 PROKARYOTES AND THEIR ENVIRONMENT

Key Notes

Prokaryotic niche diversity	Prokaryotes live in a wide variety of habitats at temperatures of around freezing to a maximum of 115°C. Their ability to use all naturally occurring carbon compounds as growth substrates allows them to grow in most places on earth. Despite their size, they are responsible for the cycling of many millions of tonnes of carbon, sulfur and nitrogen through the atmosphere annually.
Cycling of elements through the biosphere	The biogeochemical cycling of elements, such as carbon, sulfur and nitrogen, is mainly carried out by the Bacteria and Archaea, though the human population is making an increasing contribution. The turnover of elements allows the earth to function as a self-regulating entity, as proposed in Lovelock's Gaia hypothesis.
Detection of Bacteria in their natural habitats	For many years bacteriologists relied on isolation of laboratory cultures as a method of determining which Bacteria were present in a particular biotope. These methods are still used to some extent, but are now complemented by molecular methods, particularly fluorescence *in situ* hybridization (FISH) and what has recently been termed environmental genomics, (essentially PCR-based methods).
Bacterial commensualism	The Bacteria and Archaea almost always grow with other species (including higher plants and animals). The majority are very difficult to grow in pure culture in the laboratory. Bacteria and Archaea grow in communities in the environment, with the individual strains and species exchanging metabolites (often in competition with their peers) so that most organic compounds are ultimately converted into biomass, carbon dioxide and water.
Related topics	Prokaryotes and their environment (C11) Hybridization (F4) Manipulation of Cellular DNA and RNA (F16)

Prokaryotic niche diversity

Bacteria and Archaea can grow in low temperature environments (for example, the Antarctic bacterium *Flavobacterium frigidarium* isolated from marine sediments can grow at 4°C, as can an archaeon isolated nearby, *Methanococcoides burtonii*) as well as high temperature ones. The highest known temperature allowing growth of a Bacterium is 95°C (*Aquifex pyrophilus*, isolated from an Icelandic marine thermal vent), while the archaeon *Pyrolobus fumarii* will grow at 113°C. Bacteria and Archaea are found in most places in the biosphere. They have become adapted to use a wide variety of compounds as sources of energy and carbon, and may be adapted to use those anthropogenically produced chemical compounds as well. The range of metabolic diversity means that microorganisms dominate nutrient cycling processes in almost every ecosystem. This role in the cycling of elements has led Bacteria and Archaea to form very close associations with their peers as well as with plants and animals. The close nature of their commensualism, parasitism, and symbiosis mean that pure cultures of many Bacteria and Archaea are extremely difficult to grow in the laboratory. These viable but non-culturable (VBNC) microorganisms can only be detected in biotopes using visualization techniques such as **fluorescence *in-situ* hybridization** (FISH) or indirect PCR-based methods now known as **environmental genomics**.

Cycling of elements through the biosphere

James Lovelock's Gaia hypothesis portrays the earth as a self-regulating entity. The basis of this regulation is the involvement in the movement of millions of tonnes of chemical elements such as carbon, nitrogen and sulfur through the atmosphere, marine, and terrestrial environments. The interconversion between solid and gaseous compounds of the elements is mostly accomplished by microorganisms, particularly the Archaea and Bacteria. The coupling of spontaneous chemical reactions with those catalyzed by microorganisms has given the study of these earth-scale transformations the name of **biogeochemistry**. An overall scheme showing the involvement of organisms in **biogeochemical cycling** is shown in *Fig. 1*.

Sulfur cycle

The biogeochemical cycling of sulfur is extremely complex, and cannot be easily described diagrammatically. The complexity of the cycle is dictated by the number of relatively high number of oxidation states that sulfur can have in solution and in the presence of oxygen. The most significant carbon compound in the atmosphere is probably carbon dioxide, serving as a link between terrestrial and marine environments, but there is no direct sulfur equivalent. In prioritizing volatile atmospheric sulfur compounds, sulfur dioxide (SO_2), hydrogen sulphide (H_2S), carbonyl sulphide (COS), carbon disulphide CS_2), dimethyl sulphide (CH_3SCH_3), dimethyl disulphide (CH_3SSCH_3) and methane thiol (CH_3SH) may be significant, depending on the geographical location examined. The nature of sulfur's chemistry may mean that non-volatile molecules, such as methane sulphonate ($CH_3SO_3^-$) and sulphate (SO_4^-) may also be present in quantity in the atmosphere as dissolved aqueous ions.

The biogeochemical cycling of sulfur serves to illustrate regulation on a global scale. The sulfur compound **dimethylsulphonium propionate** (DMSP) is produced

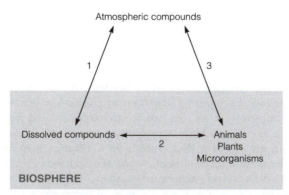

Fig. 1.   *Biogeochemical cycling. Transformation1 includes precipitation, occlusion at air/water interfaces; Transformation 2 includes growth, fixation, secretion and decay; Transformation 3 includes decay, respiration and combustion.*

intracellularly by marine photosynthetic algae. During and after algal blooms in the open oceans, this DMSP is released into solution on cell death, where it is metabolized by marine microorganisms to produce **dimethyl sulphide** (DMS). The generation of DMS is sufficiently high for some of the compound to enter the atmosphere as a gas above the ocean, where **photooxidation** by the sun's rays lead to the breakdown of DMS into two solid forms of sulfur; sulphate, and methane sulphonate. These solids act as nuclei for the condensation of water, and clouds are formed. The clouds reduce the sunlight reaching the ocean surface, and the growth of the photosynthetic algae that produced the DMSP in the first place is reduced. This regulatory cycle stops algae covering the entire open ocean, though only in regions where man has not upset this cycle by the dumping of sewerage or other compounds that the algae might feed on.

Sulfur cycling also occurs through terrestrial ecosystems, where a combination of **sulphate reducing Bacteria** (such as *Desulfobacter* and *Desulfovibrio* species), and **sulphate oxidizing Bacteria** (such as *Thiobacillus* species) interconvert elemental sulfur, hydrogen sulphide, sulphate, thiosulphate, and polythionates. Sulfur is also assimilated by all organisms into the amino acids cysteine and methionine and so ultimately into proteins. Sulphur is also assimilated for the prosthetic protein groups known as iron-sulfur clusters.

Carbon cycle

Man's geologically recent release of greenhouse gases such as carbon dioxide (CO_2) into the atmosphere are generally held to be responsible for global warming. In the absence of man, carbon dioxide serves as an atmospheric link between carbon released in the marine and terrestrial environments, with methane playing a secondary role. Outside the atmosphere, carbon forms a wide range of compounds, from simple metal cyanides to complex macromolecules such as starch, lignin, cellulose, and nucleic acids. Microorganisms take part in the ultimate conversion of all of these into carbon dioxide or methane, and these processes may be broadly divided into the aerobic and anaerobic (*Fig. 2a*). Carbon compounds are broken down aerobically either by respiration or anaerobically by fermentation. The release of carbon dioxide is balanced by the fixation of CO_2 during photosynthesis by plants and microorganisms, or by the aerobic chemolithotrophic growth of prokaryotes such as *Paracoccus denitrificans*.

Nitrogen cycle

The nitrogen cycle is the easiest to define in terms of the number of participating major nitrogenous compounds. Proteins, prosthetic groups such as chlorophyl,

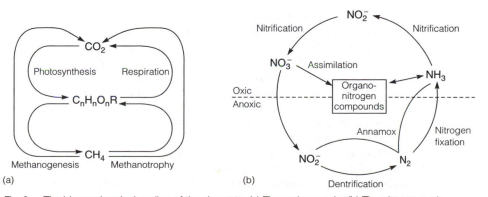

(a)　　　　　　　　　　　　　　　　　　(b)

Fig. 2.　The biogeochemical cycling of the elements. (a) The carbon cycle; (b) The nitrogen cycle.

nucleic acids, and other macromolecules can be regarded as fixed nitrogen, present in the marine or terrestrial environments. Ammonia, nitrate, and nitrite ions may participate either in the recycling of fixed nitrogen, or in the generation of atmospheric molecular nitrogen (*Fig. 2b*). The bulk of molecular nitrogen is reduced by Bacterial nitrogen fixation, while very small amounts are reduced by lightning strikes. The biologically mediated processes have been called **mineralization, nitrification, denitrification, nitrogen fixation,** and **nitrate reduction**.

Mineralization of fixed nitrogen is performed by many microorganisms during the degradation of organic matter, and results in the generation of free ammonia. This ammonia may be reassimilated, or converted into nitrate by the process known as nitrification. Well characterized nitrifiers include *Nitrosomonas europea* and is a two-step process via nitrite (*Fig. 2b*).

The conversion of nitrate to molecular nitrogen (denitrification) is performed by a taxonomically undefined group of organisms including the Bacteria *Paracoccus denitrificans* and *Pseudomonas aeruginosa*, as well as the Archaea *Haloarcula marismortui* and *Pyrobaculum aerophilum*. Recent evidence suggests that some fungi can also denitrify. Nitrate is converted sequentially into nitrite, nitric oxide, and nitrous oxide before nitrogen is released from the organism. Each stage of this process is energetically beneficial to the organism, and is generally coupled to the electron transport chain. For this reason it is also called **nitrate respiration**.

Both Archaea such as *Methanosarcina barkeri* and Bacteria such as *Rhizobium* species are also capable of nitrogen fixation or **diazotrophy**. However, the rhizobia (bacteria classified as *Rhizobium* and *Bradyrhizobium*) have the unique ability to form nodules on the roots of certain plants, allowing nitrogen to be fixed almost directly from atmosphere to plant. The majority of diazotrophs are free living, and employ various methods to protect the obligately anaerobic enzyme nitrogenase from the effects of molecular oxygen.

Detection of bacteria in their natural habitats

The isolation and laboratory cultivation of a particular Bacterium from an epitope was at one time the sole evidence that that species was present. However, many studies have shown that there are many more prokaryotes in the environment than can be cultured using laboratory media, and nucleic acid analysis has complemented classical microbial ecology. Individual cells can be labeled with a dye attached to an oligonucleotide which hybridizes to ribosomal RNA. The organisms which bind the oligonucleotide will then show up under a fluorescence microscope. Careful choice of oligonucleotide can mean that these fluorescent probes can be species or even strain specific, allowing the microbial ecologist to examine and even count various sorts of microorganism in a particular niche. Various developments of this FISH technique, have allowed the differentiation between metabolically active and inactive microorganisms.

The development of PCR means that the total ribosomal RNA (RNA should only be produced by active cells) in a microbial community can be examined. By sequencing a library of ribosomal cDNA (complementary DNA) from a particular epitope, the ecologist has some idea of the diversity of active microorganisms. Extending this idea of amplification of DNA to other genes has developed into the field of environmental genomics.

Both FISH and PCR of total genomic DNA suggest that there is a far greater diversity of Bacteria and Archaea than first thought, to the extent that the idea of discreet species barriers is beginning to break down. Although these new organisms are often held to be viable but non-culturable (VBNC), new techniques and laboratory practices are finally yielding pure and mixed cultures suitable for physiological and biochemical studies.

Bacterial commensualism and microbial communities

Bacteria are classically studied as pure cultures in the laboratory. These cultures are regarded as clonal and genetically stable. However, the situation in the environment is somewhat different: Bacteria and Archaea rarely grow as pure cultures, more often forming associations with organisms around them. These associations can range from the symbiotic (*Table 1*) to the pathogenic. The diversity of life in a particular niche often makes **functional ecology** (the study of which organisms are responsible for which activities) challenging. Many ecosystems have evolved to include many Bacteria and Archaea exchanging and competing for metabolic intermediates to a state where no single chemical compound is completely metabolized by a single organism.

Table 1. Examples of associations between bacteria and higher organisms

Bacterial/Archaeal species	Partner	Description
Rhizobium leguminosarum	Peas and other legumes	Allows plant to fix nitrogen directly from the atmosphere
Various methanogenic Archaea	Ruminant animals	The presence of Archaea in the gut results in the digestion of the cellulose content of plants
Symbiodinium species	Various coral species	Photosynthetic metabolism of the bacteria provide nutrients for the coral
Aeromonas veronii	Various leech species	Allows the digestion of blood
Vibrio fischeri	Squid	Confers a bioluminescent property on the squid

D1 MEASUREMENT OF MICROBIAL GROWTH

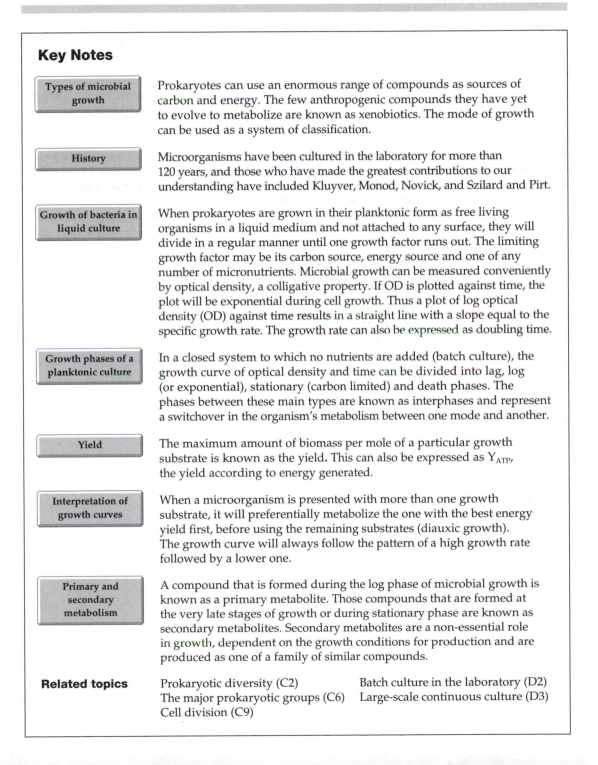

Key Notes

Types of microbial growth

Prokaryotes can use an enormous range of compounds as sources of carbon and energy. The few anthropogenic compounds they have yet to evolve to metabolize are known as xenobiotics. The mode of growth can be used as a system of classification.

History

Microorganisms have been cultured in the laboratory for more than 120 years, and those who have made the greatest contributions to our understanding have included Kluyver, Monod, Novick, and Szilard and Pirt.

Growth of bacteria in liquid culture

When prokaryotes are grown in their planktonic form as free living organisms in a liquid medium and not attached to any surface, they will divide in a regular manner until one growth factor runs out. The limiting growth factor may be its carbon source, energy source and one of any number of micronutrients. Microbial growth can be measured conveniently by optical density, a colligative property. If OD is plotted against time, the plot will be exponential during cell growth. Thus a plot of log optical density (OD) against time results in a straight line with a slope equal to the specific growth rate. The growth rate can also be expressed as doubling time.

Growth phases of a planktonic culture

In a closed system to which no nutrients are added (batch culture), the growth curve of optical density and time can be divided into lag, log (or exponential), stationary (carbon limited) and death phases. The phases between these main types are known as interphases and represent a switchover in the organism's metabolism between one mode and another.

Yield

The maximum amount of biomass per mole of a particular growth substrate is known as the yield. This can also be expressed as Y_{ATP}, the yield according to energy generated.

Interpretation of growth curves

When a microorganism is presented with more than one growth substrate, it will preferentially metabolize the one with the best energy yield first, before using the remaining substrates (diauxic growth). The growth curve will always follow the pattern of a high growth rate followed by a lower one.

Primary and secondary metabolism

A compound that is formed during the log phase of microbial growth is known as a primary metabolite. Those compounds that are formed at the very late stages of growth or during stationary phase are known as secondary metabolites. Secondary metabolites are a non-essential role in growth, dependent on the growth conditions for production and are produced as one of a family of similar compounds.

Related topics

Prokaryotic diversity (C2)
The major prokaryotic groups (C6)
Cell division (C9)

Batch culture in the laboratory (D2)
Large-scale continuous culture (D3)

Table 1.   Classification of microorganisms according to carbon and energy sources

Carbon or energy source	Description
Energy from a chemical	Chemotroph
Energy from light	Phototroph
Electrons from an organic compound	Organotroph
Electrons from an inorganic compound	Lithotroph
Carbon from carbon dioxide	Autotroph
Carbon from an organic compound	Heterotroph

Types of microbial growth

Microbes can convert an enormous range of compounds into biomass from foodstuffs humans would consider eating, right the way through to compounds, such as cyanides, that are considered highly toxic. Some man-made chemical compounds are new to the natural environment, so microbes have yet to have completed evolving pathways to deal with them. These compounds are called **xenobiotics**, but eventually microbes will evolve systems to metabolize these compounds too.

We can give a very broad classification to microbes based on how they grow (*Table 1*) but also at which temperature and pH they do it (*Fig. 1*). These terms can be combined, so *Thermus aquaticus* can be described as a thermophilic heterotroph, or *Paracoccus versutus* could be described as a mesophilic chemolitho-heterotroph (Section C2). Many other terms are used to describe the overall modes of growth of organisms (*Table 2*) and these are frequently used to describe whole populations as well as individual species.

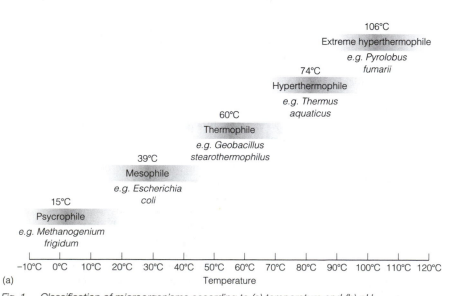

(a)

Fig. 1.   Classification of microorganisms according to (a) temperature and (b) pH.

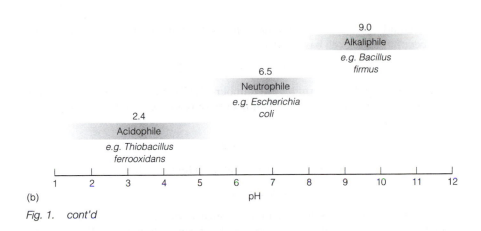

(b)

Fig. 1.   cont'd

History

Although microbes have been cultivated in pure and mixed cultures for more than 120 years, a mathematical interpretation of microbial growth is a relatively recent innovation. A summary of the timeline is shown below:

- **Kluyver (~1930)** Shake flask technique
- **Monod (1942)** Bacterial growth can be formulated in terms of growth yield, specific growth rate and the concentration of the growth limiting substrate.
- **Monod (1950); Novick and Szilard (1950)** Mathematics of continuous flow cultures
- **Pirt (1975)** Publishes seminal book *Principles of Microbe and Cell Cultivation*, summarizing the previous 40 years' work and developing these themes.

Growth of Bacteria in liquid culture

When Bacteria are grown in their **planktonic** form as free living organisms in a liquid medium and not attached to any surface, they will divide in a regular manner until one growth factor runs out. The **limiting growth factor** may be the carbon source, energy source or one of any number of micronutrients. As Bacteria (and Archaea) grow by a process of binary fission (see Section C9), at the end of a complete round of cell division there will be twice as many cells compared to the beginning of growth. If we consider a single cell, after division, this becomes two cells, those two cells become four and so on, in other words

$$1 \rightarrow 2 \rightarrow 4 \rightarrow 8 \rightarrow 16 \rightarrow 32\ldots$$

Table 2.   Other terms used to describe the growth of microorganisms

Growth property	Description
Obtain nitrogen from N_2 gas rather than 'fixed' nitrogen sources	Diazotroph
Ability to exist at low nutrient concentrations	Oligotroph
Uses reduced C1 compounds as sole source of carbon and energy	Methylotroph
Uses reduced methane as sole source of carbon and energy	Methanotroph
Able to use a mixture of heterotrophic and autotrophic growth mechanisms	Mixotroph
Only able to grow in the presence of salt	Halophile

This could be expressed mathematically as 2 to the power of n, where n is the number of times the cell has divided.

$$2^1 \rightarrow 2^2 \rightarrow 2^3 \rightarrow 2^4 \rightarrow 2^5 \rightarrow 2^6 \ldots$$

Using a method to determine the number of cells at any time (N^o) during the growth of the culture, and then the number of cells after n rounds of cell division, this becomes:

$$N = N_0 2^n$$

The equation can be rearranged to be more useful in the lab to:

$$n = 3.3(\log N - \log N_0)$$

This is not such a useful equation as such because it is difficult to count micron-sized Bacteria into a pot. Fortunately, it is easy to measure the numbers of cells in solution using a spectrophotometer. The spectrophotometer is normally set to read light at wavelengths between 500 and 600 nm – the actual figure used depending on the species being examined. As the spectrophotometer measures the amount of light scattering in a solution (the **optical density**, the more the scattering the higher the **OD**), the readings obtained reflect the number of bacteria in solution. If the OD/number of cells is plotted against time, a graph is produced similar to the one in *Fig. 2*.

A curve using the primary cell number data is **exponential** for most of the curve, reflecting $N = N_0 2^n$. Exponential curves are hard to compare, so to make better sense of how the organism is growing a more useful plot is the log OD versus time. This now means that the most of the curve is a straight line, the slope of which is the **specific growth rate, μ**. So for an organism grown under defined conditions (temperature, growth substrate, etc.), the specific growth rate can be defined as:

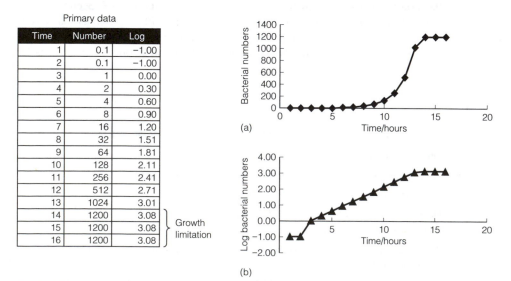

Time	Number	Log
1	0.1	−1.00
2	0.1	−1.00
3	1	0.00
4	2	0.30
5	4	0.60
6	8	0.90
7	16	1.20
8	32	1.51
9	64	1.81
10	128	2.11
11	256	2.41
12	512	2.71
13	1024	3.01
14	1200	3.08
15	1200	3.08
16	1200	3.08

Primary data

Growth limitation

Fig. 2.   The relationship between cell number and time is exponential.

$$\mu = \frac{(\log N - \log N_0)}{t} \quad \text{or}$$

$$\mu = \frac{(\log OD - \log OD)}{t}$$

It is not always convenient to plot a complete graph of the growth of an organism, so growth rate is sometimes expressed as **doubling time**. The doubling time (t_d) of a culture is the time it takes for the optical density to double, i.e. the cell numbers to multiply by two.

$$t_d = t_{2N} - t_0$$

The doubling time of *Escherichia coli* is around 20 minutes when it is growing under optimal conditions, but the doubling time of some organisms can be relatively large – the planctomycete used in the Anammox process (*Brocadia anammoxidans*) has a doubling time in excess of three weeks (see Section C6).

Since OD is related to the number of particles in solution but not directly to concentration, it is thus a **colligative property**. If two species have different cell volumes, this means that for the same OD reading the grammes biomass per ml may be different. It is possible to make a direct estimation of biomass either by taking samples and measuring the dry weight of the culture or measuring total protein or total carbon. Although these methods are accurate, none of them are rapid, so a compromise is to establish the relationship between OD and dry weight once, then extrapolate to other OD readings.

Growth phases of a planktonic culture

With the data that can be derived from optical density measurements, it is now possible to draw a complete **growth curve** of a bacterium relating biomass to the changes in the state of the culture. If no new medium or medium components are added to the container, this **batch culture** will give us information over the maximum amount of biomass that can be made using the medium components, the specific growth rate and other information.

The growth curve of a bacterium growing in a shake-flask batch culture will have the form illustrated in *Fig. 3*.

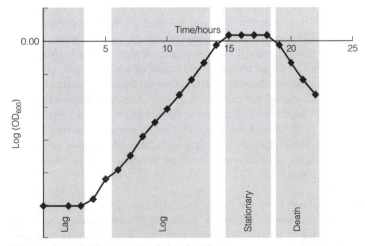

Fig. 3.   *Growth of a bacterium in batch culture.*

The phases of growth are can be divided into the **lag**, **log** (or **exponential**), **stationary**, and **death** phases. The phases in between these main types are known as **interphases** and represent a switchover in the organism's metabolism between one mode and another. If time = 0 is the point at which the culture was inoculated, the shape of the growth curve can be explained.

- **Lag phase** During this period the organism adapts to the medium into which it has been introduced. If it was grown up in a medium containing a different carbon source, of a less complex type, or grown at a different temperature, the lag phase can be long. The cells in the inoculum need to switch on new sets of genes and express new proteins to cope with the change in environment.
- **Log phase** Once all the necessary proteins are available, the cells will start dividing as quickly as they can. It is only during this phase of **exponential growth** that the equations relating to specific growth are applicable.
- **Stationary phase** Log phase cells will continue to divide, but eventually one component of the medium will run out. Media are normally designed to be **carbon limited** (i.e. the first thing to stop growth is the absence of a carbon source such as glucose or succinate), but cultures may become limited by nitrogen, phosphorous, trace elements or essential amino acids. The cells are still metabolically active but do not have the resources to divide. This stationary phase seen in laboratory cultures is indicative of the state of organisms in oligotrophic environments.
- **Death phase** After some time metabolism stops and the cells begin to die and lyse. The rate of lysis is constant over time. Optical density readings in this phase become unreliable as during lysis the cell debris now contributes to light scattering. Occasionally this is manifested in a slight increase in optical density as the number of particles in solution increases.

Yield

The amount of growth that can be obtained from a particular compound is of great interest to microbiologists, especially when the production of the maximum amount of biomass is required for the minimum cost. Substrates, such as pure glucose, are suitable for use on a small scale, but when tens or hundreds of litres of medium are to be made up the most efficient use of resources must be employed. Yield can be calculated simply by growing the same organism under the same conditions but with two different carbon and energy sources. The maximum OD at the end of each experiment signifies the best yield. However, this comparison is only valid if there is some way of normalizing the amount of carbon source added. An industrial microbiologist might look at yield in terms of grammes of biomass per unit of currency spent on the medium. Biochemically, it would be more suitable to look at grammes biomass produced per mole of carbon. If enough is known about the biochemical route by which the carbon source is utilized, it would be possible to estimate how much ATP should be generated or consumed during metabolism – for example, in comparison on the metabolism of glucose and lactose (glucose will yield more ATP). This gives the concept of Y_{ATP}, the yield according to energy generated.

Interpretation of growth curves

Apart from the times at which the various phases of growth occur and comparing the rates of growth, we can also derive other information from growth curves. If the organism is given more than one substrate to grow on, we can see which substrate it uses preferentially. Most substrates are used sequentially by Bacteria

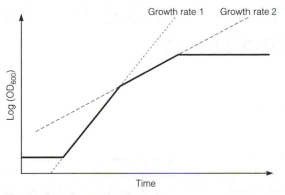

Fig. 4.   Diauxic growth of bacteria.

and not at the same time. For example if *Escherichia coli* is grown in a minimal medium containing glucose and lactose, it will use up all the glucose first (growth rate 1 in *Fig. 4*) before switching on the *lac* operon (see Section F6) and using up the lactose (growth rate 2 in *Fig. 4*). This can be explained in terms of the energetics of the compounds: the bacterium derives more ATP from the metabolism of glucose than lactose, so has a faster growth rate and uses this compound preferentially.

From this it can be see that **diauxic growth** always follows the pattern of a high growth rate followed by a lower one.

Primary and secondary metabolism

A compound that is formed during the log phase of microbial growth is known as a **primary metabolite**. Those compounds that are formed at the very late stages of growth or during stationary phase are known as secondary metabolites. As these two definitions overlap, there is some confusion over how some microbial products might be classified, but a clear primary metabolite is alcohol from *Saccharomyces cereviseae*. Here the production of alcohol is intrinsically linked to energy production, so accumulation of alcohol is parallel with the appearance of biomass.

Many industrially significant bioproducts (for example almost all non-recombinant antibiotics) are secondary rather than primary metabolites. To enhance the differences between these and the primary metabolites, secondary metabolites are also characterized as:

- Having a non-essential role in growth.
- Being dependent on the growth conditions for production.
- Being produced as one of a family of similar compounds (e.g. complex heterocycles differing in methylation).

The biochemical pathways of secondary metabolite synthesis tend to be longer than primary. Erythromycin is made by the action of 25 separate enzymes and tetracycline by more than 70.

D2 BATCH CULTURE IN THE LABORATORY

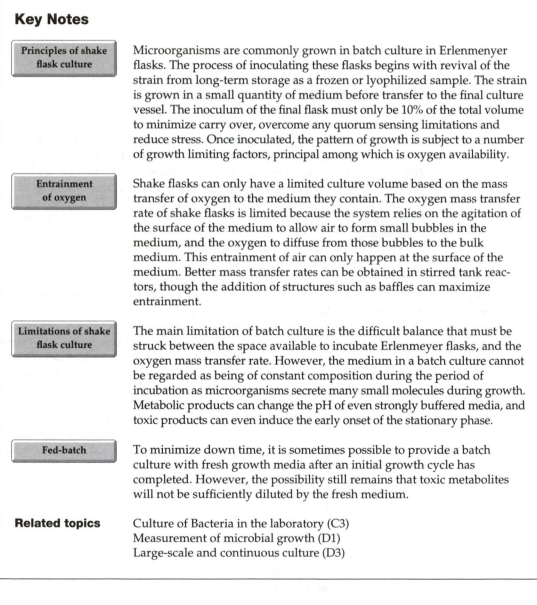

Key Notes

Principles of shake flask culture

Microorganisms are commonly grown in batch culture in Erlenmenyer flasks. The process of inoculating these flasks begins with revival of the strain from long-term storage as a frozen or lyophilized sample. The strain is grown in a small quantity of medium before transfer to the final culture vessel. The inoculum of the final flask must only be 10% of the total volume to minimize carry over, overcome any quorum sensing limitations and reduce stress. Once inoculated, the pattern of growth is subject to a number of growth limiting factors, principal among which is oxygen availability.

Entrainment of oxygen

Shake flasks can only have a limited culture volume based on the mass transfer of oxygen to the medium they contain. The oxygen mass transfer rate of shake flasks is limited because the system relies on the agitation of the surface of the medium to allow air to form small bubbles in the medium, and the oxygen to diffuse from those bubbles to the bulk medium. This entrainment of air can only happen at the surface of the medium. Better mass transfer rates can be obtained in stirred tank reactors, though the addition of structures such as baffles can maximize entrainment.

Limitations of shake flask culture

The main limitation of batch culture is the difficult balance that must be struck between the space available to incubate Erlenmeyer flasks, and the oxygen mass transfer rate. However, the medium in a batch culture cannot be regarded as being of constant composition during the period of incubation as microorganisms secrete many small molecules during growth. Metabolic products can change the pH of even strongly buffered media, and toxic products can even induce the early onset of the stationary phase.

Fed-batch

To minimize down time, it is sometimes possible to provide a batch culture with fresh growth media after an initial growth cycle has completed. However, the possibility still remains that toxic metabolites will not be sufficiently diluted by the fresh medium.

Related topics

Culture of Bacteria in the laboratory (C3)
Measurement of microbial growth (D1)
Large-scale and continuous culture (D3)

Principles of shake flask culture

The most common method of growing planktonic bacteria (i.e. those suspended in solution rather than attached to surfaces) in the laboratory is the shake or **Erlenmeyer** flask (see *Fig. 1*). This type of culture is referred to as a **batch culture** as all the components are used only for a single cycle of growth, in contrast to **continuous culture** (see Section D3) and fed-batch cultures (see below).

The inoculation process for batch cultures is not just a question of adding a few cells to the medium. Frequently the bacterial strain must be **revived** from storage, either from a **lyophilized** (freeze-dried) or frozen state. Bacterial and Archaeal cells will withstand freezing at −70°C for many years, provided the solution they are stored in contains a compound that prevents the formation of ice crystals. Crystals of ice are thought to puncture the cell membrane, rendering the frozen culture useless. However, a solution of 50% v/v glycerol or 20–50% DMSO (dimethyl sulphoxide) can prevent this from happening. To bring the cells back from their frozen state, a rich undefined medium (see Section C3) is used. This often takes the form of an agar streak, as this can simultaneously be used to check for the purity of the stored cells. An individual colony is then picked from the revival medium and inoculated into between 1 and 10 ml of the liquid medium to be used in the experiment. This first growth in a small-scale batch culture helps the strain to adapt to the medium conditions as well as to generate sufficient biomass for the next inoculation.

The 10 ml inoculum is then used to inoculate a larger, Erlenmeyer flask. Many microorganisms seem not to grow well unless the inoculum size is between 1 and 10% of the final experimental volume. The reasons for this are obscure and vary from culture to culture but may be related to one or more of the following factors:

- **Carry over** of an essential nutrient from the inoculum medium to the experimental flask.
- Some form of **quorum sensing**. Many Bacteria have mechanisms to detect the numbers of the same species in their immediate vicinity.
- Reduction in stress. On the macro level there would appear to be little difference between adding 1 ml inoculum to 100 ml of medium and adding 10 ml of inoculum to the same volume. On the microscopic scale, the inoculum and fresh medium are not perfectly and immediately mixed, so a larger inoculum may briefly form a gradient between established and new conditions, giving the individual cells slightly longer to adapt.

Once inoculated, the Erlenmeyer flask can be placed on a rotary shaker, held at constant temperature, and growth of the organisms can begin. How the organisms grows is subject to a number of **growth limiting factors**. In batch culture with Erlenmeyer flasks, the greatest limiting factor is always the concentration of oxygen, which will frequently dictate if the organism grows at all.

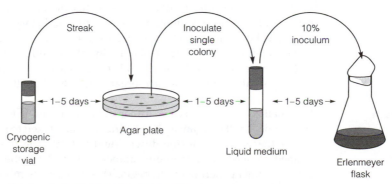

Fig. 1. A simple laboratory batch culture.

Entrainment of oxygen

Shake flasks have a limited culture volume based on the transfer of oxygen to the medium they contain. The concept of how much of a substance it is possible to move from one phase to another (in this case gaseous to aqueous) in a system is known as **mass transfer**. By trial and error it has been established that an aerobic organism will only grow optimally if the Erlenmeyer flask has a volume ten times that of the medium it contains. Thus a 250 ml Erlenmeyer flask should ideally contain only 25 ml of medium. When growing *Escherichia coli*, this volume may be raised to as much as 50 ml, but this is only possible as *E. coli* is facultatively anaerobic. Practically, the maximum size of standard Erlenmeyer flask that can be used is 5 litres, so if culture volumes of more than 500 ml are required, alternative methods such as a simple **stirred tank reactor** (see Section D3) must be used.

The oxygen mass transfer rate of shake flasks is limited because the system relies on the agitation of the surface of the medium to allow air to form small bubbles in the medium, and the oxygen to diffuse from those bubbles to the bulk medium. This **entrainment** of air can only happen at the surface of the medium, so if the level of the medium is too far up the conical flask, the surface area available is too small for mass transfer of oxygen to the medium beneath it. Entrainment of air can be increased by making the mixing of the liquid more vigorous, frequently by increasing the speed of agitation of the flask. Those organisms with a very high demand for oxygen can demand adapted flasks in which baffles in the sides of the flasks disrupt the smooth mixing of the medium further still.

Limitations of shake flask culture

As outlined above, the main limitation of batch culture is the difficult balance that must be struck between the space available to incubate Erlenmeyer flasks, and the oxygen mass transfer rate. In addition, batch culture should not be regarded as solutions of defined composition. Before the Bacteria are introduced into the media, it may be possible to define all the components of the medium even in exact terms of the elemental concentration carbon, nitrogen, phosphorous, sulfur, and so on. As soon as the organism starts to grow, the composition of the medium changes, and will continue to change until the organism stops. We assume that biomass increases and the carbon and energy sources decrease in a regular manner. However, microorganisms secrete a variety of small molecules into the medium as they grow, from protons to heterocyclic carbon compounds. Thus during batch culture, the pH as well as the concentrations of oxygen and many other compounds not only change but can fluctuate in an unpredictable manner. To reduce the effect of pH fluctuations, the growth medium is **buffered** (normally with a phosphate buffer), but even so a change of one or more units of pH during batch culture is not uncommon. The onset of stationary phase (see Section D1) is taken to be due to the lack of a suitable carbon source, but is frequently the result of the accumulation of toxic metabolites in batch culture. Despite these limitations, batch culture is a simple, quick and for the most part reproducible method of growing small quantities of microorganisms in liquid culture.

Fed-batch

For small-scale laboratory experiments, batch culture is ideal. However, relative to the time that the organism is growing, the time taken to prepare the equipment for another experiment (the **down time**) is long. The flask must be sterilized, cleaned, refilled with medium, autoclaved, and another inoculum prepared before another experiment can begin. This time can be minimized if the majority of the biomass and spent medium is poured away or removed by pumping, and fresh medium added directly to the flask. The residual biomass serves as the inoculum for the next growth cycle. Although this is attractive in terms of reducing down time, it increases the number of interventions into the experiment, and thus the possibility of contamination increases as well. In situations where the culture itself is undefined, such as during enrichment of microorganisms with a particular property, then contamination can be seen to be a smaller problem. However, the possibility still remains that toxic metabolites will not be sufficiently diluted by the fresh medium to allow exactly the same growth parameters after feeding compared to the primary culture.

D3 LARGE-SCALE AND CONTINUOUS CULTURE

Key Notes

The simple stirred tank fermenter

Some of the limitations of shake flask culture can be overcome by the construction of a vessel with pH control, agitation via an impeller and baffles, and aeration via a sparger. This increases the oxygen transfer rate (OTR) and allows the entire volume of a vessel to be filled with medium. The OTR can be monitored with an oxygen probe, and warmed and cooled *in situ*.

Other fermenter types

Depending on the application, a number of other fermenter designs can be employed in the growth of microorganisms. These include the airlift, fluidized bed and fixed bed reactors.

Continuous culture

If a stirred tank reactor is fed with fresh medium at a constant rate, and excess is allowed to flow to waste, a culture will be created with constant growth rate. This is known as steady-state culture. If limited by carbon source, this is called a chemostat, but biomass (turbidostat) and redox (potentiostat) variations are possible. The defined and constant nature of the culture allows the calculation of many parameters, including replacement time, maximum growth rate, or biomass and growth-limiting substrate concentrations.

Related topics

Measurement of microbial growth (D1)
Batch culture in the laboratory (D2)

Shake-flask batch culture provides a simple and convenient method of growing small amounts of microorganisms. However, if grammes of biomass are required for protein purification or litres of medium are needed for product recovery then the limitations of the shake flask quickly become apparent. The most significant problem is getting enough oxygen to the culture: to grow a 5 litre culture in a single flask would require a 50 litre glass Erlenmeyer if the ratio of one-tenth volume of culture to the total volume of container is maintained. Furthermore, the shake flask is a highly dynamic system. As the organism grows, it excretes primary and secondary metabolites into the medium. Some of these metabolites might actually prevent efficient use of the substrate, but more significantly there may be a change in pH by many units. If a batch culture of one litre or more is to be grown efficiently, more controlled conditions are required.

The simple stirred tank fermenter

A culture will grow in a reproducible form if it is supplied with the same medium under the same conditions of temperature, pH, and oxygen concentration. A shake flask will provide constant aeration if the medium does not exceed 500 ml, but above this volume insufficient oxygen can be entrained at the surface of the culture. The stirred tank fermenter overcomes this primary obstacle to microbial growth in larger volumes by providing agitation via an impeller, rotating at the bottom of a circular vessel. Additional mixing of the culture may be provided by one or more baffles on the sides of the vessel (see *Fig. 1*). Oxygen, normally as sterile air, enters the fermenter underneath the impeller via a sparger. The sparger breaks up the flow of air into bubbles, which are broken into yet smaller bubbles when hitting the impeller. The combination of an impeller running at >250 rpm, baffles and sparger enable a highly efficient oxygen transfer to the culture. If the speed of the impeller and the rate of flow of oxygen to the fermenter is linked via a processor to an oxygen probe in the culture, the oxygen saturation of the culture can be monitored and regulated continuously.

At high biomass loads, a vessel is subject to some cooling to the atmosphere, but may even be warmed by the metabolic energy of the culture itself. Shake-flask temperature is regulated by the air temperature of an incubator, but a fermenter is often too large both spatially and in terms of heat capacity to be regulated in this way. Instead, a jacket surrounds the fermenter vessel linked to a thermostatically controlled water supply (*Fig. 1*). This can be supplemented with warming/cooling coils in the culture itself. Finally the pH of the fermenter is kept constant with a probe linked to alkali and acid pumps.

Other fermenter types

The stirred tank fermenter provides an efficient means for the growth of most bacteria. However, as culture volumes approach hundreds of litres, the power demands of turning the impeller fast enough to ensure sufficient oxygen transfer can become too high. The **air lift reactor** (*Fig. 2*) has no internal moving parts and stirs the culture via the passage of the air itself. This design of reactor is also useful for cells which are prone to lysis by mechanical shear.

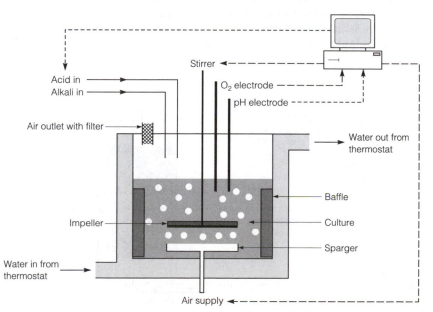

Fig. 1.   A simple fermenter with temperature, pH, and oxygen control.

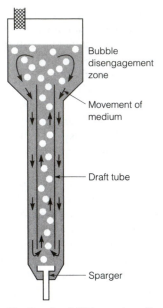

Fig. 2.   An air lift fermenter with draft tube.

When oxygenation is not so important (e.g. when using anaerobic or micoraerophilic cells) or when immobilized enzymes are used, much simpler reactors can be employed (*Fig. 3*), such as fluidised or fixed-bed reactors.

Continuous culture

A stirred tank reactor regulated for temperature, pH, and oxygen so that all conditions remain constant can be adapted to a continuous mode of action. If the continuous culture grows so that it is limited by one of the medium components, it is called a **chemostat** (*Fig. 4*), but limitations on biomass (**turbidostat**) or electron potential (**potentiostat**) can be imposed. In an industrial context, continuous culture is useful as it provides a constant production of biomaterials for downstream applications. In the research laboratory, the kinetics of the chemostat

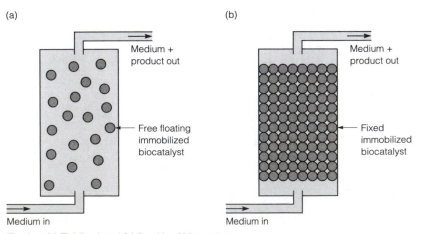

Fig. 3.   (a) Fluidized and (b) fixed bed bioreactors.

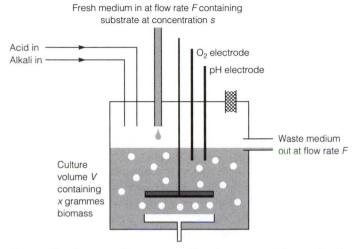

Fig. 4.   The chemostat. Temperature, pH, and oxygen control are omitted for clarity. x = biomass concentrations; s = growth limiting nutrient concentration. F = flow rate of culture V = volume.

provides additional insight into the physiology and biochemistry of pure and mixed cultures.

We assume that the action of the impeller means that the medium coming in to the culture vessel is mixed instantaneously. This is known as the **replacement time** (t_r, the time for one complete volume change of the vessel).

If V is the volume of the reactor and F is the flow rate:

$$t_r = V/F$$

If the flow rate is too high, the organism cannot divide quickly enough to maintain growth in the vessel, and eventually is diluted away (washout). The flow rate can be adjusted so that the rate of washout of biomass equals the **maximum growth rate (μ_m)**. At this point any change in substrate concentration will have a direct effect on biomass. The set of conditions under which biomass remains constant over several volume changes of the chemostat is known as a **steady state**.

The **specific growth rate** of a chemostat culture is equal to the dilution rate, D

$$D = F/V$$

This is always the case as the net increase in biomass equals growth minus output. Over an infinitely small period, this becomes:

$$V.dx = V.\mu x. \, dt - Fx.dt$$

Dividing throughout by $V.dt$

$$dx/dt = (\mu - D)x$$

So at the steady state when $dx/dt = 0$, then $\mu = D$.

We can also calculate the **biomass and growth-limiting substrate concentrations** by rewriting the balance for the growth limiting substrate (net increase in biomass = growth – output) for the case over an infinitely small period:

$$V.dx = F.s_r.dt - Fs - V\mu x. \, dt/Y$$

Where Y is the growth yield. Divide throughout by $V.dt$

$$ds/dt = D(S_r - s) - \mu xi/Y$$

In the steady state, $dx/dt = ds/dt = 0$, then the steady state values of x and s (written as x and s) are given by:

$$(\tilde{\mu} - \tilde{D})x = 0$$

and

$$D(\tilde{S}_r - \tilde{S}) - \mu x/Y = 0$$

To obtain x and s, we can use the equation for specific growth rate

$$\mu = \mu_m s/(s + K_s)$$

where K_s is the equivalent of the Michaelis Menton constant and is inversely proportional to the affinity of the organism for the substrate. As $\mu = D$, then

$$\tilde{s} = K_s D(\mu_m - D)$$

or

$$\tilde{x} = Y(s_r - \tilde{s}) = Y\{s_r - K_s D/(\mu_m - D)\}$$

The **critical dilution rate**, the dilution rate above which the culture begins to **wash out** is obtained from the maximum growth rate when $s = s_r$. Inserting this value in $\mu = \mu_m s/(s + K_s)$ results in:

$$\mu = Dc = \mu_m S_r/(S_r + K_s)$$

If s_r is very much greater than K_s, then it follows from the above equation that D_c is approximately equal to μ_m.

E1 ENZYMOLOGY

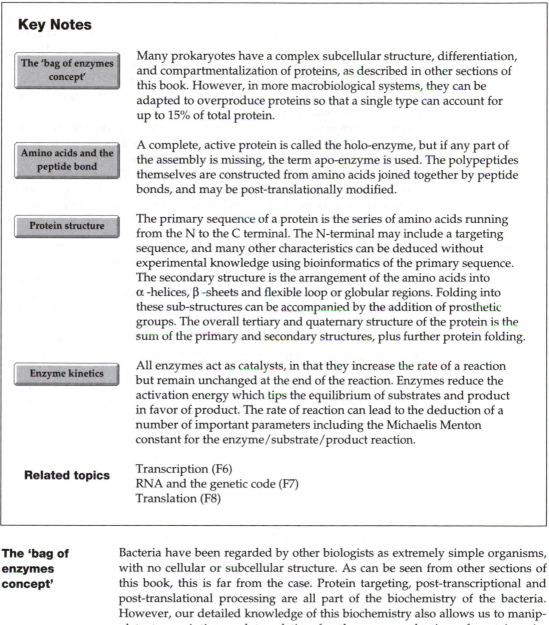

Key Notes

The 'bag of enzymes concept'

Many prokaryotes have a complex subcellular structure, differentiation, and compartmentalization of proteins, as described in other sections of this book. However, in more macrobiological systems, they can be adapted to overproduce proteins so that a single type can account for up to 15% of total protein.

Amino acids and the peptide bond

A complete, active protein is called the holo-enzyme, but if any part of the assembly is missing, the term apo-enzyme is used. The polypeptides themselves are constructed from amino acids joined together by peptide bonds, and may be post-translationally modified.

Protein structure

The primary sequence of a protein is the series of amino acids running from the N to the C terminal. The N-terminal may include a targeting sequence, and many other characteristics can be deduced without experimental knowledge using bioinformatics of the primary sequence. The secondary structure is the arrangement of the amino acids into α -helices, β -sheets and flexible loop or globular regions. Folding into these sub-structures can be accompanied by the addition of prosthetic groups. The overall tertiary and quaternary structure of the protein is the sum of the primary and secondary structures, plus further protein folding.

Enzyme kinetics

All enzymes act as catalysts, in that they increase the rate of a reaction but remain unchanged at the end of the reaction. Enzymes reduce the activation energy which tips the equilibrium of substrates and product in favor of product. The rate of reaction can lead to the deduction of a number of important parameters including the Michaelis Menton constant for the enzyme/substrate/product reaction.

Related topics

Transcription (F6)
RNA and the genetic code (F7)
Translation (F8)

The 'bag of enzymes concept'

Bacteria have been regarded by other biologists as extremely simple organisms, with no cellular or subcellular structure. As can be seen from other sections of this book, this is far from the case. Protein targeting, post-transcriptional and post-translational processing are all part of the biochemistry of the bacteria. However, our detailed knowledge of this biochemistry also allows us to manipulate transcription and translation for the over production of proteins. An *Escherichia coli* cell can be manipulated to make more than 15% of all its protein as a single recombinant form, and as such *E. coli* is the primary expression system for most protein studies from crystallography through to Michaelis Menton kinetics.

Amino acids and the peptide bond

Apart from a few ribozymes (see Section F8), all the catalysis within the Bacterial cell is carried out by proteins. In addition, the majority of the subcellular structures of the cell are proteinaceous. All proteins are made up of one or more polypeptide chains, each folded to make up a 3-dimensional shape. These polypeptides may be supplemented with other, non-protein compounds to make up the active enzyme. The complete, active protein is called the **holo-enzyme**, but if any part of the assembly is missing, the term **apo-enzyme** is used. The polypeptides themselves are constructed from **amino acids** joined together by **peptide bonds** (*Fig. 1*). The sequence of amino acids from the end of the peptide with a free amino group (the **N-terminal**) to the end with a free carboxyl group (the **C-terminal**) reflects exactly the triplet code from 5' to 3' of the mRNA that was used to translate it (see Section F8).

In most microorganisms, the same set of 20 amino acids (*Fig. 2*) are used to construct all polypeptides. A few species-specific amino acid analogues have been identified, but almost all bacteria have the ability to substitute selenocysteine for cysteine if selenium is abundant in their medium. Other **post-translational modifications** are possible, the commonest being racemization of stereoisomers between the most common L-form and the rarer D-forms. In this respect bacteria are unusual in the use of D-forms which are rarely used in eukaryotes.

Protein structure

The **primary sequence** of the polypeptides is the first point at which the function of the protein begins to emerge. Proteins buried in membranes tend to have a higher percentage of hydrophobic amino acids. Conversely, cytoplasmic or periplasmic proteins consist of more hydrophilic amino acids. The primary sequence may also encode **N-terminal targeting sequences**, for example to export the protein via the *tat* pathway to the periplasm, or the *sec* system to the outside world. Function and primary sequence are linked to the extent that the science of **bioinformatics** relies on peptide sequence to give the first clues as to the function of an unknown protein.

A long chain of amino acids joined to form a polypeptide would have no function. All proteins are folded to a greater or lesser extent. Amino acids will spontaneously arrange themselves into one of three **secondary structures.**

- **α-helix** – a spiral of amino acids sometimes represented as a cylinder in diagrammatic representations of proteins (*Fig. 3*).

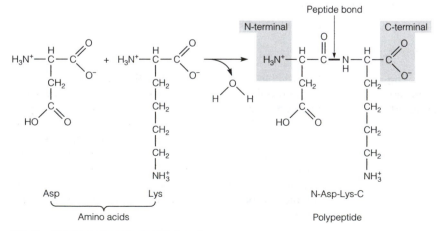

Fig. 1.   The formation of a peptide bond.

- **β sheet** – a flat arrangement, formed by a zig-zag of amino acids. Represented by a flat arrow in diagrammatic representations of proteins (*Fig. 3*).
- **Loop** – a part of a polypeptide that forms a flexible bridge between sheets and/or helices.

The final structure of the protein is in some measure determined by the primary sequence and secondary structure, and is classified according to the relative amounts of α-helix and β -sheet. However, amino acids may be further modified (by glycosylation, acetylation, hydroxylation, methylation or the addition of nucleotides), the polypeptide may be cut and other polypeptides may be joined to it, covalently or by other forms of protein–protein interaction. Finally there may be one or more **prosthetic groups** added, such as cofactors, haem or iron-sulfur clusters.

Aromatic side chains

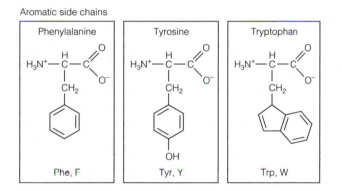

Non-polar aliphatic side chains

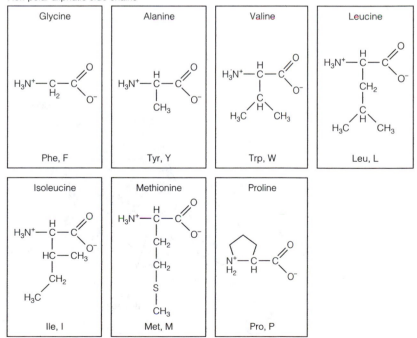

Fig. 2.   *The 20 most common amino acids with their three-letter and single-letter abbreviations.*

Continued

Polar uncharged side chains

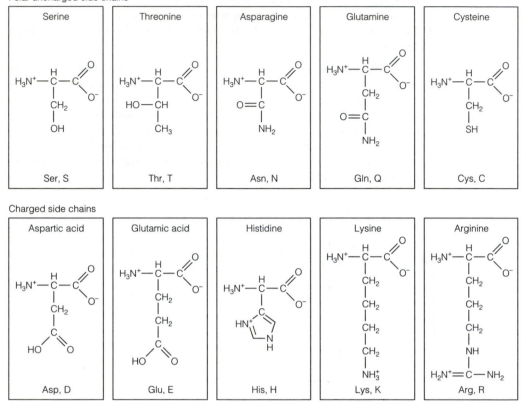

Charged side chains

Fig. 2. cont'd

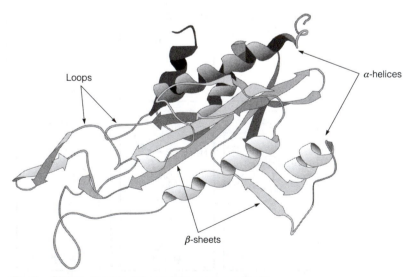

Fig. 3. Ribbon diagram of the restriction enzyme EcoRI.

The overall **tertiary and quaternary structure** of the protein is the sum of the primary and secondary structures, plus further protein folding. The distinction between tertiary and quaternary structure is blurred, with tertiary structure defined as the shape of a single polypeptide, and quaternary structure being the shape of all the peptides in the protein. However, many polypeptides do not adopt a final shape until they have interacted with their neighbors. Proteins will adopt a quaternary structure spontaneously, but many proteins are actively folded by the cellular apparatus. A class of proteins known as chaperonins aid the folding of polypeptides into the correct active form, and minimize the occurrence of improperly folded protein. The importance of correct folding is illustrated by **prions** in human disease: the presence of a correctly folded prion protein in the brain causes no disease, however should another fold-form of the same protein dominate, then spongiform encephalitis will arise.

Tertiary and quaternary structure is stabilized by a number of intra-protein bonds apart from the peptide bonds holding the amino acids together. Among these stabilizing bonds between amino acids on the same or different polypeptide chains are:

- Hydrophobic interactions between areas of hydrophobicity in the protein.
- Ionic bonds between areas of hyrophilicity in the protein.
- Hydrogen bonds.
- Weaker bonds such as van der Waals forces.

In addition, proteins translocated to the periplasm can form sulfur bridges (**disulphide bonds**) between cysteine residues.

Enzyme kinetics   All enzymes act as catalysts, in that they increase the rate of a reaction but remain unchanged at the end of the reaction. Enzymes reduce the activation energy which tips the equilibrium of substrates and product in favor of product:

$$A + B \rightleftharpoons C$$

$$A + B + enzyme \rightleftharpoons C + enzyme$$

In reality the enzyme often forms unstable intermediates in which the substrates and products are transiently bound to the peptide chain of prosthetic

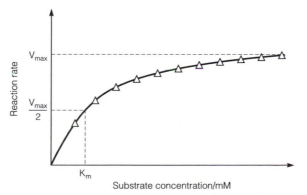

Fig. 4.   The Michaelis Menton constant inferred from the maximum rate of reaction.

groups. Enzymes are extremely specific in the reactions they can perform in comparison to chemical catalysts such as platinum, and tend to perform only one type of reaction at their active site. Many proteins have more than one active site and so appear to perform more than one reaction. The study of an enzyme's kinetics gives an insight into the mechanism by which the reaction is performed, and the effect of inhibitors.

The estimation of the rate of a reaction is valuable to microbiologists because it allows the calculation of the affinity of the enzyme for particular substrates. Affinity is an important concept as not only does it give a measure of how efficient one enzyme performs a reaction over another, it also allows the deduction of the cellular concentration of substrates. The affinity of an enzyme is approximately equal to the K_m (**Michaelis Menton constant**) of the enzyme. The K_m is equal to have the maximum reaction rate (V_{max}) and can be deduced from plotting the reaction rate with increasing substrate concentration under defined conditions (*Fig. 4.*).

E2 HETEROTROPHIC PATHWAYS

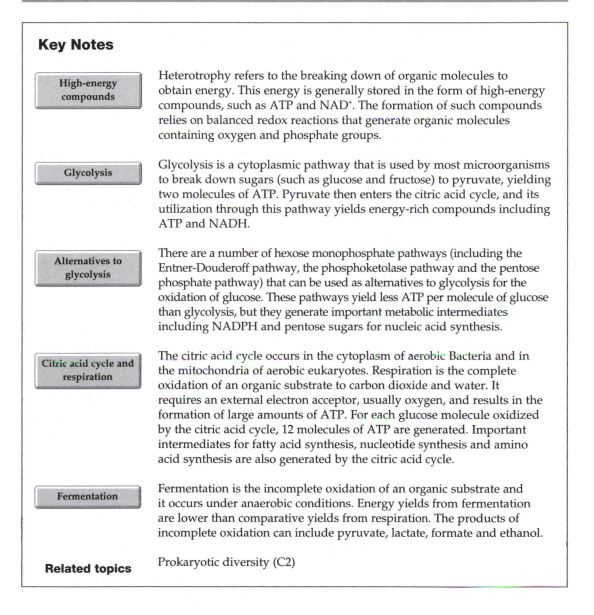

Key Notes

High-energy compounds

Heterotrophy refers to the breaking down of organic molecules to obtain energy. This energy is generally stored in the form of high-energy compounds, such as ATP and NAD^+. The formation of such compounds relies on balanced redox reactions that generate organic molecules containing oxygen and phosphate groups.

Glycolysis

Glycolysis is a cytoplasmic pathway that is used by most microorganisms to break down sugars (such as glucose and fructose) to pyruvate, yielding two molecules of ATP. Pyruvate then enters the citric acid cycle, and its utilization through this pathway yields energy-rich compounds including ATP and NADH.

Alternatives to glycolysis

There are a number of hexose monophosphate pathways (including the Entner-Douderoff pathway, the phosphoketolase pathway and the pentose phosphate pathway) that can be used as alternatives to glycolysis for the oxidation of glucose. These pathways yield less ATP per molecule of glucose than glycolysis, but they generate important metabolic intermediates including NADPH and pentose sugars for nucleic acid synthesis.

Citric acid cycle and respiration

The citric acid cycle occurs in the cytoplasm of aerobic Bacteria and in the mitochondria of aerobic eukaryotes. Respiration is the complete oxidation of an organic substrate to carbon dioxide and water. It requires an external electron acceptor, usually oxygen, and results in the formation of large amounts of ATP. For each glucose molecule oxidized by the citric acid cycle, 12 molecules of ATP are generated. Important intermediates for fatty acid synthesis, nucleotide synthesis and amino acid synthesis are also generated by the citric acid cycle.

Fermentation

Fermentation is the incomplete oxidation of an organic substrate and it occurs under anaerobic conditions. Energy yields from fermentation are lower than comparative yields from respiration. The products of incomplete oxidation can include pyruvate, lactate, formate and ethanol.

Related topics

Prokaryotic diversity (C2)

High-energy compounds

The ability to produce high-energy compounds for metabolism and storage is a prerequisite for cell survival. Energy is acquired by cells through a series of balanced **oxidation-reduction** (**redox**) reactions from organic or inorganic substrates. The simplest redox reaction can be seen in the reaction below

$$H_2 + \frac{1}{2} O_2 \rightarrow H_2O$$

H_2 = reductant (electron donor) that becomes oxidized
O_2 = oxidant (electron acceptor) that becomes reduced

The energy that is released in redox reactions is stored in a variety of organic molecules that contain oxygen atoms and phosphate groups. **ATP**, adenosine triphosphate, is a **high-energy compound** found in almost all living organisms. It is synthesized in catabolic reactions, where substrates are oxidized, and utilized in anabolic, biosynthetic reactions. Intermediates called **carriers** participate in the flow of energy from the electron donor to the terminal electron acceptor. The co-enzyme **nicotinamide adenine nucleotide** (NAD^+) is a freely diffusable carrier that transfers two electrons and a proton, and a second proton from water, to the next carrier in the chain.

$$NAD^+ + 2H^+ + 2e^- \rightleftharpoons NADH + H^+$$

The reactions for the phosphorylated derivative ($NADP^+$) are similar. NAD^+ is usually used in energy-generating reactions and $NADP^+$ in biosynthetic reactions.

All protozoa, all fungi and most prokaryotes synthesize ATP by oxidizing organic molecules. This can be either via **respiration** or by **fermentation**. Respiration requires a terminal electron acceptor. This is usually oxygen, but nitrate or sulfate are among the compounds used in anoxic conditions. Fermentation requires an organic terminal oxygen acceptor.

Microorganisms can be grouped according to the source of energy they use, and by the source of carbon which may either be an organic molecule or from CO_2 (carbon dioxide fixation) (*Table 1*).

Glycolysis (Embden-Meyerhof-Parnas)

The reactions termed **glycolysis** take place in the cytoplasm of all prokaryotes and eukaryotes. The pathway generates two ATP molecules per molecule of glucose degraded, and feeds substrates into subsequent metabolic pathways.

The steps in glycolysis are shown in *Fig. 1* but the overall net reaction can be summarized as follows:

$$\text{Glucose} + 2ADP + 2P_i + 2NAD^+ \rightarrow 2 \text{ pyruvate} + 2ATP + 2NADH + 2H^+$$

The reactions at the beginning of the pathway require two ATP molecules, but the gross yield of ATP per glucose molecule is four, giving a net gain of two ATP per glucose.

Table 1.   Classification of microorganisms by energy and carbon source utilized

	Type	Electron donor	Energy source	Carbon source	Examples
Organotrophs	Chemo-organotroph	Organic compounds	Redox reactions of organic compounds	Organic compounds	All fungi, all protists, most terrestrial bacteria
	Photo-organotroph	Organic compounds	Light	Carbon dioxide and organic compounds	Nonsulphur Bacteria
Lithotrophs	Chemo-lithotrophs	Inorganic compounds	Redox reactions of inorganic compounds	CO_2	*Thiobacillus, Nitrosomonas, Nitrobacter, Hydrogenomonas, Beggiotia*
	Photolithotrophs	Inorganic compounds	Light	CO_2	Photosynthetic green and purple Bacteria, photosynthetic protista

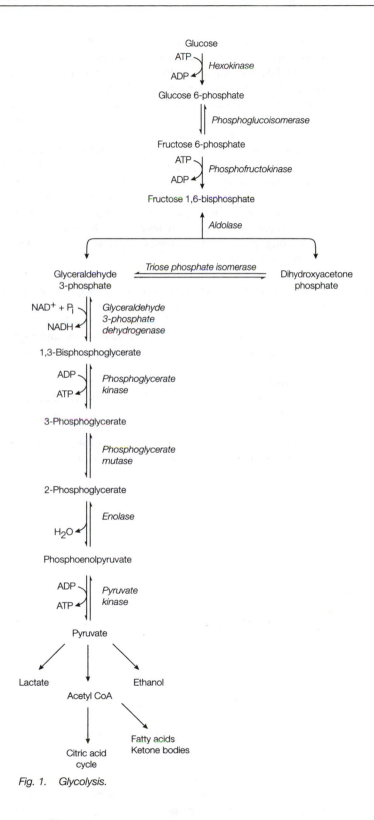

Fig. 1.   Glycolysis.

The initial reactions at the beginning of the pathway transform the 6 carbon sugar glucose into glucose 1,6-bisphosphate, via two phosphorylation reactions. There follows a near symmetrical split into two 3C phosphorylated compounds (glyceraldehyde-3-phosphate and dihydroxyacetone phosphate (DHA)). These compounds will interconvert as an equilibrium reaction via the enzyme triose phosphate isomerase. Glyceraldehyde-3-phosphate is the substrate for subsequent reactions of glycolysis.

Further energy is added to the glyceraldehyde-3-phosphate by the addition of a second high-energy phosphate group, from NADPH to the aldehyde group. There then follow two reactions where the high-energy phosphate groups of 1,3-bisphosphoglycerate are used to form ATP from ADP, mediated by two kinase enzymes, phosphoglycerate kinase and pyruvate kinase. These reactions are termed **substrate level phosphorylations**.

The final product of glycolysis is pyruvate, which feeds into respiration in aerobic conditions.

Alternatives to glycolysis

Some important groups of Bacteria, for example some Gram-negative rods, do not use glycolysis to oxidize glucose. They use a different mechanism, the **Entner-Douderoff** (*Fig. 2*), which yields one mole of ATP, NADPH and NADH from every mole of glucose. This is a **hexose monophosphate pathway** (HMP), and in this pathway only one molecule of ATP is produced per molecule of glucose metabolized.

Another HMP is the **phosphoketolase** pathway, which is another method for glucose breakdown found in *Lactobacillus* and *Leuconostoc* spp. when grown on 5-carbon sugars (pentoses). The pathway produces lactic acid, CO_2 and either ethanol or acetate (*Fig. 3*).

An important HMP is the **pentose phosphate pathway** (PPP), which often operates in conjunction with glycolysis or other HMP pathways. The PPP is

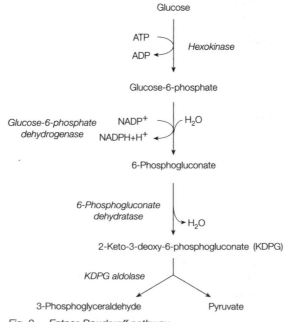

Fig. 2.   *Entner-Douderoff pathway.*

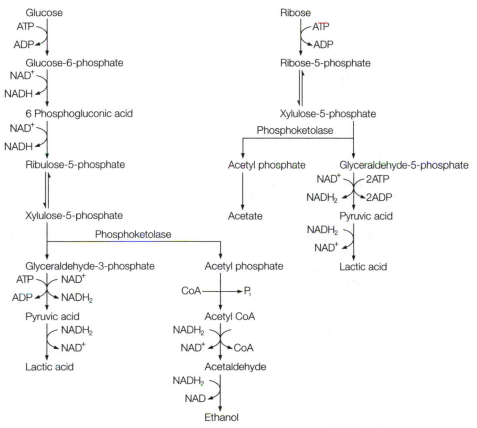

Fig. 3. The phosphoketolase pathway.

an important provider of intermediates that serve as substrates for other bio-synthetic pathways. This pathway yields NADPH/+H+ and pentoses which are used in the synthesis of nucleotides including, FAD, ATP and coenzyme A (CoA).

The reactions can be summarized as

$$\text{Glucose-6-phosphate} + 2\,NADP^+ + water \rightarrow \text{Ribose-5-phosphate} + 2\,NADPH + 2H^+ + CO_2$$

There are three important stages to this pathway:

1. Glucose-6-phosphate is converted to ribulose-5-phosphate, generating two NADPH + 2H+

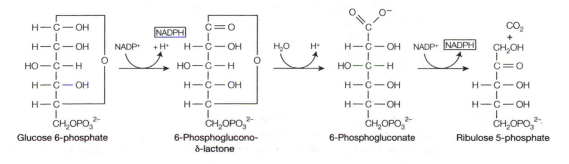

2.   Ribulose-5-phosphate isomerises to ribose-5-phosphate

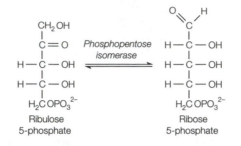

3.   Excess ribose-5-phosphate is converted to fructose-6-phosphate and glyceraldehyde, via a series of reactions, to enter glycolysis.

The citric acid cycle and respiration

The **citric acid cycle** is found in the cytosol of aerobic prokaryotes, and the mitochondria of eukaryotes. Anaerobic organisms have incomplete cycles whilst facultative aerobic organisms only have a functional citric acid cycle in the presence of O_2.

Complete oxidation of organic substrates to CO_2 and water via the citric acid cycle requires an external electron acceptor; the best studied are oxygen, nitrate or sulfate. This process yields large amounts of energy stored as ATP. The product of glycolysis, pyruvate, can be completely oxidized using enzymes of the citric acid cycle.

In summary, during the operation of the citric acid cycle three carbon atoms of pyruvate are completely oxidized to CO_2, and four hydrogen atoms reduce NAD^+ and FAD (*Fig. 4*, reactions 1–10).

The cycle begins with an oxidative decarboxylation (reaction 1), where CO_2 is released and NADH is formed. The resulting 2C (acetyl) unit is linked to CoA (reaction 2), and this high energy compound couples with oxaloacetate (4C) to form a 6C unit, citric acid. The acetyl group of the citric acid is then further metabolized, with 2 decarboxylation reactions releasing CO_2 (reactions 15 and 16) and 4 more coenzyme molecules are reduced. At the end of the cycle oxaloacetate is regenerated as the acceptor of further acetyl units (reaction 10).

The reduced co-enzymes are then oxidized by a respiratory electron transport chain which may use oxygen, nitrate or sulfate as terminal electron acceptors. This allows for NAD^+ regeneration and the synthesis of ATP, a process known as oxidative phosphorylation. NAD^+ regeneration is essential as levels are limited in cells.

Each turn of the citric acid cycle yields three NADH molecules and one $FADH_2$ molecule which via oxidative phosphorylation generates ATP molecules. Including the single ATP molecule that is formed in the conversion of succinyl CoA to succinate, each molecule of glucose oxidized by the citric acid cycle produces 12 ATP molecules.

The cycle produces intermediates for many other biosynthetic pathways, including fatty acid biosynthesis (citrate), amino acid synthesis (α-ketoglutarate), and nucleotide synthesis (α-ketoglutarate and oxaloacetate).

Fermentation

Fermentation is an **incomplete oxidation** of an organic substrate. During fermentations an electron donor becomes reduced, and energy is trapped by **substrate level phosphorylation**. Fermentation products include pyruvate if

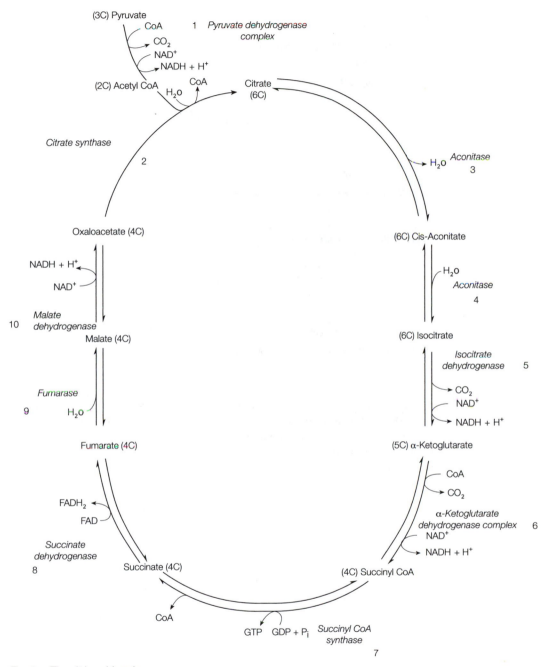

Fig. 4.   The citric acid cycle.

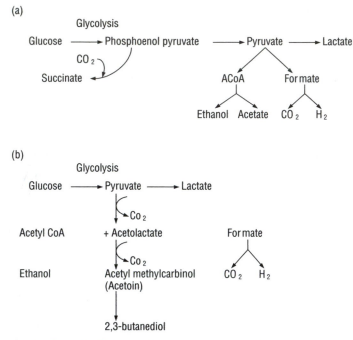

Fig. 5.   Products of fermentation. (a) Mixed acid. (b) Butanediol.

the **glycolytic** pathway is used, or lactate, formate, 2,3-butanediol, and ethanol from **the butanediol pathway** (used by *Klebsiella, Erwinia, Enterobacter* and *Serratia* spp.) or succinate, ethanol, acetate and formate from a **mixed acid fermentation** (found in *Escherichia, Salmonella* and *Shigella* spp.). See *Fig. 5* for details.

E3 ELECTRON TRANSPORT, OXIDATIVE PHOSPHORYLATION AND β-OXIDATION OF FATTY ACIDS

Key notes

Electron transport	Electron transport is used to create a proton motive force (PMF) across membranes. This PMF is used by all microorganisms to generate ATP via a membrane bound ATPase. The Archaea and Bacteria also use PMF to drive the movement of flagellae, allow transport of charged substrates across membranes and maintain their osmotic potential. In eukaryotic microbes, PMF is established across the inner membrane of the mitochondrion. In the electron transport chain, a series of balanced oxidation and reduction reactions drives the movement of electrons through the carrier series from NADH to oxygen. During this process energy is released and ATP is synthesized.
Anapleurotic pathways	Lost intermediates from glycolysis and the citric acid cycle are replenished by anapleurotic reactions, where carbon dioxide is fixed into three-carbon compounds by carboxylation reactions.
Glyoxalate cycle	Some substrates that microbes can utilize as carbon sources, for example the two-carbon compound acetate, can lead to the depletion of citric acid cycle intermediates. Reactions that result in the loss of CO_2 during the cycle can be avoided by using the glyoxalate cycle.
Fatty acid oxidation	Fatty acids can be used as substrates by microorganisms through the fatty acid or beta oxidation pathway. This is located in the mitochondria of eukaryotes and the cytoplasm of prokaryotes.
Anaerobic respiration	Many microbes live in low or no oxygen environments. Alternative electron acceptors, such as nitrate and sulfate, can be utilized instead of oxygen by these organisms to complete the electron transport chain.
Methanogenesis	Archaea have many biochemical pathways in common with the Bacteria. A unique property is the possession of a pathway that converts C1 compounds into methane with the release of energy.
Related topics	Prokaryotes and their environment (C11) Eukaryotic cell structure (L2)

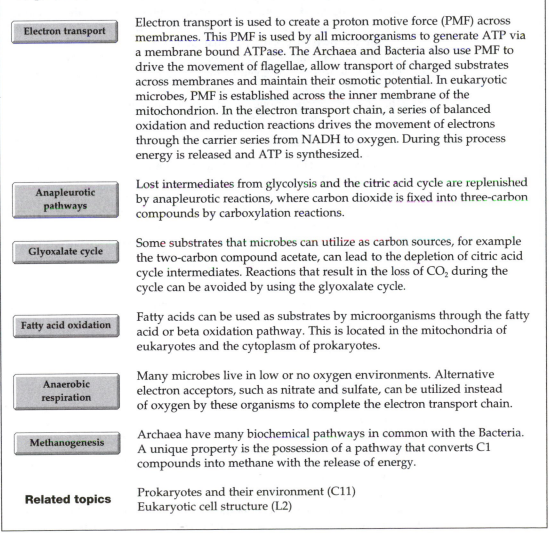

Electron transport

Peter Mitchell theorized that the generation of ATP only occurred because mitochondria and Bacteria could pump protons across a membrane. These primary pumps lead to a generation of a charge across the membrane, known as the **proton motive force** (PMF). As the protons try to move back across the membrane, the energy of their movement can be harnessed in a number of ways. The most important use of the PMF in many aerobic organisms is the **generation of ATP**, normally using the enzyme f_1f_0 ATPase. This ATP-generating enzyme is found in the cytoplasmic membrane of Bacteria and the inner membrane of the mitochondria, and in the absence of the PMF actually cleaves ATP into ADP and phosphate. However when PMF is applied, the ATPase works essentially in reverse, generating ATP from ADP and phosphate. It is thus known as a **secondary pump** (*Fig. 1*). For all this to happen efficiently, the organism or organelle must conform in several ways to the Mitchell hypotheses which are:

1. Protons are pumped across mitochondrial and bacterial membranes in such a way as to generate an **electric potential** across the membrane. A membrane bound-enzyme (ATPase) couples synthesis of ATP to the flow of protons down the electric potential gradient.
2. Solutes can accumulate against a concentration gradient by the coupling of proton flow to the movement of the solute by a **transmembrane protein**. These cotransporters may act as **symports, antiports or uniports.**
3. The flow of protons through the flagellar transmembrane proteins rotates components of the flagellum and allows a prokaryote to move.

There are five types of component molecule:

1. Enzymes that catalyze transfer of hydrogen atoms from reduced NAD^+ to flavoproteins (NADH dehydrogenases).
2. Flavoproteins. Flavin mononucleolides (FMN) and flavin adenine dinucleotide (FAD). The flavins are reduced by accepting a hydrogen atom from NADH and oxidized by losing an electron.
3. Electron carriers, cytochromes. Cytochromes are porphyrin containing proteins each of which can be reduced or oxidized by the loss of a single electron:

$$\text{Cytochrome-Fe}^{2+} \rightleftarrows \text{Cytochrome-Fe}^{3+} + \varepsilon\text{-}$$

4. Iron sulfur proteins. These are carriers of electrons with a range of reduction potentials
5. Quinones. These are lipid-soluble carriers that can diffuse through membranes carrying electrons from iron-sulfur proteins to cytochromes.

The current view of electron transport is summarized in *Fig. 1*.

Protons for the final reduction in the transfer chain are supplied by the disassociation of water, providing the build up of hydroxyl ions on the inside of the membrane.

Protons flow back into the mitochondrial matrix or bacterial cell through the enzyme ATP synthase, driving ATP synthesis. This enzyme is in two parts, one localized on the Bacterial cytoplasmic or mitochondrial matrical side and the other which spans the membrane to the outside of the bacterial cell or the inter-membrane space of the mitochondrion.

The rate of **oxidative phosphorylation** is set by the availability of ADP, electrons only flow down the chain when ATP is needed. When there are high levels of ATP and energy-rich compounds like $NADH^+$ and $FADH_2$ accumulation of citric acid inhibits the citric acid cycle and glycolysis.

However, an alternative theory, the **conformational change hypothesis**, proposes ATP synthesis occurs because of conformational changes created in

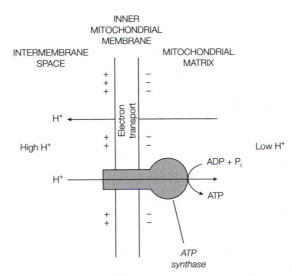

Fig. 1.   The role of ATPase as a secondary pump. From Hames and Hooper, Instant Notes in Biochemistry, *2nd edn. © BIOS Scientific Publishers 2000.*

the ATP synthase enzyme caused by electron transport. This theory is currently being intensively researched.

Anapleurotic pathways

The intermediates of glycolysis and the citric acid cycle are used as precursors of biosynthetic pathways. To maintain the energy-yielding processes of glycolysis and the citric acid cycle these lost intermediates must be replenished **by anapleurotic reactions**. Three carbon compounds are carboxylated to form oxaloacetate. Both pyruvate and phosphoenol pyruvate (PEP) can be used in these reactions, e.g.

$$\text{PEP} + \text{HCO}_3^- \xrightarrow{\text{PEP carboxylase}} \text{oxaloacetate} + \text{P}_i$$

Glyoxalate cycle

A number of organic acids can be used by microorganisms as electron donors and carbon sources. Those that are common to the citric acid cycle, citrate, malate, fumarate and succinate for example, can be metabolized using the enzymes of the citric acid cycle. However, utilization of acetate via the citric acid cycle will cause the depletion of oxaloacetate. If this occurs the citric acid cycle could not operate. To compensate for the loss of oxaloacetate the **glyoxalate shunt occurs** (*Fig. 2*).

In this pathway, which shares many of the reactions of the citric acid cycle, reactions that give rise to CO_2 evolution are bypassed, and isocitrate is split into succinate and glyoxalate by the enzyme isocitrate lyase. Succinate can be used in biosynthetic reactions, whilst glyoxalate is combined with Acetyl CoA via malate synthase to form malate, which enters the citric acid cycle.

Fatty acid oxidation

Fatty acids can be used as substrates for microbial metabolism. The metabolic process is called **beta oxidation**, and it occurs in the mitochondria of eukaryotes and the cytoplasm of prokaryotes. Two carbon units are removed from the fatty acid to yield their acyl CoA. The pathway is outlined in *Fig. 3*.

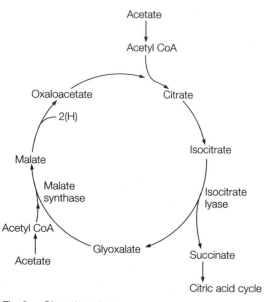

Fig. 2.   Glyoxalate shunt.

The pathway begins with the activation of the fatty acid by CoA. There then follow two separate dehydrogenation reactions where electrons are transferred to FAD and NAD⁺, finally yielding CoA and an activated fatty acid to restart the cycle.

Anaerobic respiration

Anaerobic respiration occurs in prokaryotes that are unable to use oxygen as a terminal electron acceptor. These organisms are termed **obligate anaerobes**. Other prokaryotes can use anaerobic respiration facultatively, if oxygen happens to be unavailable. Less energy is generated during anaerobic respiration than in aerobic respiration. Nitrate, sulfate and carbon dioxide can be used as **alternative electron acceptors**.

Nitrate respiration uses the most common alternative electron acceptor, nitrate (*Fig. 4*). The first step of the reaction is catalyzed by the enzyme nitrate reductase, an enzyme which is only synthesized under anaerobic conditions. The product is nitrite, which is excreted by most staphylococci and enterobacteriaceae, but other Bacteria will reduce nitrite further to ammonia or nitrogen gas. This reaction, and the enzyme that catalyzes it, is termed **dissimilatory** because nitrogen is reduced during the biological breakdown of organic compounds. This type of respiration leads to **denitrification.**

Methanogenesis

Energy metabolism in the Archaea has many similarities to either the Bacteria or the Eukarya, subject to slight modifications (e.g. the Entner-Doudoroff pathway for glucose catabolism). Pathways cannot be generalized because of the metabolic diversity of the Archaea, but the metabolism of carbon sources resulting in the release of methane (methanogenesis) is unique. Substrates such as carbon monoxide, formate and carbon dioxide are metabolised anaerobically in the presence of hydrogen:

$$CO_2 + 4H_2 \rightarrow CH_4 + 2H_2O$$

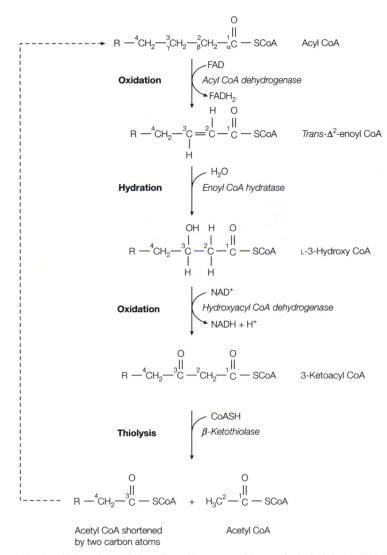

Fig. 3.   Fatty acid oxidation. From Hames and Hooper, Instant Notes in Biochemistry, 2nd edn. © BIOS Scientific Publishers 2000.

C1 compounds such as methanol, methylamine and dimethylsulfide are used when hydrogen is an electron donor in the following way:

$$CH_3OH + H_2 \rightarrow CH_4 + H_2O$$

However some methanogens can use methanol in the absence of hydrogen:

$$4CH_3OH + CH_4 \rightarrow CO_2 + 2H_2O$$

Compound such as acetate are cleaved in an acetotrophic process:

$$CH_3COO^- + H_2O \rightarrow CH_4 + HCO_3^-$$

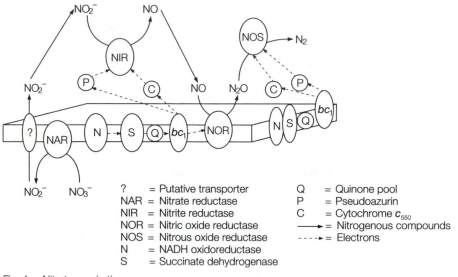

?	= Putative transporter	Q	= Quinone pool
NAR	= Nitrate reductase	P	= Pseudoazurin
NIR	= Nitrite reductase	C	= Cytochrome c_{550}
NOR	= Nitric oxide reductase	——▶	= Nitrogenous compounds
NOS	= Nitrous oxide reductase	----▶	= Electrons
N	= NADH oxidoreductase		
S	= Succinate dehydrogenase		

Fig. 4.   Nitrate respiration.

All these reaction are chemiosmotically linked to ATP synthesis. The acetotrophic reaction yields the least amount of energy per mole substrate, whereas the CO_2-type substrates yield the most.

E4 AUTOTROPHIC REACTIONS

Key Notes

Chemolithotrophy

Autotrophic microorganisms can survive in the absence of organic carbon sources by fixing atmospheric or dissolved CO_2 to form carbohydrates. Chemolithotrophs have the ability to fix CO_2 using the Calvin cycle, and the energy required to drive the reactions comes from the oxidation of inorganic substrates such as ammonia.

Photosynthesis

Photosynthesis can be divided into two sets of reactions, those that are light-dependent (light reactions) and those that are light-independent (dark reactions). The light reactions convert light into chemical energy through the synthesis of ATP, which is then used to drive the Calvin cycle (dark reactions). Photosynthesis may be described as oxygenic if oxygen is generated (as in the cyanobacteria and the photosynthetic eukaryotes) or as anoxygenic if it is not (as in the green and purple Bacteria). The light reaction can be driven by photosystem I and II in eukaryotes, but may only be driven by photosystem I in some prokaryotes.

Light reactions in bacterial photosynthesis

Photosynthetic green and purple Bacteria contain chlorophyll A and B, and carry out anoxygenic photosynthesis that utilizes only photosystem I.

Light reactions in eukaryotic photosynthesis

In eukaryotes, photosynthesis occurs in the chloroplasts, and involves photosystems I and II. The light-dependent reactions generate $NADPH + H^+$, and the resulting proton gradient is used to generate ATP by non-cyclic phosphorylation.

Dark reactions in eukaryotic photosynthesis

The dark (light-independent) reactions of photosynthesis are called the Calvin cycle and use the energy generated from light-dependent reactions to synthesize carbohydrates from CO_2 and H_2O.

Related topics

The mojor prokaryotic groups (C6)
Fungal nutrition (J2)
Nutrition and metabolism (K2)

Chemolithotrophy   Chemolithotrophy is found in a limited number of microorganisms. Chemolithotrophs obtain their energy by the oxidation of inorganic substrates and their carbon from CO_2. However, these reactions yield less energy than oxidation of glucose to CO_2, so large quantities of substrates have to be oxidized to generate enough energy for sufficient ATP and NADH generation. The process whereby ammonia is oxidized to nitrate is termed nitrification, and ATP can be generated via this reaction. However, electrons cannot be donated directly for NADH production from ammonia or nitrate because they have a more positive redox potential than NAD^+, a process termed reversed electron flow allows ATP to be used to generate small but sufficient amounts of NADH (*Fig. 1*).

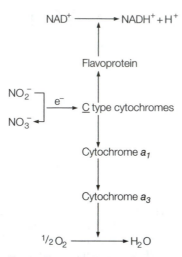

Fig. 1. *Reversed electron flow.*

Photosynthesis   A large number of microorganisms have the ability to use sunlight to generate
                       ATP by photophosphorylation. This process may not generate oxygen, a reaction
                       termed anoxygenic, as found in the green and purple Bacteria, or it may generate
                       oxygen, termed oxygenic, by the photolysis of water, as found in the blue green
                       Bacteria and algae (*Table 1*). The reactions are complex but can be divided into two
                       sets of reactions, the light reactions where light energy is converted into chemical
                       energy (ATP), and the dark reactions where ATP is used to synthesize glucose.

The light reaction   Photosynthetic Bacteria contain **bacteriochlorophylls** a and b, with absorption
in bacterial   maxima of 775 and 790 nm respectively. These pigments are contained within
photosynthesis   sac-like extensions of the plasma membrane called **chlorosomes** in green
                           sulfur and non-sulfur bacteria and **intracytoplasmic** vesicles in purple Bacteria.

Table 1.   *Classification of photosynthetic microorganisms according to hydrogen (reductant) and carbon source*

Nutritional classification	Examples	Carbon source	Hydrogen source	Oxygen evolution
Primarily photolithotrophs	Green sulfur Bacteria (Chlorobiaceae)	CO_2, acetate, butyrate	H_2, H_2S, $S_2O_3^{2+}$	Negative
	Purple sulfur Bacteria, (Chromatiaceae)	CO_2, acetate, butyrate	H_2, H_2S, $S_2O_3^{2+}$	Negative
Photo-organotrophs	*Purple non-sulfur Bacteria (Rhodospirrillaceae)	Organic (CO_2)	H_2, organic	Negative
	*Green gliding Bacteria (Chloroflexaceae)	Organic (CO_2)	H_2, organic	Negative
	Halobacteria (Archaea)	Organic	Organic	–
Photolithotrophs	Blue green Bacteria (Cyanobacteria)	CO_2	H_2O	Positive
Photolithotrophs	Photosynthetic protista	CO_2	H_2O	Positive

* Can grow as chemoorganotrophs aerobically in the dark.

Bacterial photosynthesis is an **anoxygenic** (non-oxygen producing) photosynthesis which relies on photosystem I only, and is termed a cyclic phosphorylation (*Fig. 2*).

There is no net change in the numbers of electrons in the system. ATP synthesis occurs during the generation of a protein motive force during photosynthesis, which allows ATP synthase to synthesize ATP. The electrons expelled from the reaction center return to the bacteriochlorophyll via the electron transport chain. The photosynthetic apparatus consists of four membrane-bound pigment–protein complexes, plus an ATP synthase. For NADPH synthesis, Bacteria must use electron donors like hydrogen, H_2S, sulphur and organic compounds with a more negative reduction potential than water. In this case direct transfer can occur via ferrodoxin.

The purple sulfur Bacteria cannot synthesize NADPH directly by photosynthetic electron transport. This is because their acceptor molecules are more positive than the $NADP^+/NADPH$ couple (−0.32 volts). In this case electrons enter at the cytochromes from the electron donors, and ATP from the light reactions is used to reduce $NADP^+$ to $NADPH^+$, a process termed energy-dependent reverse electron flow (*Fig. 3*).

The light reaction in eukaryotic photosynthesis

Photosynthesis occurs within **antenna complexes** and **reaction centers** in the thylakoid membrane of chloroplasts in eukaryotic microorganisms. Antenna complexes are formed from several hundred chlorophyll molecules plus accessory pigments. Light excitation of the chlorophyll molecule results in an electron in a chlorophyll molecule being excited to a higher orbit, and this

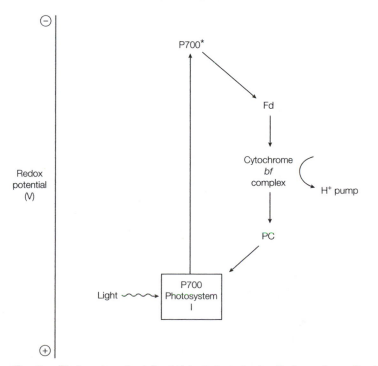

Fig. 2. Photosystem I of bacterial photosynthesis. Redrawn from Brock, Biology of Microorganisms, Madigan et al., Prentice Hall Inc., USA, 1997.

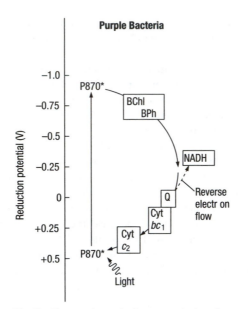

Fig. 3.   Energy-dependent reverse electron flow.

energy is transferred between chlorophylls until it is channeled into the chloro-phyll molecules of the reaction center.

The reaction center contains two photosystems, called **photosystem I (PS I)** and **photosystem II (PS II)**, with different light-energy absorption maxima. PS I absorbs at 700 nm, and PS II at 680 nm. The reaction centers are linked by

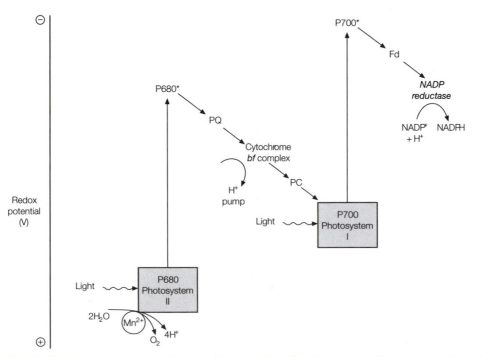

Fig. 4.   The Z scheme for noncyclic photophosphorylation. From Hames and Hooper, Instant Notes in Biochemistry, 2nd edn. © BIOS Scientific Publishers 2000.

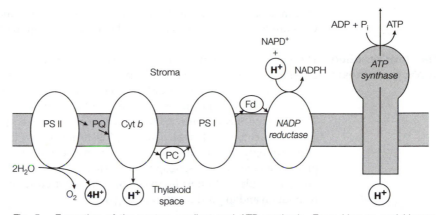

Fig. 5.   Formation of the proton gradient and ATP synthesis. From Hames and Hooper,
Instant Notes in Biochemistry, *2nd edn. © BIOS Scientific Publishers 2000.*

other electron carriers, and if the components are arranged by their redox potentials they assume a Z shape, so the scheme is called the **Z scheme** (*Fig. 4*).

The reactions of the Z scheme generate NADPH from NADP. ATP is generated by non-cyclic phosphorylation reactions because of the creation of a **proton gradient** between the thylakoid space and the stroma by the reactions of PS I and PS II. An ATP synthase is present in the thylakoid membrane, and H$^+$ is pumped from the stroma into the thylakoid space, generating ATP (*Fig. 5*).

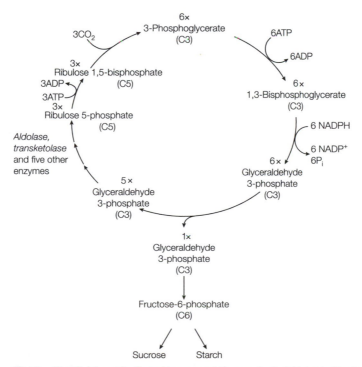

Fig. 6.   The Calvin cycle. From Hames and Hooper, Instant Notes in Biochemistry, *2nd edn.*
© BIOS Scientific Publishers 2000.

PS I may operate without PS II in some circumstances, and in this reaction no O_2 is produced; only ATP is produced via the proton gradient.

The dark reaction of photosynthesis The light-independent reactions of photosynthesis use the NADPH and ATP generated from the light reactions to synthesize carbohydrates from CO_2 and water. This is called the **Calvin cycle**. A key enzyme in this cycle is **rubisco** (*Fig. 6*) (ribulose bisphosphate carboxylase), a large, multi-component enzyme. This enzyme incorporates CO_2 into ribulose 1,5-bisphosphate to form first a six-carbon compound, which then splits to form two three-carbon molecules (3-phosphoglycerate). Subsequent reactions regenerate ribulose 1,5-bisphosphate from one of the 3-phosphoglycerate molecules, to continue the cycle. The other molecule of 3-phosphoglycerate is transported to the cytosol and used in respiration and to produce storage sugars.

E5 BIOSYNTHETIC PATHWAYS

Key Notes

Carbohydrates	Carbohydrates are synthesized from precursors by a process termed gluconeogenesis. This pathway is almost the reverse of glycolysis, except that three irreversible glycolytic enzymes are replaced by three synthetic enzymes specific to this pathway.
Amino acids	Some Bacteria can fix atmospheric nitrogen when fixed nitrogen sources (nitrate, nitrite or ammonia) are not available. All other microorganisms must use fixed nitrogen. Only ammonia can be incorporated directly to form amino acids; nitrate and nitrite must be reduced to ammonia first. There are five amino acid families from which 20 different amino acids are synthesized.
Nucleic acids	Nucleic acids are made of nucleotides, which are cyclic nitrogen-containing compounds. Purines are bicylic, while pyrimidines have a single ring. When linked to a phosphorylated pentose sugar they are termed nucleosides. They are the building blocks of DNA and RNA.
Lipids	Lipids are synthesized from fatty acids. They are long chain molecules of around 18 carbon atoms, and they may be saturated (contain no double bonds) or unsaturated (contain one or more double bonds).
Related topics	Fungal nutrition (J2) Nutrition and metabolism (K2) Prokaryotes and their envirnonment (C11) Structure of nucleic acids (F3) Enzymology (E1) DNA-The primary information macromolecule (F1) Structure of nucleic acids (F3)

Carbohydrates   New cellular material has to be produced by microorganisms in order for them to grow and reproduce. This process is called **biosynthesis**. Small molecules are produced initially from building blocks obtained from the environment, larger molecules being synthesized from these basic units.

Small molecules include glucose, amino acids and nucleic acids. Larger macromolecules include complex carbohydrates, cellulose, lipids and proteins.

All heterotrophic species of organisms must synthesize glucose from sources other than carbon dioxide. The process is termed **gluconeogenesis** (*Fig. 1*) and essentially it is a pathway that reverses the process of glycolysis.

There are several points at which irreversible glycolysis reactions are substituted with gluconeogenic reactions.

- Phosphoenolpyruvate kinase converts pyruvate to phosohoenolpyruvate.
- Fructose-6-phosphate is synthesized by fructose bisphosphatase from fructose 1-6 phosphate.
- Glucose synthesis by glucose-6-phosphatase from glucose-6-phosphate.

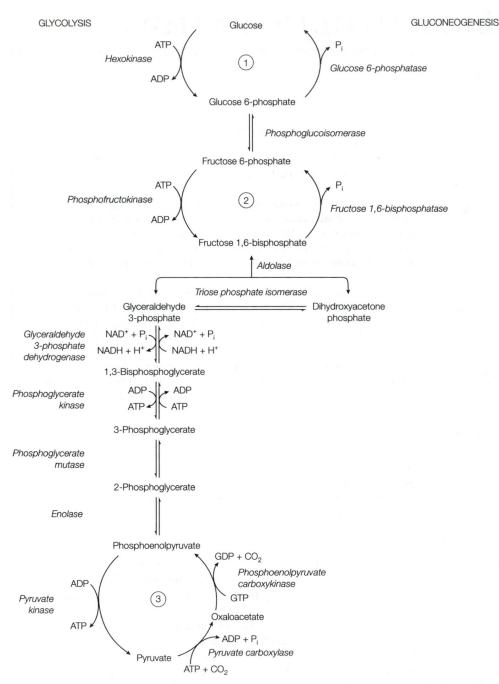

Amino acids

Assimilation of nitrogen in microbes is very variable. Only one group can utilize atmospheric nitrogen, in a process called **nitrogen fixation**. This reaction is only seen in proteobacteria such as *Azotobacter* or *Rhizobium* as well as in Gram positive Bacteria such as some species of *Clostridium*.

Nitrogen fixation is mediated by an oxygen-sensitive enzyme called **nitrogenase**

$$N_2 + 6H^+ + 12\ ATP + 12\ H_2O \rightarrow 2NH_3^+ + 12\ ADP + 12\ P_i$$

This reaction is an extremely energy expensive one.

All other members of the microbial world utilize nitrate, nitrite or ammonia. Nitrate and nitrite must be reduced to ammonia before assimilation via nitrate and nitrite reductase.

There are 20 different amino acids found in **proteins**. However there are not 20 different biosynthetic pathways because there are five amino acid families that share parts of their biosynthesis. The five families are based on glutamate, aspartate, pyruvate, serine or chorismate (*Fig. 2*).

NADPH glutamate dehydrogenase will synthesize glutamate from α-ketoglutarate by the amination of the 2 carbon organic acid. A series of **transaminations** to other organic acid intermediates of the TCA cycle synthesizes aspartate. Glutamine is synthesized from glutamate by the enzyme glutamine

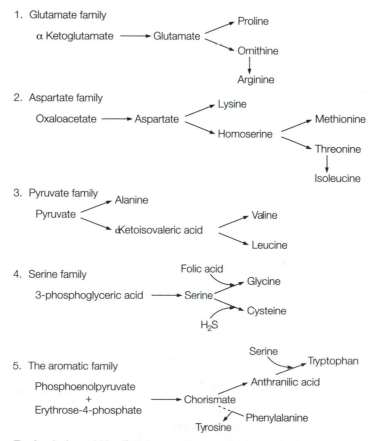

1. Glutamate family

α Ketoglutamate ⟶ Glutamate ⟨ Proline / Ornithine → Arginine

2. Aspartate family

Oxaloacetate ⟶ Aspartate ⟨ Lysine / Homoserine ⟨ Methionine / Threonine → Isoleucine

3. Pyruvate family

Pyruvate ⟨ Alanine / αKetoisovaleric acid ⟨ Valine / Leucine

4. Serine family

Folic acid
3-phosphoglyceric acid ⟶ Serine ⟨ Glycine / Cysteine
H_2S

5. The aromatic family

Phosphoenolpyruvate + Erythrose-4-phosphate ⟶ Chorismate
Serine → Tryptophan / Anthranilic acid / Phenylalanine / Tyrosine

Fig. 2. Amino acid families share common pathways.

synthetase. Alanine, aspartate and asparagine are formed by transamination of glutamate to pyruvate or oxaloacetate respectively.

Nucleic acids

Nucleic acids are formed of **purines** and **pyrimidines**, cyclic nitrogen containing compounds called **nucleotides** (See Section F3). The purines, adenine and guanine are two-ringed structures synthesized from the precursor ribose-5-phosphate, but six other molecules contribute to its formation (*Fig. 3a*). Pyrimidines, uracil, cytosine and thymine are synthesized by the carbamation of aspartic acid via the enzyme aspartate carbamyl transferase, cyclization forming the single ring structure (*Fig. 3b*).

Nucleosides are formed from a purine or pyrimidines linked to a pentaose sugar and if the sugar is phosphorylated it is termed a nucleotide.

Lipids

Fatty acid synthesis occurs in the cytosol using NADPH as the reductant (*Fig. 4*). An **acyl carrier protein** (ACP) is central to the synthesis of fatty acids. Acetyl CoA is carboxylated to malonyl CoA and then both acetyl CoA and malonyl CoA are formed by the transacylase enzymes. AcetylACP and MalonylACP condense to form acetoacetyl ACP. The acetoacetyl ACP undergoes a reduction, dehydration and a further reduction to produce 4 carbon butytryl ACP.

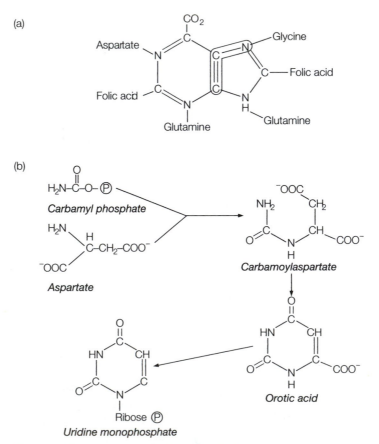

Fig. 3.   (a) Purine structure. (b) Pyrimidine structure.

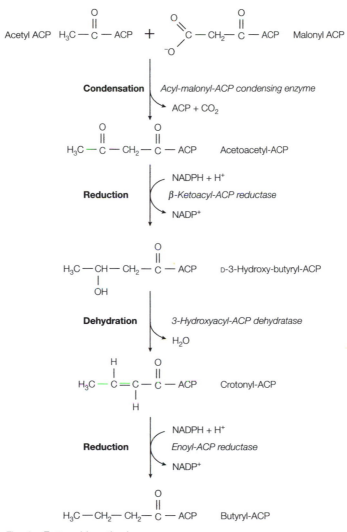

Fig. 4.   Fatty acid synthesis.

This cycle repeats to synthesize the long chain fatty acids needed for **membrane synthesis**. Lipids are synthesized from long chain saturated fatty acids, containing on average 18 fatty acids with a single unsaturated bond. Some may be branched. In aerobic Bacteria and eukaryotic microbes double bonds are introduced in the saturated fatty acid chain by the action of desaturase enzymes. Facultative and anaerobic Bacteria form double bonds during the synthesis of the fatty acid.

F1 DNA – THE PRIMARY INFORMATIONAL MACROMOLECULE

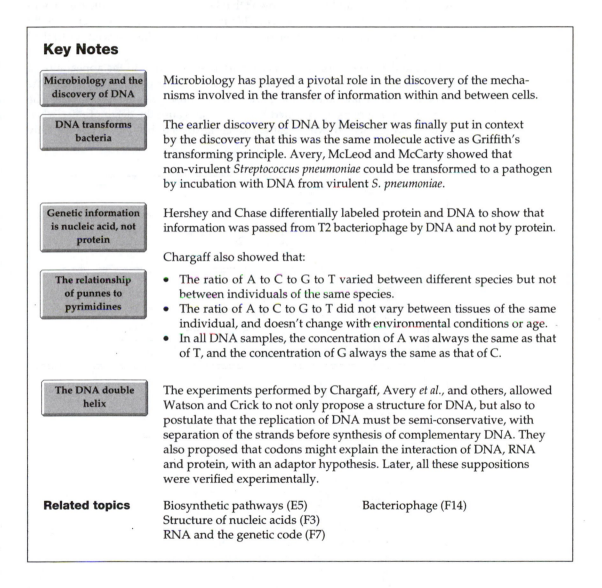

Key Notes

Microbiology and the discovery of DNA	Microbiology has played a pivotal role in the discovery of the mechanisms involved in the transfer of information within and between cells.
DNA transforms bacteria	The earlier discovery of DNA by Meischer was finally put in context by the discovery that this was the same molecule active as Griffith's transforming principle. Avery, McLeod and McCarty showed that non-virulent *Streptococcus pneumoniae* could be transformed to a pathogen by incubation with DNA from virulent *S. pneumoniae*.
Genetic information is nucleic acid, not protein	Hershey and Chase differentially labeled protein and DNA to show that information was passed from T2 bacteriophage by DNA and not by protein. Chargaff also showed that:
The relationship of punnes to pyrimidines	• The ratio of A to C to G to T varied between different species but not between individuals of the same species. • The ratio of A to C to G to T did not vary between tissues of the same individual, and doesn't change with environmental conditions or age. • In all DNA samples, the concentration of A was always the same as that of T, and the concentration of G always the same as that of C.
The DNA double helix	The experiments performed by Chargaff, Avery *et al.*, and others, allowed Watson and Crick to not only propose a structure for DNA, but also to postulate that the replication of DNA must be semi-conservative, with separation of the strands before synthesis of complementary DNA. They also proposed that codons might explain the interaction of DNA, RNA and protein, with an adaptor hypothesis. Later, all these suppositions were verified experimentally.
Related topics	Biosynthetic pathways (E5) Bacteriophage (F14) Structure of nucleic acids (F3) RNA and the genetic code (F7)

Microbiology and the discovery of DNA

Microbiology has played a pivotal role in the discovery of the mechanisms involved in the transfer of information within and between cells. The biggest milestone in molecular genetics is the elucidation of the structure of DNA by Watson, Crick, Franklin, and Wilkins in the early 1950s, but the cellular role for this macromolecule had already begun to be established.

DNA transforms bacteria

Although biochemistry was well advanced by 1930, the identity of the molecule responsible for the transfer of genetic information between mother and daughter cell was not agreed upon. Protein was held to be the most likely candidate as it held sufficient diversity in sequence to code for the many cellular variations. DNA had been purified as early as 1868 (by **Freidrich Meischer**, from pus in surgical bandages) and many of its characteristics determined, but had not been explicitly linked to information flow. A breakthrough came in 1944 when **Avery, MacLeod, and McCarty** showed that DNA could confer virulence on non-pathogenic *Streptococcus pneumoniae*. This finding was built on earlier work by **Griffith**, who had shown that the same effect could be achieved with heat-killed bacteria (*Fig. 1*).

Griffith described virulent *S. pneumoniae* as smooth due to the appearance of the colonies on agar plates. The virulence factor is a capsule which causes the smooth colonies and allows the Bacterium to cause pneumonia in mice. The finding that live non-encapsulated Bacteria could cause the death of mice when mixed and incubated with heat-killed encapsulated bacteria was not fully

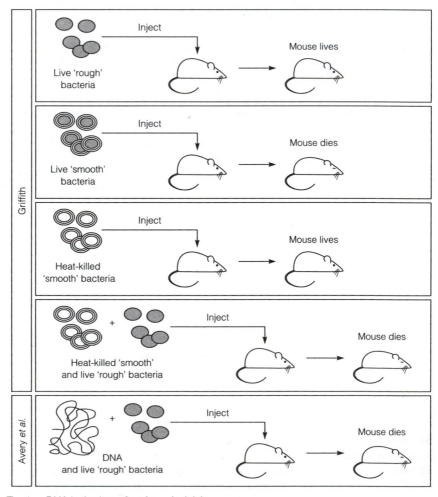

Fig. 1.   DNA is the 'transforming principle'.

explained at the time. Griffith did not fractionate *S. pneumoniae* but did suggest that a subcomponent of the encapsulated Bacteria was responsible, a component he called the **'transforming principle'**. The fractionation was performed by Avery, MacLeod, and McCarty, who went on to not only find that the principle was DNA, but also propose that DNA was a carrier of genetic information. In many ways these experiments were a series of fortunate events, in that Griffith could have chosen many other Bacteria to study, but chose one that was naturally competent (see Section F12) under the conditions he used. Furthermore, *S. pneumoniae* can be transformed by linear fragments of dsDNA, and had Avery, MacLeod, and McCarty carried out their experiments with *E. coli*, they would not have obtained such easily explained results. Meanwhile, the debate over whether DNA was the only repository for genetic information continued.

Genetic information is nucleic acid, not protein

Another microbiological experiment was used to determine if DNA alone could be responsible for the persistence of genetic information in a daughter cell. The experiment supposed that labeled DNA would be detectable in all progeny of a cell, and conversely that labeled protein would also persist, if either were the primary informational macromolecule. This sounds simple in theory, but practically bacteriophage (see Section F14) were the only way to manufacture a genetic entity with either DNA labeling only or protein labeling only. **Hershey and Chase** ^{32}P-radiolabeled T2 bacteriophage DNA, and then allowed the self-assembly of virions (see Section F14) with unlabeled protein. Concurrently they ^{35}S-radiolabeled T2 bacteriophage coat protein, and made virions with unlabeled DNA (*Fig. 2*).

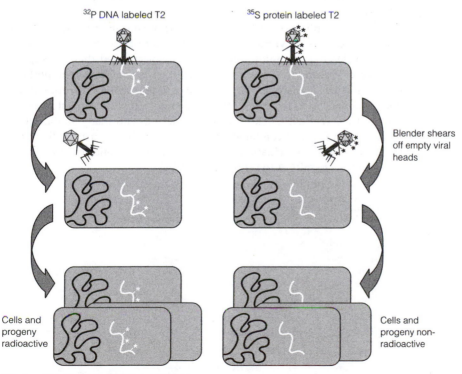

^{32}P DNA labeled T2 ^{35}S protein labeled T2

Blender shears off empty viral heads

Cells and progeny radioactive

Cells and progeny non-radioactive

Fig. 2. The Hershey–Chase experiment.

By interrupting the infection of *E. coli* by T2, by using a blender, they were able to separate the heads from the cells by centrifugation. In this way they demonstrated that the DNA entered the cell and caused a change but not the T2 protein, conclusively demonstrating that DNA carries genetic information.

The relationship of purines to pyrimidines

The composition of DNA had been rigorously examined by **Chargaff** and colleagues in the late 1940s. They established that DNA had a set of properties, commonly known as **Chargaff's rules**. These were that:

- The ratio of A to C to G to T varied between different species but not between individuals of the same species.
- The ratio of A to C to G to T did not vary between tissues of the same individual, and does not change with environmental conditions or age.
- In all DNA samples, the concentration of A was always the same as that of T, and the concentration of G always the same as that of C, commonly summarized as **A + T = G + C**.

The structure of DNA could not have been worked out without these rules, and the implications of the structure would not have been fully appreciated without the work on the transforming principle and interrupted T2 infection.

The DNA double helix

The process by which the structure of DNA was found to be a double helix has been well documented elsewhere. The salient points of the structure are summarized in Section F3, but it is important to see the structure in the context of the previous chemical and biological evidence – for example, the double helix was only regular in the way that X-ray diffraction experiments suggested if G was paired with C and T with A. More remarkably, Watson and Crick were able to suggest much more about the biochemistry of DNA than the mere shape of the molecule alone. They proposed that the replication of DNA must be **semi-conservative**, with separation of the strands before synthesis of **complementary** DNA. Subsequently, the very nature of DNA as an informational macromolecule led them to suppose the existence of **codons** (see Section F7) and thus to an **adaptor hypothesis**. In conjunction with the discovery of mRNA (see Section F6) they proposed that, in order for amino acids to reflect DNA sequence, a molecule must intervene between the two. This led Crick to draw a molecule with many similarities to tRNA (see Section F7). The combination of these hypotheses surrounding DNA structure has led to the elucidation of most of our understanding of molecular biology, from gene to mRNA to protein (via tRNA).

F2 GENOMES

Key Notes

Comparative genomics	The information available from completely and partially sequenced genomes allows comparison of gene function across phylum boundaries. The relatively small size of microbial genomes means that many more species have complete sequences compared to higher organisms.
Generalized structure of the bacterial genome	The singular circular model of a Bacterial genome is applicable to *Escherichia coli*, but other bacteria have multiple chromosomes some of which are linear. Genome size seems to be related to the organism's ecological niche, with obligate intracellular pathogens having the smallest genomes, and free-living, ubiquitous organisms the largest. A prokaryote's genome may include one or more plasmids, which can provide specialized genetic information for activities not normally associated with cell growth (e.g. antibiotic production). Eukaryotes and prokaryotes differ in the arrangement of genes. Relatively speaking, Bacterial genomes are composed mainly of genes, whereas regulatory systems for genes comprise most plant and animal genomes.
Generalized structure of Archaeal genome	A typical Archaeal genome is very similar to that of a Bacterial one. Generally the chromosome is single and circular, of a similar size to the bacteria and may be complemented by the presence of plasmids. However, Archaea have many genes and genetic systems in common with eukaryotes as well as Bacteria.
Eukaryotic genomes	Eukaryotes have many more chromosomes than prokaryotes (between 4 and 105 per cell) and do not generally have plasmids. Extracellular DNA does appear in the form of the mitochondrial and chloroplast genomes, which are held outside the cell nucleus.
Related topics	Prokaryotic Diversity (C2) The major prokaryotic groups (C6) Composition of a typical prokaryotic cell (C7) Structure of a nucleic acids (F3)

Comparative genomics

With genomic data, via DNA sequencing, becoming more easily available, it is becoming increasingly relevant to consider the whole of a microorganism rather than its individual genes. In this way subtle differences (for example, what makes *Bacillus anthracis* the causative agent of anthrax compared to the genetically very similar *Bacillus cereus*) can be examined. The arrangement of genes relative to one another, their presence or absence, the sequence of the genes, and intergenic regions are all used to compare not only species of Bacteria, but also to examine more distant relationships, e.g. between higher animals and microorganisms. The size of higher organisms' genomes (the human genome is 3200 Mbp) has made widespread whole genome sequencing a task performed by a few consortia of public and private laboratories worldwide. However, the relatively small size of Bacterial and Archaeal genomes (3 to 8 Mbp) has led to the release of new complete genome sequences on an almost weekly basis. The way in which DNA sequences are obtained and assembled into genome-sized pieces is beyond the scope of this text, but a primary consideration in approaching genomics is how the microorganism's genome is arranged. The structure of the viral genome is considered in Section L4.

Generalized structure of the bacterial genome

The Bacterial genome is often portrayed as a stable, single, circular molecule. However, the genome of most Bacteria are fluid (constantly changing in response to external stimuli) and composed of several molecules including extra chromosomes, megaplasmids and plasmids.

The model organism for molecular biology, *Escherichia coli*, is considered to be the paradigm for all Bacterial and Archaeal genomes. However, its single haploid circular chromosome, consisting of around 4.6 million bp, is rather unusual compared to other genera, but is by far the best studied. Other Bacterial genomes comprise several chromosomes, some of which are circular and some of which are linear (*Table 1*).

The size of a Bacterial genome is related to the ecological niche in which the organisms live. Obligate pathogens, such as the causative agent of epidemic typhus (*Rickettsia prowasekii*), seem to have minimized their genomes to such an extent that they rely on host proteins and metabolites in order to replicate. This is taken to the extreme in the smallest known genome, that of *Mycoplasma pneumoniae* which is composed of less than 0.9 million base pairs of DNA. In comparison, free living organisms, such as the metabolically versatile *Pseuodomoans aeruginosa* and *Streptomyces coelicolor*, have to cope with changes in temperature over tens of degrees, varying carbon and energy sources in the space of minutes and other environmental challenges. As a consequence they have a larger complement of genes regulated by a more complex sensing apparatus, and thus a larger genome.

Another strategy used by microorganisms to cope with transient environmental change is the acquisition of plasmids. Plasmids are small circular extrachromosomal pieces of DNA, which replicate independently of the genome. In contrast to

Table 1.   Chromosomal structures of Bacterial genomes*

Organism (Total size bp)	Number of circular chromosomes	Number of linear chromosomes	Extrachromosomal DNA
Escherichia coli (4 639 221 bp)	1	0	Plasmids may be present
Mycoplasma pneumoniae (816 394 bp)	1	0	Some species in the genus have a single plasmid
Rickettsia prowasekii (1 111 523 bp)	1	0	No known plasmids
Paracoccus denitrificans (~3 740 000 bp)	3 (2 + 1.1 + 0.64 Mbp)	0*	Plasmids and megaplasmids may be present
Ralstonia eutropha (5 810 922 bp)	2	0	Plasmids may be present
Deinococcus radiodurans	2	0	Plasmids may be present
Streptomyces coelicolor (6 667 507 bp)	0	1	Linear and circular plasmids may be present

*An additional linear chromosome has been found in the closely related species Paracoccus pantotrophus

the singular genome, there may be between ten and 100 000 complete copies of a plasmid in a Bacterial cell. Plasmids may carry genes that allow the microorganism to become pathogenic (one of the main differences between species of *Salmonella* is the presence of plasmid(s) carrying pathogenicity factors), resist antibiotics (resistance to kanamycin, streptomycin, and many other antibiotics is carried on plasmids) or metabolize a particular set of compounds (for example, the proteins making up the *xyl* pathway used by *Pseudomonas putida* for the degradation of toluene). Occasionally these plasmids are integrated into the genome and only exist as extrachromosomal DNA in the presence of certain physiological stimuli. While the plasmids that are used in molecular biology are in the range of 2.5 to 10 K bp, naturally occurring plasmids can be many hundreds of thousands of base pairs in size, bringing into question the philosophical difference between these megaplasmids and the chromosomes themselves.

The characteristics that distinguish Bacterial genomes from the eukaryotes lie mainly in how the genetic information is arranged. Relatively speaking, the Bacterial genome is information rich, containing many regions coding for proteins and RNA but comparatively few regions involved with the regulation of expression. Genes of similar function tend to be clustered together, and often genes in a single metabolic pathway or all involved in the synthesis of a complex multi-subunit protein are found in operons (see Section F6). Genes in an operon are sometimes so tightly packed together they overlap. The fluidity of the Bacterial genome is reflected in gene order found in different Bacterial genera: there is no similarity in the arrangement of genes among the major phyla, and often gene order is very different in species of the same genus.

Different Bacterial genomes have varying composition in terms of nucleotides. The G+C content of the Bacteria ranges from 25 to 75%, and this is often reflected in the more frequent use of certain codons (see Section F7) for certain amino acids (termed codon usage). While Bacterial genomes do contain repeating elements, they are often long repeats of >10 bp and may be associated with pathogenicity islands, insertion sequences or the remnants of excised lysogenic bacteriophage.

Generalized structure of Archaeal genome

A typical Archaeal genome is very similar to that of a Bacterial one (see *Table 2*). Generally the chromosome is single and circular, of a similar size to the bacteria and may be complemented by the presence of plasmids (see Section F15). The main differences are in the fine structure of the arrangement of genes and the proteins that associate with the genomic DNA.

While the Archaea have operons and tend to exhibit clustering of genes according to function, the arrangement of the genes have elements in common with both the eukaryotes and the Bacteria. An Archaeal operon may contain genes

Table 2. Chromosomal structures of Archaeal genomes

Organism (total size bp)	Number of circular chromosomes	Number of linear chromosomes	Extrachromosomal DNA
Methanococcus jannaschii (1 664 970 bp)	1	0	Species characteristic-plasmids of 58,407 bp and 16,550 bp
Aeropyrum pernix (1 669 695 bp)	1	0	No known plasmids
Pyrococcus horikoshii (1 738 505 bp)	1	0	Plasmids may be found

that have closest relatives in both the other kingdoms, and rarely the genes themselves may be made up of domains which may have origins in different Kingdoms.
However, about a third of the genes in any archeon is unique to this kingdom.

The replication origin of the Archaeal genome has many features in common
with the eukaryotes and this similarity in the gross chromosomal features is
apparent through the use of histone-like proteins to stabilize the chromosomal
tertiary structure.

**Eukaryotic
genomes**

The smallest eukaryotic genome is that of the parasite *Encephalitozoon cuniculi*
(2.5 million bp), and many eukaryotic microorganisms have smaller genomes than
the larger more differentiated organisms. Eukaryote genomes are characterized by
having a large number of chromosomes (between 4 and 105 in the haploid state,
Table 3) but do not generally have stable extrachromosomal DNA as plasmids.
As well as the nuclear chromosomes, some of the cell organelles (mitochondrion,
chloroplast) have their own chromosomes which code for proteins specific for the
function of the organelle. For detailed information on the human genome, consult
Genomes 2 (Brown, BIOS).

Fungal genomes are characterized by their lack of introns (only 43% of *S. pombe*
genes contain introns of a total of 4730). These introns are small, being only
50–200 bp in size compared to the introns of >10 kb in mammals. Although the
genes are not as tightly packed as in Bacteria or Archaea, fungi are information
rich and contain little repetitive DNA (50–60% of the *S. cerevisae* nuclear genome
is transcribed, compared to 33% of *Schizophyllum commune* and only 1% in *Homo
sapiens*).

Table 3.   Features of eukaryotic genomes

Organism	Description	Genome size (bp)	Haploid number of chromosomes
Encephalitozoon cuniculi	Human pathogen	2 507 519	11
Saccharomyces cerevisiae	Budding yeast	12 495 682	4
Cyanidioschyzon merolae	A unicellular red alga	16 520 305	10
Plasmodium falciparum	Causes malaria	22 853 764	7
Neurospora crassa	Fungus	38 639 769	7
Caenorhabditis elegans	Microscopic worm	100 258 171	6
Arabidopsis thaliana	Flowering plant	115 409 949	5
Drosophila melanogaster	Fruit fly	122 653 977	4
Homo sapiens	Human	$\sim 3.3 \times 10^9$	23
Fugu rubripes	Pufferfish	$\sim 3.65 \times 10^8$	21
Oryza sativa	Rice	$\sim 4.3 \times 10^8$	18
Amphibians	Various	$10^9 - 10^{11}$	
Equisetum arvense	Horsetail	$>10^{11}$	105

F3 STRUCTURE OF NUCLEIC ACIDS

Key Notes

Nucleotides and nucleosides

The monomeric unit of DNA and RNA polymers is the nucleotide. Each nucleotide is made up of base (which can either be a purine or a pyrimidine) joined together via a glycosidic bond to ribose sugar. The polymer is formed via a phopshodiester bond between the ribose units. DNA and RNA contain the purine bases adenine and guanine, and the pyrimidine cytosine. However, RNA alone has the pyrimidine uracil and DNA alone thymine.

DNA and RNA

DNA and RNA differ in the hydroxylation of the ribose (deoxyribonucleic acid compared to ribonucleic acid). Most cellular RNA is single stranded but DNA is double stranded, with two polymeric molecules joined by hydrogen bonding between bases. Bases are always aligned so that A=T and G≡C. DNA polymers are antiparallel, so that the ends are the 3′ hydroxyl group on one polymeric molecule, and the 5′ phosphate group of its partner. The strands are arranged in a double helix with a major and a minor groove, and are destabilized by heat and/or high salt concentrations, dissociating into single strands. The complementary nature of DNA means that the sequence of only one strand of DNA is written from 5′ to 3′, as the other can be deduced from the first.

Related topics

Biosynthetic pathways (E5)
DNA - The primary information macromolecule (F1)
Manipulation of cellular DNA and RNA (F16)

Nucleotides and nucleosides

Both RNA and DNA are polymers. The monomeric unit is the nucleotide, which consists of three main parts: a nitrogenous **base**; a sugar and a phosphate. A nucleoside consists of the base and sugar parts alone. The nitrogenous base can either be a **purine** or a **pyrimidine** (*Fig. 1*). The sugar part is a **ribose** with five carbons, and as the sugar is held (chemically speaking) to be a side-group of the nitrogenous base, these carbons are labeled 1-prime (1′) to 5-prime (5′). The purine or pyrimidine is linked at the 1′ position, the phosphate at the 5′. Both DNA and RNA contain the purines adenine and guanine, normally abbreviated to A and G. In nucleotides, the purine is linked to the ribose via a glycosidic bond joining ribose C1′ to purine C9.

In contrast, DNA contains nucleotides with pyrimidines cytosine (C) and thymine (T), but RNA cytosine and uracil (U). Cytosine, uracil and thymine are joined via their C1 position to the C1′ of ribose. The nucleotide monomers are joined together to form DNA or RNA via the phosphate and sugar groups, so that the 3′ carbon of one sugar is joined via the phosphate to 5′ carbon of another. This concept of 5′ and 3′ is important when assigning a direction to DNA and RNA polymers.

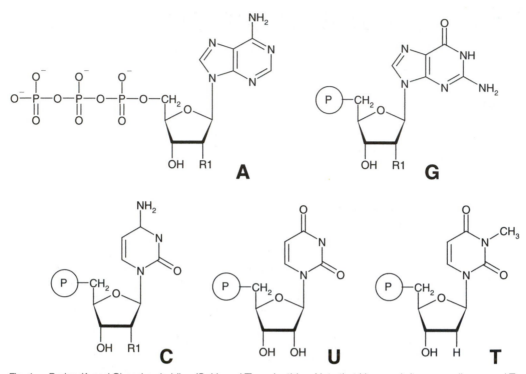

Fig. 1.   *Purine (A and G) and pyrimidine (C, U, and T) nucleotides. Note that U can only be an oxyribose, and T can only be deoxyribose. A, G, and C can be deoxy- or oxy-ribose, signified by R1. The full chemical structure of phosphate is shown for A, but omitted for clarity in the remaining nucleotides.*

DNA and RNA

Ribonucleic acid (RNA) is made up of the nucleotides A, C, G, and U in which the ribose has a hydroxyl (OH) group at the 2' position. It occurs as a single strand of polynucleotides in the cell (*Fig. 2*), whereas DNA (**deoxyribonucleic acid**) is double stranded (*Fig. 3*) and is made up of the nucleotides A, C, G, and T with only a hydrogen at the 2' position. The absence of the oxygen molecule at the 2' position gives DNA the 'deoxy' part of its name. The strands of DNA are held together by hydrogen bonds (*Fig. 3*) between the bases A and T (two hydrogen bonds) or G and C (three hydrogen bonds).

This property of hydrogen bonding only between A = T or G ≡ C, along with data from **X-ray diffraction** showing that the spacing of the nucleotides was regular, were some of the factors that allowed Watson and Crick to deduce a 3-dimensional structure for DNA. The model proposed that the two strands of DNA ran **anti-parallel** to one another, so that the backbone of each strand of phosphate and ribose were identical to one another, retaining the 5' to 3' linkages (*Fig. 3*). This orientation means that the charged phosphate groups are on the outside, rendering the molecule soluble, while the more hydrophobic bases are in the interior. The X-ray diffraction patterns also gave a clear indication that the two antiparallel strands were arranged in a right-handed **double helix** (*Fig. 4*).

The double helical structure means that there is a **minor groove** and a **major groove** on the exposed part of the polymer. Different types of molecules (other biomolecules or small chemical compounds) preferentially bind to either the major or minor grooves. The interaction of other molecules, plus variation in salt concentration or temperature, can cause the double helix to become

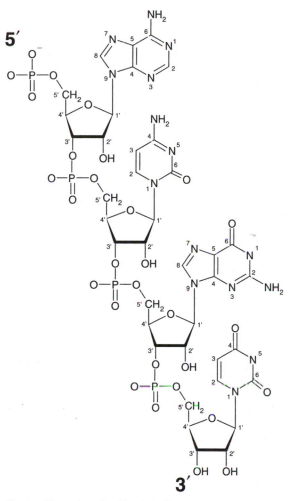

Fig. 2.   The polynucleotide chain. Illustrated is a single-stranded RNA molecule.

partly unwound, overwound or even to **dissociate** into two single strands completely.

Rather than drawing a full chemical structure of DNA and RNA each time, by convention, both DNA and RNA sequences are written from 5′ to 3′ using single initials for their bases only, e.g. 5′ -ACGTCTTAGCTAGC-3′ or 5′-AUCGACU-UGCAGC-3′. The presence of U in RNA only and T in DNA only provides a convenient marker showing whether the sequence is deoxyribonucleic acid or ribonucleic acid. Only one strand of DNA is written down, as the nature of the base-pairing means that the **complementary sequence** can be written down quickly and easily, with A matched to T and C matched to G.

5′ -ACGTCTTAGCTAGC-3′
3′ -TGCAGAATCGATCG-5′

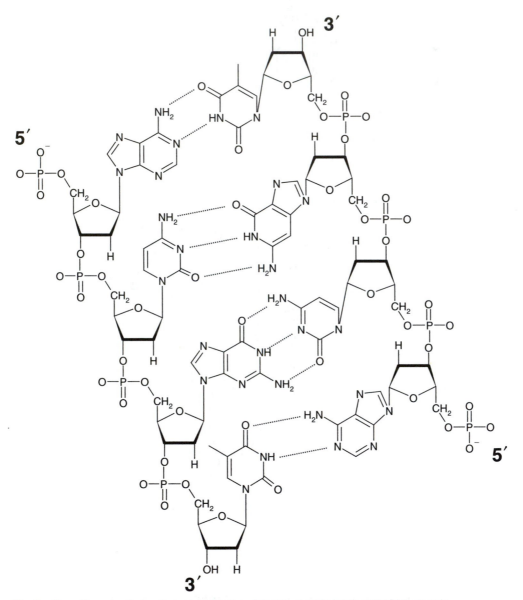

Fig. 3.   Two-dimensional chemical representation of double-stranded antiparallel DNA strands.

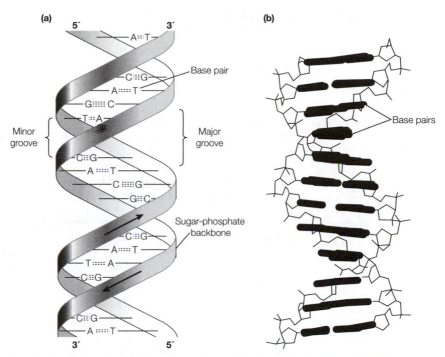

Fig. 4.   The DNA double helix (a) schematic view of the structure; (b) a more detailed structure, highlighting the stacking of the base pairs. From Instant Notes in Molecular Biology, *2nd Edn., L1, Fig. 5, p. 35.*

F4 HYBRIDIZATION

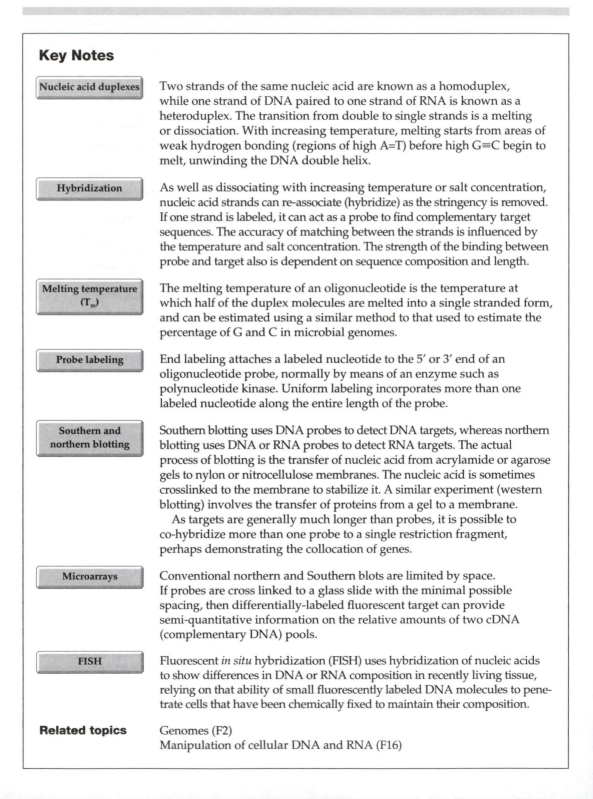

Key Notes

Nucleic acid duplexes	Two strands of the same nucleic acid are known as a homoduplex, while one strand of DNA paired to one strand of RNA is known as a heteroduplex. The transition from double to single strands is a melting or dissociation. With increasing temperature, melting starts from areas of weak hydrogen bonding (regions of high A=T) before high G≡C begin to melt, unwinding the DNA double helix.
Hybridization	As well as dissociating with increasing temperature or salt concentration, nucleic acid strands can re-associate (hybridize) as the stringency is removed. If one strand is labeled, it can act as a probe to find complementary target sequences. The accuracy of matching between the strands is influenced by the temperature and salt concentration. The strength of the binding between probe and target also is dependent on sequence composition and length.
Melting temperature (T_m)	The melting temperature of an oligonucleotide is the temperature at which half of the duplex molecules are melted into a single stranded form, and can be estimated using a similar method to that used to estimate the percentage of G and C in microbial genomes.
Probe labeling	End labeling attaches a labeled nucleotide to the 5′ or 3′ end of an oligonucleotide probe, normally by means of an enzyme such as polynucleotide kinase. Uniform labeling incorporates more than one labeled nucleotide along the entire length of the probe.
Southern and northern blotting	Southern blotting uses DNA probes to detect DNA targets, whereas northern blotting uses DNA or RNA probes to detect RNA targets. The actual process of blotting is the transfer of nucleic acid from acrylamide or agarose gels to nylon or nitrocellulose membranes. The nucleic acid is sometimes crosslinked to the membrane to stabilize it. A similar experiment (western blotting) involves the transfer of proteins from a gel to a membrane. As targets are generally much longer than probes, it is possible to co-hybridize more than one probe to a single restriction fragment, perhaps demonstrating the collocation of genes.
Microarrays	Conventional northern and Southern blots are limited by space. If probes are cross linked to a glass slide with the minimal possible spacing, then differentially-labeled fluorescent target can provide semi-quantitative information on the relative amounts of two cDNA (complementary DNA) pools.
FISH	Fluorescent *in situ* hybridization (FISH) uses hybridization of nucleic acids to show differences in DNA or RNA composition in recently living tissue, relying on that ability of small fluorescently labeled DNA molecules to penetrate cells that have been chemically fixed to maintain their composition.
Related topics	Genomes (F2) Manipulation of cellular DNA and RNA (F16)

Nucleic acid duplexes

The elegant model of the DNA double helix proposed by Watson and Crick is one example of a nucleic acid **duplex**, in that it is composed of two chemically non-identical polynucleotides. For example, the sequence of one strand may be 5'-AATCGTAGCTGATCGG-3' while the other will be the complementary sequence 5'-CCGATCAGCTACGATT-3' (see Section F3). However, both strands are DNA so this is known as a **homoduplex**. Nucleic acids will also form other duplexes, both DNA:RNA (a **heteroduplex**) and RNA:RNA (a rare homoduplex in cell biology), The strands form double helices with similar dimensions to the Watson–Crick DNA double helix, and are stabilized by hydrogen bonds between base-pairs. Both homo- and heteroduplexes can be split into their component single strands by the action of heat, certain chemicals or a combination of the two. The transformation from duplex to single strands is known as **melting** or **disassociation.** The process starts in regions of the duplex with less hydrogen bonding, where there are more A=T than G≡C base pairs so the helix is slightly weaker. The formation of single strands is then completed at higher G≡C regions, so it can appear that the helix **unwinds**.

Hybridization

Two strands of nucleic acid can reform a double helix once the heat or chemical melting agent is removed (a process known as **re-association** or **hybridization**). An imperfect double helix can even form if the match in sequence between the two strands is not exact. This unusual ability of DNA and RNA to melt and then reform helices is exploited in many molecular biology based microbiology experiments, where a labeled polynucleotide with some form of marker attached (normally known as the **probe**) is used to find similar nucleic acid sequences (the target). The conditions under which the hybridization of target and probe is allowed to take place change according to how specific the reassociation is. Under conditions of high heat and/or salt, only a perfect match of target and probe can occur, as a conventional double helix with the highest chemical stability is reformed (*Fig. 1*). As the heat and or salt concentration is lowered, the probe can hybridize to other sequences on the target that may not be a perfect match (i.e. a mismatch in base pairing is allowed). The conditions which influence the amount of non-specific hybridization are collectively known as **stringency**.

The stringency required to allow a particular hybridization to occur is influenced by several sequence-specific factors. A greater number of G or C in the probe will mean that the stability of the duplexes is higher (3 hydrogen bonds per base pair) so stringency can be lower. The length of the probe will also have an effect – longer probes are more stable but can tolerate a greater number of mismatches. A convenient way of describing the hybridization properties of an oligonucleotide is via the melting temperature.

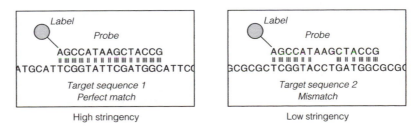

Fig. 1.   *Hybridization of a DNA probe to a DNA target.*

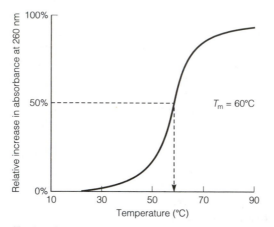

Fig. 2.   Determination of the melting temperature of oligonucleotides.

Melting temper-
ature (T_m)

The melting temperature of an oligonucleotide is defined as the temperature at which half of the duplex molecules are melted into a single-stranded form. This would be an extremely difficult experiment to perform, but fortunately the absorbance of double-stranded and single-stranded nucleotides is slightly different. At A_{260nm} single-stranded nucleotides have a higher absorbance, so if the absorbance of a duplex is measured with time in a defined solution, an estimate of T_m can be made (*Fig. 2*).

Knowing the T_m of an oligonucleotide can allow the researcher to predict at what temperature stable hybridization will occur, usually 15 to 25°C below the melting temperature, when a probe is only a few tens of base pairs in length. An experiment similar to determining the T_m of short oligonucleotides is used to estimate the percentage of G and C in microbial genomes an important character in microbial classification.

Probe labeling

The probe can be visualized in a number of ways by attaching either radioactive or fluorescent molecules to the oligonucleotide. These markers can then be detected by techniques, such as autoradiography, scintillation counting or scanning densitometry, which are outside the scope of this text. Radioactive labels include ^{32}P, ^{33}P and ^{35}S, normally included by substituting an existing atom of one or more nucleotides in the probe with the radioactive equivalent. Bulkier fluorescent molecules are attached covalently to oligonucleotides by chemical means. In both cases there are two approaches to labeling. **End labeling** (*Fig. 3*) attaches a labeled nucleotide to the 5′ or 3′ end of an oligonucleotide probe, normally by means of an enzyme such as polynucleotide kinase. **Uniform labeling** (*Fig. 4*) incorporates more than one labeled nucleotide along the entire length of the probe, normally by making copies of the probe with DNA polymerase I or making a PCR product with labeled nucleotides.

Polynucleotide kinase
plus [$\gamma$$^{32}P$]ATP catalyses
phosphate exchange or
phosphorylation of free
5′ hydroxyl group

'Cold' probe

AGCCATAAGCTACCG

 AGCCATAAGCTACCG

Labeled probe

Fig. 3.   Probe end labeling.

Random hexamers anneal to 'cold' probe

CCGTTG   AGCCAT   CCGTAA
GCGGCAAGCTAGCATGCATTCGGTATTCGATGGCATTCGTCGA

DNA polymerase I
with cold dCTP, dGTP
and dTTP but with
$[\gamma^{32}P]$dATP synthesize
complementary strand

CCGTTGCATCGTACGTAAGCCATAAGCTACCGTAAGCAGCT
GCGGCAAGCTAGCATGCATTCGGTATTCGATGGCATTCGTCGA

Many copies made, complementary labeled
probe copy can be used after denaturation

Fig. 4.   *Probe uniform labeling.*

Southern and northern blotting

Some of the most commonly used applications for hybridization in microbiology are the processes named after Professor E. Southern, in which hybridization is carried out on a membrane. The process of transferring nucleic acid to the membrane is known as blotting, so detection of DNA targets using a DNA probe is known as **Southern blotting**, whereas, a sort of opposite, **northern blotting** is the detection of RNA targets using a DNA or RNA probe. An analogous technique in which proteins are blotted onto a membrane and then detected using antibodies is known as **western blotting**. Note that the capitalization only applies to the surname of the inventor.

The target DNA in Southern blotting is usually genomic DNA treated with one or more restriction enzymes and separated by molecular weight using agarose gel electrophoresis (see Section F16). The DNA is denatured (normally using an alkaline solution such as NaOH) and transferred to a nitrocellulose or nylon membrane by capillary action or the application of a vacuum (*Fig. 5*). This target DNA is sometimes **cross-linked** to the membrane using ultraviolet light to

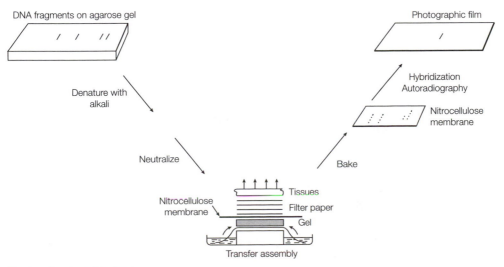

Fig. 5.   *Southern blotting.*

improve attachment. The procedure for northern blotting is identical, except that total RNA is used and this generally isn't digested in Bacterial molecular biology.

The membrane and its target DNA or RNA are then treated with the probe and, after hybridization is judged to have been completed, the membrane is washed thoroughly to remove any unbound probe. The temperature at which this washing is carried out and the composition of the wash buffer provide the stringency, determining whether the probe binds only to exactly complementary sequences, or whether it can also bind to sequences with some mismatch as well. Southern blotting can be used to address many microbiological problems. In a binary fashion it can be used to determine whether a gene interruption experiment has occurred successfully if the probe hybridizes to the cassette inserted into a gene, or to determine the presence or absence of a gene in a particular species or strain of bacterium. In a more deductive manner, it can be used for gene mapping experiments, both to show that two probes **cohybridize** to the same restriction fragment, indicating their proximity to one another, or for restriction mapping as outlined in *Fig. 6*.

The detection of mRNA using a northern blotting procedure is used to determine the length of a particular transcript, which is useful when determining if the gene of interest is transcribed monocystronically or as part of an operon. Northern blotting can be used to show that a particular gene is switched on under certain conditions, by running RNA isolated from two differently incubated cultures, side by side. Careful use of scanning densitometry, in conjunction with northern blotting can even give an indication of the level of expression of a particular gene, but this is being replaced by easier techniques such as QRT-PCR (see Section F16).

Microarrays

The disadvantage of using agarose gel electrophoresis coupled to Southern or northern blotting is that only a few experiments can be performed. The advent of machines capable of spotting defined amounts of oligonucleotide in very small areas, and the development of new techniques to cross-link these oligonucleotides, to surfaces such as glass, have led to the development of microarray techniques.

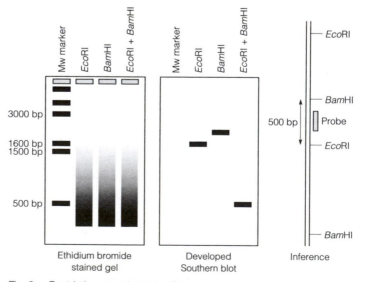

Fig. 6. Restriction mapping using Southern blotting.

The principle is much the same as Southern blotting, except that the probe is now immobilized and the target is labeled. Oligonucleotides spanning open reading frames from a genome of known sequence are amplified by PCR, denatured, and spotted in equal amounts onto a glass slide by robot in specific positions, at densities of up to 1000 spots per cm^2. The slide is hybridized with dye-labeled cDNA from cultures grown under two different conditions. One pool of cDNA is labeled with a fluorescent dye, such as Cy3, the other with Cy5. Using scanning fluorescence densitometry of the slide, the signals produced are proportional to the amounts of cDNA present and allow identification of cell- or condition-specific changes in transcript level over a limited dynamic range. This process can be further automated and, if factory prepared silicon high-density arrays are used, millions of oligos spanning an entire eukaryotic genome and representing the entire transcriptome of an organism can be analyzed.

FISH

Fluorescent *in situ* hybridization (FISH) uses hybridization of nucleic acids to show differences in DNA or RNA composition in recently living tissue, relying on the ability of small fluorescently labeled DNA molecules to penetrate cells that have been suitably treated. Normally this **fixing** process leads to the death of the cell and can be considered a destructive procedure. However, as long as the sample can be sectioned thinly enough to allow transmission of fluorescence, chromosomes can be 'painted' different colors depending on their gene content. Alternatively, different types of prokaryotes in a mixed colony can be highlighted using carefully designed species-specific probes, with the results collected using a fluorescence microscope. The latter technique is used frequently in modern microbial ecology.

F5 DNA REPLICATION

Key Notes

DNA replication

DNA replication in all organisms proceeds via similar mechanisms. Although the speed of replication can vary, it proceeds by common rules:

- replication is semi-conservative;
- all DNA and RNA polymerases synthesise DNA in a 5′ to 3′ direction;
- all cellular DNA and RNA polymerases initiate synthesis with an RNA primer;
- sll DNA and RNA polymerases require magnesium ions to function.

Synthesis of nucleic acids by polymerases

All the bacterial nucleic acid polymerases share a common hand-like structure, with the template polynucleotide binding to the palm (the active site), and nucleotides entering between the thumb and fore-finger. Bacteria have several DNA polymerases, but only one core RNA polymerase, whereas the eukaryotes have many specialized variants of both.

Initiation of DNA replication

DNA inside the cell is highly supercoiled, so must be unwound and rewound by topioisomerase I or DNA gyrase as replication proceeds. Bacterial DNA is always replicated from the origin of replication. Two primosomes are formed here, and DNA replication occurs bidirectionally.

DNA replication fork

The DNA replication fork is formed at the origin of replication by the action of DNA helicase, and re-annealing of the single strands is prevented by single-stranded binding protein. The opened helix is termed the replication bubble, in which the primosome forms. DNA primase adds a few complementary RNA nucleotides to the template strand, which acts as a primer for the main DNA replication complex, DNA polymerase III. The leading strand is synthesized continuously, but the lagging strand is made in short lengths that are eventually joined together. Primase adds short RNA fragments (Okazaki fragments) to the lagging strand, allowing DNA polymerase III to extend from 5′ to 3′ until it meets the next RNA primer, where extension stops. The short stretches of RNA are then removed by DNA polymerase I, and any breaks in the phosphodiester bonds mended by DNA ligase.

Proofreading

DNA polymerases I and III both have a separate active site that allows these enzymes to check for perfect complementarity as DNA is synthesized.

Termination of synthesis and resolution of replicated circular genomes

The movement of the two replisomes is stopped at 180° from *ori* by the binding of terminator utilization substrate (*Tus*). This leaves the two complete circular chromosomes inter-twined, a situation that is resolved by the action of XerC and XerD at the *dif* site.

Linear genomes

The full details of the replication of linear genomes has yet to be elucidated but, in *Borrelia burgdorferi*, a central *ori* is replicated bidirectionally and terminated by unknown mechanisms. In the bacteriophage σ29 replication is initiated at either end of the chromosome and is terminated in the center by the collision of the two replisomes.

Related topics	DNA-The primary information macromolecule (F1)
	Structure of nucleic acids (F3)
	Mutation (F10)
	DNA repair (F11)
	Recombination (F13)
	Manipulation of cellular DNA and RNA (F16)

DNA replication in all organisms takes place via similar mechanisms, and much of the detail has been gained from our understanding of the molecular biology of the bacterium *Escherichia coli*. In all bacteria, DNA replication is triggered by cell mass, but the speed of genomic duplication varies from species to species. In *E. coli*, new bases are added at approximately 1000 nucleotides per second (nt s^{-1}), but in *Mycoplasma capricolum* this rate falls to only 100 s^{-1}. This means that, although the mycoplasma has a considerably smaller genome, both bacteria can completely replicate their genomes in around 45 minutes.

There are two rules that govern how DNA replication can occur. The first is that replication is **semi-conservative**, in that double-stranded DNA is separated and each strand acts as a template for synthesis of a new strand composed of bases exactly complementary to the template. The second is that all nucleic acid polymerases (DNA and RNA polymerases) synthesize DNA in a 5′ to 3′ direction (see Section F3) using a primer to initiate synthesis. This primer is RNA, whether or not DNA or RNA is to be synthesized. Nucleotides (also known as bases) are added to the exposed 3′-hydroxyl group of the primer so that they are exactly complementary to the template strand (A matched with T or U; C matched with G) (see Section F3).

Synthesis of nucleic acids by polymerases

All enzymatic extension of polynucleic acids requires a template and the synthesis of nucleic acids proceeds in a 5′ to 3′ direction. The RNA and DNA polymerases, in common with many other enzymes associated with nucleic acid metabolism, also require Mg^{2+} or a similar divalent ion to function. This ion is not strictly a prosthetic group for these enzymes, but helps in the binding of nucleotides to the active site. All the bacterial nucleic acid polymerases share a common structure, in that the protein around the active site folds to form a structure that could be said to be like a hand. The template polynucleotide binds to the palm (the active site), and nucleotides enter between the thumb and forefinger.

Bacterial cells have a variety of DNA polymerases (*Table 1*) each with specialized functions, but most only have one RNA polymerase. In contrast most eukaryotes have a plethora of both specialized DNA and RNA polymerases. Unfortunately our knowledge of the cell biology of the Archaea is not sufficiently developed to be able to draw many inferences on this topic, but we do know that their primary replicative polymerase resembles one class of eukaryotic DNA polymerase more than any of the bacterial types.

Table 1. Properties of bacterial and Archaeal DNA polymerases

Polymerase	Function	Origin	Direction of synthesis
DNA polymerase I	Aids in both replication and repair	Bacteria	5′ → 3′
DNA polymerase II	Primary repair complex	Bacteria	5′ → 3′
DNA polymerase III	Primary replication complex	Bacteria	5′ → 3′
Family B polymerase	Primary replication complex	Archaea	5′ → 3′

**Initiation of
DNA replication**

Our knowledge of Bacterial chromosomal replication is mostly limited to circular chromosomes composed of double-stranded DNA. Rather than being arranged as a simple circle, the DNA is **supercoiled** – twisted and folded in on itself. This is to compact the DNA into the very small volume of the cell. The model microorganism, *Escherichia coli*, is only 2 or 3 μm long, yet its 4.6 million bp chromosome is about 1 mm in circumference. The number of supercoils in the genome is determined by the competitive action of two winding and unwinding enzymes, topoisomerase I and DNA gyrase. The equilibrium between the two is established according to the physiological state of the organism, so is influenced by external environmental factors. In addition, topoisomerase and gyrase have to act locally to allow DNA replication to begin. The supercoiled DNA must first be relaxed to allow replication proteins access to the template strands.

Bacterial DNA is always replicated from a single location on the chromosome, known as the **origin of replication**. This origin (abbreviated as *ori*) is a sequence of around 300 base pairs, and here a group of enzymes collectively called the **primosome** acts (*Fig. 1*). Two primosomes are formed at the origin of replication of circular genomes, so DNA replication occurs **bidirectionally.**

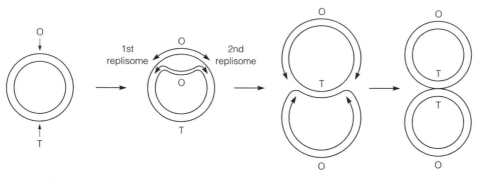

Fig. 1.   Resolution of circular bacterial genomes.

**The DNA
replication fork**

Once DNA replication has been initiated, a large number of enzymes must function to allow DNA polymerase access to the template strand. The DNA is separated into single strands by **DNA helicase**, and spontaneous re-annealing is prevented by **single-stranded binding protein** (SSB). The primosome is then formed in the developing replication bubble, allowing the binding of **DNA primase**. This primase adds a few complementary RNA nucleotides to the template strand, which acts as a primer for the main DNA replication complex, DNA polymerase III (*Fig. 2*). The extension of DNA complementary to the template strand means that the replication bubble can be extended quickly. However it was less clear for many years how the other strand (often called the lagging strand) was replicated at the same time and yet still synthesized from 5′ to 3′.

An elegant explanation of the extension of the replication bubble was that it was discontinuous, again using DNA polymerase III extending from RNA primers. It was found that primase regularly adds short RNA fragments (**Okazaki fragments**) to the lagging strand, allowing DNA polymerase III to extend from 5′ to 3′ until it meets the next RNA primer, where extension stops. The lagging strand is, for a short time, a DNA:RNA heteroduplex, before DNA polymerase I functions in its repair mode and removes RNA with an **exonuclease** activity, replacing it with complementary DNA nucleotides. Any gaps in the newly synthesized lagging strand are then resolved by DNA ligase. In this way, both the template and lagging strands of the replication bubble are quickly and accurately replicated (*Fig. 2*).

Proofreading

At each stage of DNA replication, the sequence of the newly-synthesized DNA is checked to see that it exactly matches the template strand, a function known as proofreading. First the incorporation of each nucleotide is dictated so that it exactly matches the complementary template base just by the shape of the DNA polymerase III active site. A separate active site in the enzyme (in the epsilon sub-unit) then checks the base-pairing again after incorporation as the polymerase moves

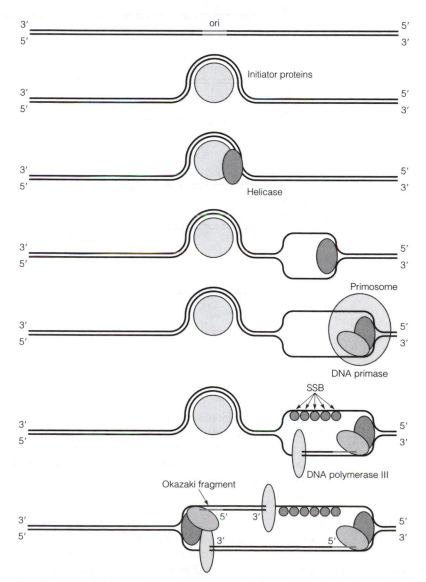

Fig. 2.   DNA replication. Adapted from Instant Notes in Molecular Biology, 2nd Edn, E1, Fig. 2, p. 74.

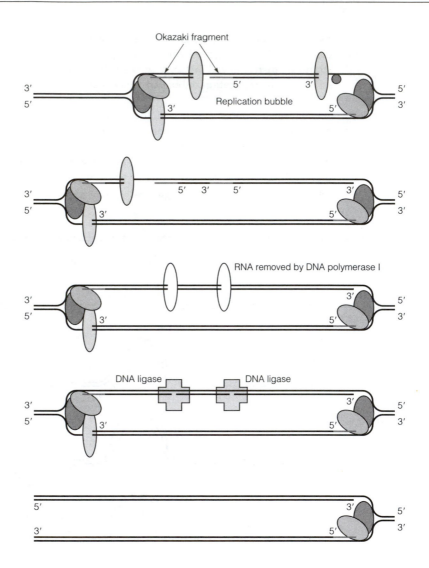

Fig. 2.   DNA replication (continued).

down the template strand. Similarly, DNA polymerase I has a proofreading domain in many organisms (though not in *Escherichia coli*) to ensure that the replacement of RNA with DNA is accomplished without error (*Fig. 3*). DNA polymerase I also participates in the continuous checking of the replicated double stranded genome as part of the mutation repair systems described in Section F11.

Termination of synthesis and resolution of replicated circular genomes

The synthesis of a copy of a circular Bacterial genome is signaled by a region coded within the DNA. This site is at a 6 o'clock position on the genome, if the *ori* is at 12 o'clock. A terminator utilization substrate (*Tus*) binds at this site, stopping the movement of the two replisomes. This might appear to be the end of the replication process, but the cell is now faced with the challenge of separating the two

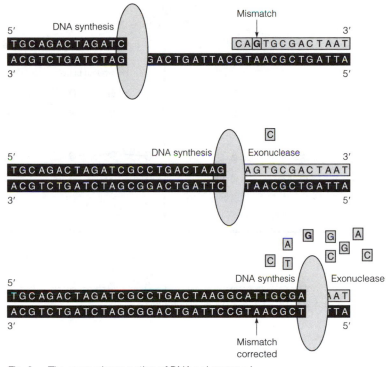

Fig. 3.   The exonuclease action of DNA polymerase I.

complete chromosomes from one another. Without an effective means of resolving the two chromosomes, they will remain intertwined with one another. Two intertwined genomes close to one another are prone to generalized recombination (see Section F13) and must be separated as quickly as possible. A short DNA sequence known as *dif* is found in the middle of the termination region, and here the proteins XerC and XerD bind both double helices and, via a transient DNA/protein complex known as the Holliday junction, cut and resplice the DNA to form two separate circular chromosomes. Thus a process of site-specific recombination (see Section F13) stops more widespread generalized recombination.

Linear genomes   A Bacterial chromosome is typically presented as a single circle of double-stranded DNA. In reality, the situation is far more complex. Most importantly Bacteria may have one or more chromosomes (see Section F2), each of which may be linear or circular. However, the full details of the mechanisms of bacterial linear DNA replication have yet to be fully elucidated. The linear chromosome of *Borrelia burgdorferi* has an origin of replication in the center, and presumably replicates bi-directionally (Figure 4), though the mechanisms of termination are unclear. More unusual mechanisms of linear chromosome replication are exhibited by the bacteriophage, such as the *Bacillus* 'phage σ29, where replication is initiated at either end of the chromosome and is terminated in the center by the collision of the two replisomes. Similar systems may exist in other Bacteria and Archaea.

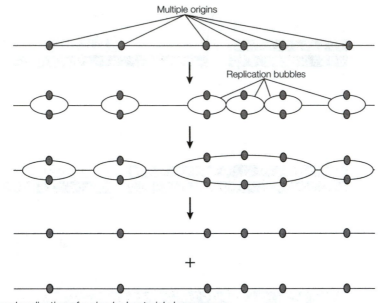

Fig. 4.   Bidirectional replication of a circular bacterial chromosome.

F6 TRANSCRIPTION

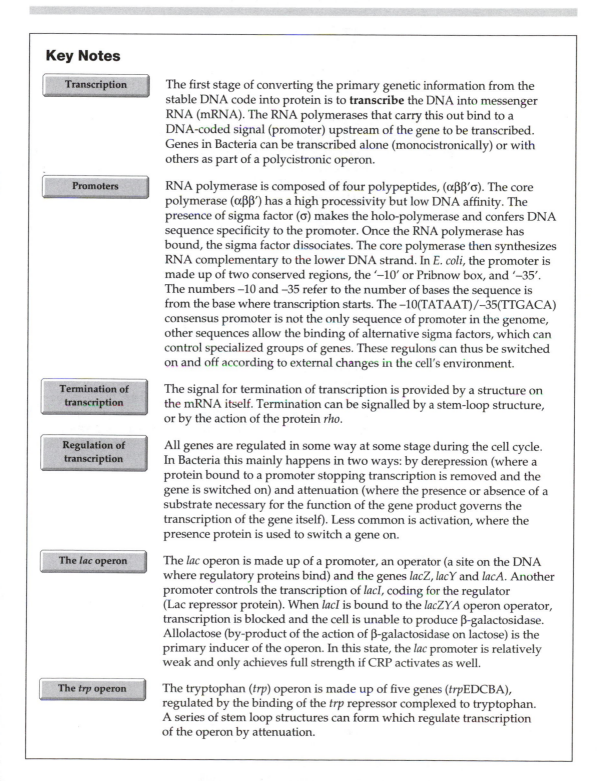

Key Notes

Transcription

The first stage of converting the primary genetic information from the stable DNA code into protein is to **transcribe** the DNA into messenger RNA (mRNA). The RNA polymerases that carry this out bind to a DNA-coded signal (promoter) upstream of the gene to be transcribed. Genes in Bacteria can be transcribed alone (monocistronically) or with others as part of a polycistronic operon.

Promoters

RNA polymerase is composed of four polypeptides, ($\alpha\beta\beta'\sigma$). The core polymerase ($\alpha\beta\beta'$) has a high processivity but low DNA affinity. The presence of sigma factor (σ) makes the holo-polymerase and confers DNA sequence specificity to the promoter. Once the RNA polymerase has bound, the sigma factor dissociates. The core polymerase then synthesizes RNA complementary to the lower DNA strand. In *E. coli*, the promoter is made up of two conserved regions, the '–10' or Pribnow box, and '–35'. The numbers –10 and –35 refer to the number of bases the sequence is from the base where transcription starts. The –10(TATAAT)/–35(TTGACA) consensus promoter is not the only sequence of promoter in the genome, other sequences allow the binding of alternative sigma factors, which can control specialized groups of genes. These regulons can thus be switched on and off according to external changes in the cell's environment.

Termination of transcription

The signal for termination of transcription is provided by a structure on the mRNA itself. Termination can be signalled by a stem-loop structure, or by the action of the protein *rho*.

Regulation of transcription

All genes are regulated in some way at some stage during the cell cycle. In Bacteria this mainly happens in two ways: by derepression (where a protein bound to a promoter stopping transcription is removed and the gene is switched on) and attenuation (where the presence or absence of a substrate necessary for the function of the gene product governs the transcription of the gene itself). Less common is activation, where the presence protein is used to switch a gene on.

The *lac* operon

The *lac* operon is made up of a promoter, an operator (a site on the DNA where regulatory proteins bind) and the genes *lacZ*, *lacY* and *lacA*. Another promoter controls the transcription of *lacI*, coding for the regulator (Lac repressor protein). When *lacI* is bound to the *lacZYA* operon operator, transcription is blocked and the cell is unable to produce β-galactosidase. Allolactose (by-product of the action of β-galactosidase on lactose) is the primary inducer of the operon. In this state, the *lac* promoter is relatively weak and only achieves full strength if CRP activates as well.

The *trp* operon

The tryptophan (*trp*) operon is made up of five genes (*trpEDCBA*), regulated by the binding of the *trp* repressor complexed to tryptophan. A series of stem loop structures can form which regulate transcription of the operon by attenuation.

Related topics	Electron transport, oxidative phospharylation and β-oxidation of fatty acids (E3)
	Structure of nucleic acids (F3)
	RNA and the genetic code (F7)
	Translation (F8)
	Signal transduction and environmental sensing (F9)

Transcription

The first stage of converting the primary genetic information from the stable DNA code into protein is to **transcribe** the DNA into messenger RNA (**mRNA**). This process is carried out by RNA polymerase, which has many common features with the DNA polymerases. The structure of the RNA polymerase is a little similar to the DNA polymerases in the active site region, and also requires Mg^{2+} to function and synthesizes nucleic acid in a 5′ to 3′ direction. However, the RNA polymerases do not require a primer to initiate synthesis. Instead, their signal for the initiation of transcription is a specific sequence on the DNA, known as the **promoter** region.

A gene's promoter is said to be 'upstream', in that the promoter region is situated to the 5′ end of the coding region. The promoter region allows the RNA polymerase to bind and begin transcription so that the resulting mRNA contains not only the coding region itself, but also all the signals to start and stop the synthesis of the polypeptide. How RNA polymerase works is intrinsic to the concept that one gene makes one polypeptide. In eukaryotes, genes are arranged so that the promoter region is in such a position so that when transcription occurs a single mRNA molecule is produced which can be used to code for a single polypeptide (**monocistronic**). Genes which code for polypeptides that have a common purpose (such as the manufacture of a multi-polypeptide protein) are placed in many different parts of the genome, frequently on different chromosomes. In prokaryotes (both the Bacteria and the Archaea) genes are more likely to be arranged so that the coding regions for enzymes involved in a single pathway are clustered together. Furthermore, several genes may be arranged so close to one another that they are transcribed from a single promoter (*Fig. 1*). This **polycistronic** arrangement is called an **operon** (see the *lac* operon below).

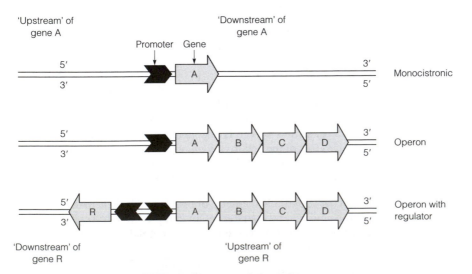

Fig. 1.   *The arrangement of genes and promoters in bacteria.*

Promoters   RNA polymerase is composed of four different polypeptides, called beta, beta prime, alpha and the **sigma factor** ($\alpha\beta\beta'\sigma$). The **core polymerase** ($\alpha\beta\beta'$) has a high processivity (RNA polymerizing activity) but low DNA affinity. The addition of the sigma factor to make **holo-polymerase** confers DNA sequence specificity, forcing the RNA polymerase to bind at the promoter region. Once the RNA polymerase has bound, the sigma factor dissociates. The core polymerase then synthesizes RNA complementary to the lower DNA strand. The promoter region tends to have a slightly higher A and T content than the surrounding DNA, and the most abundant form of promoter in *Escherichia coli* has two fairly well conserved sequences, TATAAT (the **Pribnow** box or –10 sequence) and TTGACA (the – 35 sequence). The higher A/T content means that there are fewer hydrogen bonds to hold the double stranded DNA together and thus the double helix is easier to force apart, a process termed melting. Melting of the promoter region allows access of the RNA polymerase, which specifically targets the –10 and –35 regions by the use of the sigma factor. The numbers –10 and –35 refer to relative position in relation to the number of bases the sequence is from the base where transcription starts (known as the transcription start site, *Fig. 2*).

The –10(TATAAT)/–35(TTGACA) **consensus promoter** is not the only sequence of promoter in the genome. There are many promoters with a sequence very similar to –10/–35 and the greater the difference in the base sequence is to this consensus, the weaker the promoter. A strong promoter (such as that of the *lac* operon) forms a very tight bond with the sigma factor, and transcription is very likely to be initiated from such a promoter. A weak promoter binds the sigma factor by only a few bases, and is concomitantly less likely to initiate RNA synthesis. Other promoters have completely different sequences that bind **alternative sigma factors**. A good example of this is the alternative sigma factor produced in response to low oxygen in *Escherichia coli* and some other bacteria. The sigma factor still allows RNA polymerase to recognize a –10 site, but instead will only allow the holo-polymerase to bind where there is an additional specific sequence (FNR site) at around 42 bases upstream of the transcription start site. The use of alternative sigma factors allows a whole group of genes and operons, known as a **regulon**, to be switched on and off according to external changes in the cell's environment. Induction of the FNR regulon allows the cell to induce all the genes which are useful to cope with low oxygen, principally alternative electron acceptors (see Section E3).

Termination of transcription   Confusingly, the signal for termination of transcription is provided by a structure on the mRNA itself. After the last gene in the operon has been completely trans-cribed, the RNA polymerase continues transcription past the last gene's termination codon. This part of the mRNA folds up into a **stem-loop** structure (*Fig. 3*) that causes the RNA polymerase to pause and cease transcription. The RNA polymerase then dissociates from the DNA and the mRNA it has generated, leaving the complete transcript ready for the ribosomes to translate it into protein (see Section F8).

Regulation of transcription   The strength of a promoter and alternative sigma factors represent two ways in which the cell can adjust the rate of mRNA transcription from a particular promoter, and thus the amount of individual proteins in the cell. There are some genes that seem to be switched on most of the time, particularly those genes involved in 'housekeeping' functions of the cell, such as central metabolic path-ways, and those that account for gene products, such as the ribosomal components However, we have come to recognize that all genes are **regulated** in some way at some stage during the cell cycle, and the bacterial cell has a variety of means of altering the flow of information from the genome to the proteome.

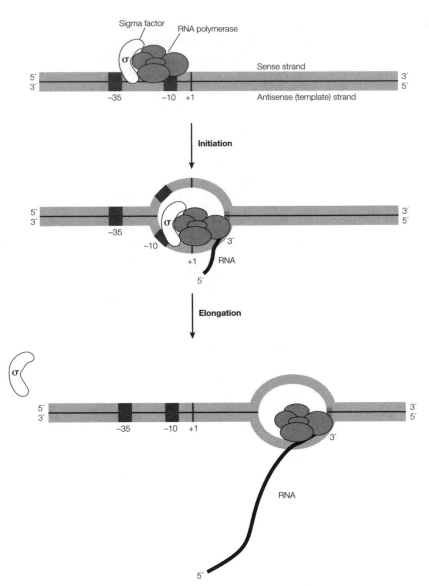

Fig. 2.   Formation of the transcription complex. From Instant Notes in Molecular Biology, *2nd Edn, K4, Fig. 1, p. 193.*

The two most common forms of regulation cited in the study of bacteria are **derepression** (where a protein bound to a promoter stopping transcription is removed and the gene is switched on) and **attenuation** (where the presence or absence of a substrate necessary for the function of the gene product governs the transcription of the gene itself). The former is exemplified by the *E. coli lac* operon, the latter by the *trp* operon from the same organism. It should be noted that **activation** (where the presence of a protein is used to switch a gene on) is used much less frequently in Bacteria than derepression. The contrary is true in the Archaea and in eukaryotes, where transcription factors are the main regulators in activating gene expression.

5′ **... NNNNAAGCGCCGNNNNCCGGCGCUU**UUUU – OH 3′

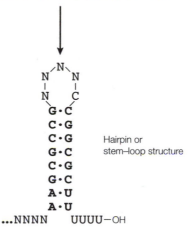

Hairpin or
stem–loop structure

...NNNN   UUUU–OH

Fig. 3.   RNA hairpin structure. From Instant Notes in Molecular Biology, *2nd Edn, K1, Fig. 3, p. 185.*

The *lac* operon

The most commonly used model for transcription and the regulation of transcription is the *lac* operon of *Escherichia coli*. The operon comprises a promoter, an **operator** (a site on the DNA where regulatory proteins bind) and three genes. These three genes allow the Bacterium to use lactose instead of glucose as a carbon source (*Fig. 4*), *lacZ* coding for the enzyme β-galactosidase (the gene product LacZ, (note the difference in italicization and capitalization between the gene and the protein it makes), *lacY* coding for a permease (gene product LacZ) that allows lactose through the membrane, and *lacA*, a gene of poorly understood function that codes for a transacetylase.

Upstream of the *lacZYA* promoter is another gene called *lacI* which has its own promoter (*Fig. 5*). The product of this gene, LacI, is the primary regulator of the *lac* operon and is sometimes called the **Lac repressor**. When LacI is bound to the *lacZYA* operon operator, transcription is blocked, the cell is unable to produce β-galactosidase and thus is unable to use lactose as a carbon source.

The functions of all the genes were found as a result of analysis of mutations. Changes in the DNA sequence in the genes themselves can knockout individual genes (*Table 1*).

The finding of a mutation that meant an inactive LacI was produced, and this allowed constitutive expression of *lacZ* (i.e. in the presence and **absence** of lactose) was significant and was one finding that allowed Jacob and Monod to begin to elucidate the mechanism of LacI repression. They proposed that LacI was

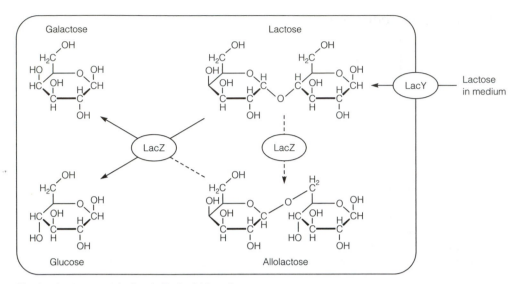

Fig. 4.   Lactose metabolism in Escherichia coli.

always produced from its own promoter, but was structurally altered by the presence of lactose itself which stopped it binding to the *lacZYA* promoter (*Fig. 6*). It
was found that the molecule that caused this change and induced the *lac* operon
was allolactose (*Fig. 4*). Allolactose is the primary **inducer** but there are other
molecules such as XGAL that can also induce the operon.

The model was backed up with studies in the non-structural parts of the operon,
in the promoters of both *lacZYA* and *lacI*, and the operator of *lacZYA* (*Table 2*).

The *lac* promoter itself is relatively weak, even when induced by allolactose.
High levels of transcription are only achieved when not only is the promoter
de-repressed and when it is activated by the **cAMP receptor protein** (**CRP**). This
is sometimes called **catabolite activator protein** (**CAP**). If the cell is rich in energy
from other sources, it may not need to utilize lactose at all. The precursor of ATP,
AMP, is in equilibrium with a cyclic form, cAMP (*Fig. 7*). When glucose is low,
there is an abundance of AMP, which in turn means there is a lot of cAMP. The
cAMP binds to CRP which can then activate the *lac* operon by as much as 40-fold.

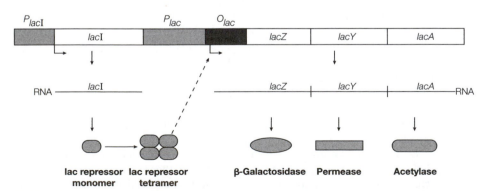

Fig. 5.   Structure of the lactose operon. From Instant Notes in Molecular Biology, *2nd Edn, L1, Fig. 1, p. 198.*

Table 1.   Effect of mutations in lac operon structural genes

Mutation	Genotype	Phenotype	Notes
Inactivation of LacI gene	lacI⁻ lacZ⁺ lacY⁺ lacA⁺	In the presence of lactose: β-galactosidase⁺ In the absence of lactose: β-galactosidase⁺	LacZ LacY LacA produced in both the presence and absence of lactose (constitutively).
Inactivation of LacZ gene	lacI⁺ lacZ⁻ lacY⁺ lacA⁺	In the presence of lactose: β-galactosidase⁻ In the absence of lactose: β-galactosidase⁻	No β-galactosidase (LacZ), so lactose cannot be metabolized
Inactivation of LacY gene	lacI⁺ lacZ⁺ lacY⁻ lacA⁺	In the presence of lactose: β-galactosidase⁺ In the absence of lactose: β-galactosidase⁻	No permease (LacY), so lactose cannot enter the cell to be metabolized
Inactivation of LacA gene	lacI⁺ lacZ⁺ lacY⁺ lacA⁻	In the presence of lactose: β-galactosidase⁺ In the absence of lactose: β-galactosidase⁻	No transacetylase, but no effect on lactose metabolism

The *trp* operon

The tryptophan operon comprises five genes (*trp*EDCBA), which, when transcribed and, translated, enable the cell to synthesize tryptophan from glutamine and chorismate (*Fig. 8 (a)*). Tryptophan itself inhibits the first step of the pathway at the protein level, but the cell also imposes regulation during both transcription and translation.

It is important to remember that the cell does not need the tryptophan synthesis enzymes when there is an abundance of tryptophan itself, a contrast to the *lac* operon where lactose abundance requires that the cell induces a metabolism system. The transcription of the *trp* operon is thus stopped by the binding of the *trp* repressor complexed to tryptophan. As the levels of tryptophan fall, the operon is

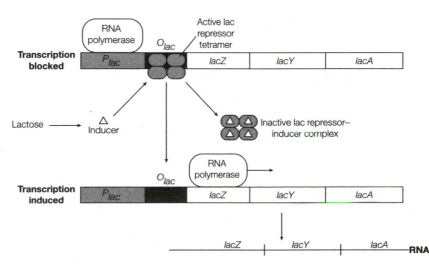

Fig. 6.   Binding of the inducer activates the lac repressor. From Instant Notes in Molecular Biology, 2nd Edn, L1, Fig. 1, p. 199.

Table 2.   Effect of mutations in lac *operon promoters and operators*

Mutation	Genotype	Phenotype	Notes
Inactivation of *lacZ* promoter e.g. mutation in –10 region	LacI⁺ LacZ⁻ LacY⁻ LacA⁻	In the presence of lactose: β-galactosidase⁻ In the absence of lactose: β-galactosidase⁻	With no promoter and no permease or β-galactosidase, lactose cannot be metabolized
Change in *lacZYA* operator means LacI can no longer bind	LacI⁺ LacZ⁺ LacY⁺ LacA⁺	In the presence of lactose: β-galactosidase⁺ In the absence of lactose: β-galactosidase⁺	β-galactosidase and permease produced constitutively
Change in *lacI* promoter to make it stronger	LacI⁺ LacZ⁺ LacY⁺ LacA⁺	In the presence of lactose: β-galactosidase⁻ In the absence of lactose: β-galactosidase⁻	As too much LacI is produced, all the allolactose is titrated out Thus there is always some LacI bound to the *lacZYA* operator, and β-galactosidase is never produced

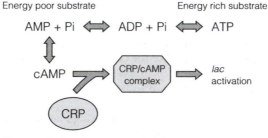

Fig. 7.   CRP responds to cAMP.

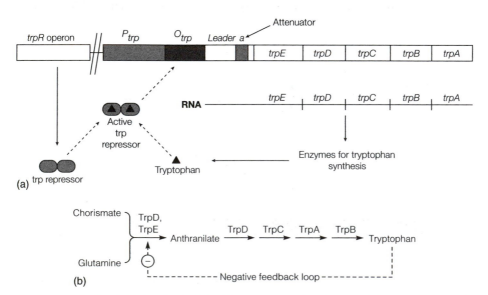

Fig. 8.   (a) Structure if the trp *operon and function of the* trp *repressor.* Instant Notes in Molecular Biology, *2nd Edn, L2, Fig. 1, p. 20; (b) The pathway of tryptophan synthesis in* E. coli.

de-repressed. The difference between repression and de-repression is about 70-fold, much less than that of the *lac* operon in its 'on' and 'off' states, and this gave a clue that there was more than one mechanism regulating *trp*EDCBA expression.

The number of bases between the *trp* promoter/operator and triplet 1 of *trpE* is unusually long. The 162 nt form a rho-independent terminator site. When transcribed, consisting of a short GC-rich palindrome followed by eight uracil bases. If this palindrome can hybridize to itself, it forms a stem-loop structure (or hairpin) which is a highly efficient transcriptional terminator, allowing the RNA polymerase to synthesize only the first 140 bp of the operon. The mechanism by which these short transcripts are made is termed **attenuation.**

In bacteria, transcription by RNA polymerase and translation by ribosomes often occur in close proximity, with the ribosome binding to its mRNA binding

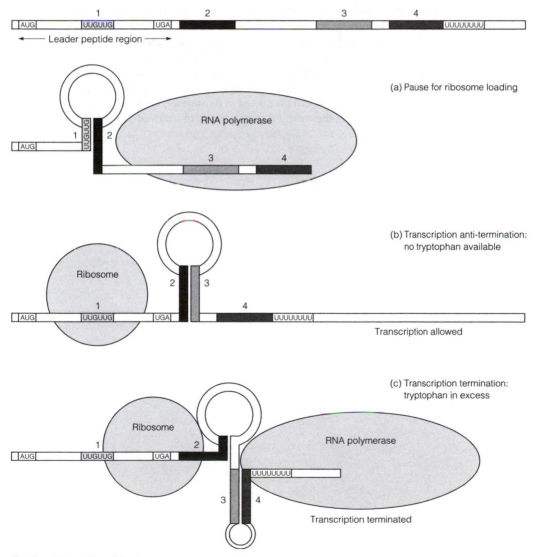

Fig. 9.   Attenuation of the trp operon.

site as soon as the RNA polymerase has transcribed it. To ensure that the RNA polymerase and ribosome are as close to one another as possible, the initial transcription of the first hundred bases results in the formation of a stem loop between regions 1 and 2 of the mRNA (*Fig. 9*). This causes the RNA polymerase to pause, and while this takes place, a ribosome loads itself onto the mRNA at the 5′ end.

When tryptophan is absent or at very low levels, the ribosome begins to translate the 5′ end of the *trp* mRNA, until it encounters two codons coding for tryptophan itself (UUGUUG, region 1 in *Fig. 9*). As tryptophan is at such low levels, there are few tRNA-Trp molecules around for the ribosome to incorporate into the growing peptide and, while the ribosome is waiting, a hairpin between regions 2 and 3 forms which reinforces this pause and allows the RNA polymerase to transcribe the entire operon. Contrast this with the situation when tryptophan is abundant: again the RNA polymerase begins transcription, pauses and the ribosome loads, but now there is no barrier to the ribosome incorporating Trp, so it can move forward, unfolding the hairpin between regions 2 and 3. This causes region 3 to be available to form another secondary structure, this time with region 4 (*Fig. 9*). However, the combination of region 3 and 4 coupled to the presence of poly U a little further downstream is a rho-independent transcription termination signal. The RNA polymerase falls off after only having manufactured the first 140 nt of the *trp* mRNA. In this way, the transcription of the *trp* operon is attenuated in the presence of tryptophan, but permitted in its absence, allowing the cell to conserve energy by avoiding the wasteful transcription of un-needed mRNA.

F7 RNA AND THE GENETIC CODE

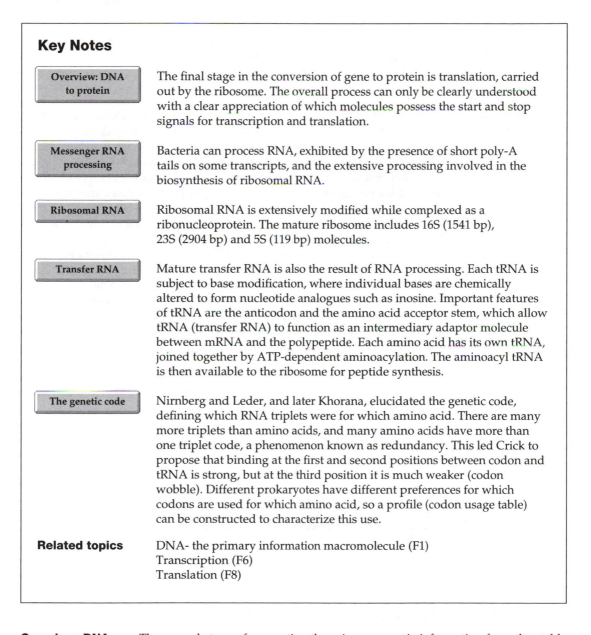

Key Notes

Overview: DNA to protein

The final stage in the conversion of gene to protein is translation, carried out by the ribosome. The overall process can only be clearly understood with a clear appreciation of which molecules possess the start and stop signals for transcription and translation.

Messenger RNA processing

Bacteria can process RNA, exhibited by the presence of short poly-A tails on some transcripts, and the extensive processing involved in the biosynthesis of ribosomal RNA.

Ribosomal RNA

Ribosomal RNA is extensively modified while complexed as a ribonucleoprotein. The mature ribosome includes 16S (1541 bp), 23S (2904 bp) and 5S (119 bp) molecules.

Transfer RNA

Mature transfer RNA is also the result of RNA processing. Each tRNA is subject to base modification, where individual bases are chemically altered to form nucleotide analogues such as inosine. Important features of tRNA are the anticodon and the amino acid acceptor stem, which allow tRNA (transfer RNA) to function as an intermediary adaptor molecule between mRNA and the polypeptide. Each amino acid has its own tRNA, joined together by ATP-dependent aminoacylation. The aminoacyl tRNA is then available to the ribosome for peptide synthesis.

The genetic code

Nirnberg and Leder, and later Khorana, elucidated the genetic code, defining which RNA triplets were for which amino acid. There are many more triplets than amino acids, and many amino acids have more than one triplet code, a phenomenon known as redundancy. This led Crick to propose that binding at the first and second positions between codon and tRNA is strong, but at the third position it is much weaker (codon wobble). Different prokaryotes have different preferences for which codons are used for which amino acid, so a profile (codon usage table) can be constructed to characterize this use.

Related topics

DNA- the primary information macromolecule (F1)
Transcription (F6)
Translation (F8)

Overview: DNA to protein

The second stage of converting the primary genetic information from the stable DNA code into protein is to **translate** the messenger RNA into a polypeptide. This process is carried out by the ribosome. The ribosome can only bind to messenger RNA, at sites to the 5' of the AUG start codon. Confusion often arises over the location of the promoter, the ribosome binding site (RBS) and the transcript start site, as these are often all marked on schematics showing the features of a gene in its genomic location (*Table 1 and Fig. 1*).

Table1.   *Location of nucleic acid signals for transcription and translation*

Conversion	Start signal	Stop signal	Notes
DNA to RNA	Found on DNA	Found on mRNA	See Section F6
TRANSCRIPTION	PROMOTER	TERMINATOR	
mRNA to protein	Found on mRNA	Found on mRNA	See Section F8
TRANSLATION	RIBOSOME BINDING SITE	TERMINATION CODON	

Messenger RNA processing

A common misconception about prokaryotes is that they are unable to perform any processing on their RNA, a false distinction often drawn between them and the eukaryotes. Although transcript processing is less common in prokaryotes, it still does occur. There is evidence that some bacterial transcripts have short poly-A tails added, and there is evidence of intron/exon-like processing in a very few genes. The Archaea possess many of the transcript processing features associated with eukaryotes, including extensive intron/exon structures, but neither the bacteria nor the Archaea rely heavily on mRNA splicing to generate protein diversity. Instead they rely on population diversity to cope with environmental change.

One function in which extensive RNA processing is always present in bacteria is in the manufacture of the ribosomal RNA. In *E. coli* there are seven different operons for ribosomal RNA (*rrn operons*), scattered throughout the genome. The number, amount of duplication location, and order of the genes within these operons varies in different species. At the *E. coli* K12 *rrn*A genomic locus the 16S rRNA gene (*rrs*A) is separated from the 23S rRNA gene (*rrl*A) by genes coding transfer RNA (*ile*T and *ala*T). Downstream of *ala*T at the end of the operon is the gene for 5S rRNA (*rrf*A) (*Fig. 1*). The operon is transcribed as a single polycistronic RNA of about 5000 nucleotides but is quickly processed by RNase III, plus other site-specific RNA-hydrolyzing enzymes called M5, M16 and M23.

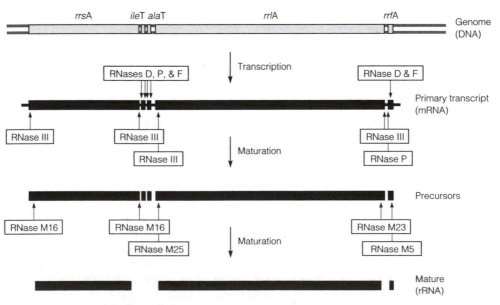

Fig. 1.   Processing of the E. coli *K12 rrn A primary transcript.*

The processing takes place in the order outlined in *Fig. 1*, and begins just as the unprocessed RNA emerges from the RNA polymerase, i.e. the 5′ end of the operon is processed before the 3′.

Ribosomal RNA

The immature *rrn* transcript can form stable secondary structures made up of stems of RNA:RNA complementary duplexes, and loops of single-stranded RNA. The structure allows the protein to bind to the rRNA, initially to form a **ribonucleoprotein (RNP)** complex. Some of these proteins remain to become part of the mature ribosome (see Section F8), but some are lost as 24 base-specific methylations of the rRNA take place. It is also at this stage that the first RNase III stage of maturation occurs. Further processing of the precursors by M5, M16, and M23 RNases generates three ribosomal RNA species, named 16S, 23S, and 5S. The 'S' stands for Svedburg units and refers to the sedimentation coefficient – an older measure of molecular weight and size. With the advent of sequencing, the sizes of the *rrn* genes are equally well known, 16S being 1541 bp, 23S 2904 bp, and the smallest, 5S is only 119bp in *E. coli*.

Transfer RNA

Mature transfer RNA is also the result of RNA processing, resulting in a molecule 60 to 95 nt long. *Fig. 1* shows the involvement of RNases D, P, and F in the maturation of tRNA^{Ala} from the *ala*T primary transcript. These RNases, along with RNase E, are also responsible for the maturation of most tRNA molecules. In addition, each tRNA is subject to **base modification**, where individual bases are chemically altered to form nucleotide analogues, such as inosine. The number of these modifications and their position in the mature tRNA alters according to which amino acid is attached.

The most important features of any tRNA molecule is its **anticodon**, found in approximately the middle of the molecule, and the **amino acid acceptor stem** at the 3′ end. These features are necessary for the central role of tRNA as the **adaptor molecule,** transporting amino acids to the ribosome and holding them in a position so that the relevant codon and anticodon are complementary, and the amino acid is in a suitable orientation for a peptide bond to form in the nascent peptide chain (see Section F8). Like rRNA, tRNA has a secondary structure composed of stems of RNA duplex and loops of single-stranded RNA. This is generally said to be a cloverleaf structure, as three loops are formed. The cloverleaf frequently appears as a 2-dimensional structure in this and other books, but it is worth bearing in mind that extensive hydrogen bonding throughout the molecule confers a 3-dimensional shape.

The loading of an amino acid onto its specific tRNA is termed **aminoacylation**, and is performed at the expense of ATP (*Fig. 2*). Aminoacyl-tRNA-synthetase first binds both the amino acid and ATP as distinct molecules, before converting the ATP to AMP and then attaching the mononucleotide to the amino acid. The uncharged tRNA then enters the enzyme's active site and where the amino acid is transferred from AMP to the tRNA's acceptor stem. This catalyzes the release of the **aminoacyl tRNA** which is then available to the ribosome.

The genetic code

The task of finding out which codon corresponded to which amino acid was performed by **Nirnberg and Leder** in 1964, and further refined by Khorana at about the same time. The code was elucidated by incubating mRNA of defined sequence with *E. coli* RNA polymerase, and all possible tRNAs and amino acids. They then examined either which tRNA bound to the polymerase, or what the sequence of the resulting polypeptide was. This gave us the familiar dictionary of

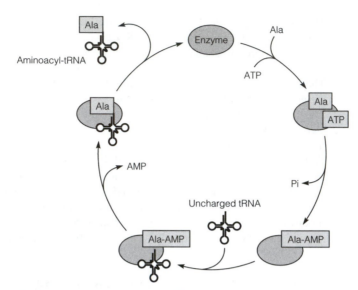

Fig. 2.   Biosynthesis of aminoacyl tRNA.

codons and the amino acids that they are translated into (*Table 2*). The table shows that many amino acids are coded for by more than one triplet, for example serine has six possible codons. This allows all the 3^4 permutations to represent the 20 most common amino acids, plus three stop codons. This phenomenon of **redundancy** led Crick (see Section F1) to propose that binding at the first and second positions between codon and tRNA is strong but at the third position it is

Table 2.   The universal genetic code. From Instant Notes in Molecular Biology, *2nd Ed, P1, Table 1, p257*

First position (5′ end)	Second position							Third position (3′ end)
	U		C		A		G	
	Phe	UUU	Ser	UCU	Tyr	UAU	Cys	UGU	U
	Phe	UUC	Ser	UCC	Tyr	UAC	Cys	UGC	C
U	Leu	UUA	Ser	UCA	**Stop**	UAA	**Stop**	UGA	A
	Leu	UUG	Ser	UCG	**Stop**	UAG	Trp	UGG	G
	Leu	CUU	Pro	CCU	His	CAU	Arg	CGU	U
	Leu	CUC	Pro	CCC	His	CAC	Arg	CGC	C
C	Leu	CUA	Pro	CCA	Gln	CAA	Arg	CGA	A
	Leu	CUG	Pro	CCG	Gln	CAG	Arg	CGG	G
	Ile	AUU	Thr	ACU	Asn	AAU	Ser	AGU	U
	Ile	AUC	Thr	ACC	Asn	AAC	Ser	AGC	C
A	Ile	AUA	Thr	ACA	Lys	AAA	Arg	AGA	A
	Met	AUG	Thr	ACG	Lys	AAG	Arg	AGG	G
	Val	GUU	Ala	GCU	Asp	GAU	Gly	GGU	U
	Val	GUC	Ala	GCC	Asp	GAC	Gly	GGC	C
G	Val	GUA	Ala	GCA	Glu	GAA	Gly	GGA	A
	Val	GUG	Ala	GCG	Glu	GAG	Gly	GGG	G

much weaker, a concept known as **codon wobble**. Physiologically it is taken to mean that a codon, such as CCU (proline), may be paired with a GGA, GGU, GGC or GGG anticodon on the tRNA. This may be the case in eukaryotic organisms with a full complement of tRNAs. In the Bacterial world, it is hypothesized that carrying every single tRNA genomically is energetically wasteful – most micro-organisms attempt to carry the minimum number of genes necessary for their selected lifestyle. Thus most bacteria carry a limited set of tRNAs, with a minimum of redundancy. This minimization can be seen when trying to express genes from one bacterium in another. Frequently the gene will not be made heterologously because the host bacterium does not carry one or more of the tRNAs required, and translation stutters or stops.

As different organisms have different genomic %G+C compositions the set of tRNA molecules a species keeps tends to reflect this bias. G+C-rich organisms such as *Paracoccus denitrificans* will have more codons in their genes ending in G or C compared to an A-T rich organism such as *Campylobacter jejuni*. If we look at all the codons in all the genes of a particular organism, we can compile a **codon usage table**, showing which the commonest codons are.

The codon usage table also reveals that the 'universal genetic code' portrayed in *Table 2* has some exceptions in the bacteria. The use of AUU (Ile in *Table 2*), GUG (Val) or UUG (Leu) has been noted as a start codon, still resulting with the inclusion of fMet at the protein N-terminal. In *Mycoplasma*, UGA codes for Trp rather than acting as a termination signal.

F8 TRANSLATION

Key Notes

The ribosome	The ribosome is a riboprotein, with three RNA molecules (5S, 16S, and 23S) bound to polypeptides. The function of the ribosome is to translate an mRNA into a polypeptide. The mRNA contains a ribosome binding site (Shine Delgarno sequence) at its 5′ end, which is followed by an initiation signal sequence, or start codon. In Bacteria this codon is normally AUG (formylmethionine). The termination codon is also found as an mRNA-encoded signal.
Initiation	Before encountering an mRNA molecule, the components (30S and 50S subunits) of the ribosome float free in the cytoplasm. Initiation factors, GTP, tRNA^{fMet} cooperate with the binding of the subunits to the start codon, and after GTP hydrolysis and the dissociation of IF3, then 2 and 1, the 70S initiation complex is formed.
Elongation	The 70S initiation complex has three main sites: the P site holds either tRNA^{fMet} or a tRNA-bound polypeptide; the A site holds incoming charged tRNA and the E site binds uncharged tRNA. The translocation of the ribosome from 5′ to 3′ makes these sites available in sequence, catalyzed by elongation factors (EF-Tu and EFG-GTP). In this way a nascent polypeptide emerges from the ribosome.
Termination	When the ribosome arrives at a stop codon, a protein release factor causes the cleavage of the completed polypeptide and dissociation of the ribosome back into free 30S and 50S subunits.
The polysome	In Bacteria and Archaea, transcription and translation is tightly coupled, with mRNA emerging from the RNA polymerase to be translated almost immediately by the ribosome.
Related topics	Transcription (F6) RNA and the genetic code (F7)

The ribosome   The second stage of converting the primary genetic information from the stable DNA code into protein is to **translate** the messenger RNA into a polypeptide. The agent for this conversion is the ribosome. There is frequent confusion between the binding sites for transcription and translation. This confusion is enhanced by the highlighting of the promoter/operator, ribosome binding sites and the various control and termination signals on DNA sequences. However, the ribosome binds at a **ribosome binding site** (RBS, sometimes referred to as the **Shine–Dalgarno** sequence) which is only found on mRNA. The initiation signal sequence for ribosomal translation is AUG, again on the mRNA, making up the first codon of the gene (*Fig. 1*). The termination signal of the ribosome is also on the mRNA, at the stop codon. Contrast this with RNA polymerase, whose binding site (the promoter) is DNA, yet confusingly the termination signal is intrinsic to the mRNA it produces.

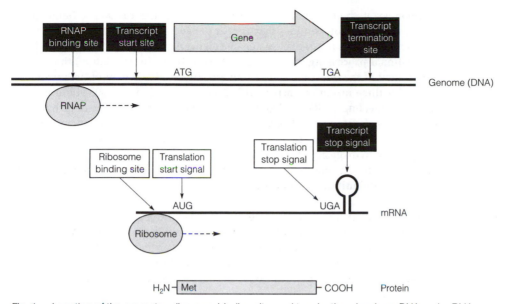

Fig. 1.   Location of the promoter, ribosome binding site, and termination signals on DNA and mRNA.

The ribosome itself is an unusual protein structure. It is composed of many polypeptides, but also includes three rRNA molecules per ribosome. The 5S, 16S, and 23S rRNA have important catalytic functions, so the ribsome can be considered more as a ribozyme rather than a conventional enzyme. In the cytoplasm, the ribosome exists as two separate RNA/protein complexes, known as 30S and 50S. The 'S' designation stands for Svedburg units, a measure of the degree of sedimentation during centrifugation. The relationship between S and molecular weight is non-linear, but gives an indication of size. The signal of the formation of the intact ribosome is the presence of mRNA in the cytoplasm. Ribosome size varies between kingdoms. In the Bacteria complete ribosome is a 70S riboprotein, containing 5S, 16S and 23S rRNA. In the Archaea, 5S and 23S rRNA is present, but the third rRNA is 18S. In the eukaryotes, the situation is much more complex and, indeed, the mechanisms of initiation, elongation and termination are different, involving many more ancillary proteins.

Initiation

The signal for the formation of a holo-ribosome, capable of converting mRNA into a polypeptide, is the presence of a mRNA-encoded ribosome binding site (RBS) at a suitable distance from a start codon, AUG. The consensus binding site for the ribosome is the sequence AGGAGG, 5 to 9 nt from the start codon, but the actual sequence is subject to considerable variation between genes. In *Escherichia coli* the RBS is often easy to spot by eye within the mRNA sequence, but in other bacteria the situation is more confused. It is often only possible to estimate the position of the RBS based on knowledge gained of the position of the promoter, transcription start site and start codon – it is assumed that the RBS is between the TSS and the start codon.

The start codon is normally AUG in Bacteria, which codes for methionine. Other start codons are used more rarely, for example GTG sometimes appears. Although GTG codes for valine, its position as the initiator codon means that here, as with all start codons, the tRNA added will be charged with **formylmethionine** (fMet). This means that all completed Bacterial polypeptides have methionine at their N terminal when they leave the ribosome. Some holo-enzymes contain polypeptides that appear to have other N terminal amino acids, but this is only as a result of cleavage of a signal peptide or another form of **post-translational processing** by specific proteases. To begin translation, the smaller 30S subunit of the ribosome is first activated by the attachment of the nucleotide GTP, plus three **initiation factors** (IF1, 2, and 3). The charged 30S subunit stabilizes the annealing of tRNA^{fMet} to the start codon, which causes the detachment of IF3. The 30S/ tRNA^{fMet}/mRNA complex stimulates the attachment of the 50S ribosomal subunit, an energetic process which results in the departure of IF1 and IF2, plus the hydrolysis of GTP to GDP and inorganic phosphate. The ribosome is now ready to add amino acids to the tRNA bound formylmethionine, and is termed the **70S initiation complex**.

Elongation

The 70S initiation complex has three sites in which tRNA can be bound: the **P site** holds either tRNA^{fMet} or a tRNA-bound polypeptide (*Fig. 2*); the **A site** holds incoming charged tRNA and lastly the **E site** binds uncharged tRNA. In addition, the ribosome has binding sites to hold the mRNA template and a catalytic site to enable the formation of peptide bonds between adjacent charged tRNAs.

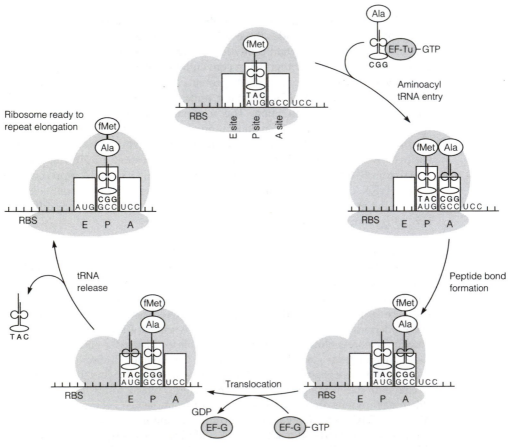

Fig. 2.   Translation elongation.

The ribosome initially holds tRNA^{fMet} in the P site annealed to the complementary start codon, and the first step in elongation is to allow the entry of a charged aminoacyl tRNA to match the next codon on the mRNA. The aminoacyl tRNA is first bound to an elongation factor (**EF-Tu**) This elongation factor has the nucleotide GTP attached to it. Entry of tRNA-EF-Tu-GTP is an energetic process, leading to the hydrolysis of the complexed GTP to GDP and results in the complementary base pairing of the tRNA anticodon to the 2nd codon of the mRNA. The example given in *Fig. 2* is the entry of tRNA^{Ala} , which shows how this aminoacyl tRNA is held in the A site so that the amino acids of tRNA^{Ala} and tRNA^{fMet} are held in such a position as to allow the easy formation of a peptide bond. The peptide bond is formed concomitantly with the cleavage of the bond between tRNA and fMet, so that the **nascent polypeptide** (or growing polypeptide) is bound to the tRNA in the A site and the P site contains an uncharged tRNA. To allow the next codon to be utilized, the ribosome must move (or **translocate**) three nucleotides to the 5′ of the mRNA.

The translocation of the ribosome is a process that requires energy, provided by a second, nucleotide-complexed elongation factor, **EFG-GTP**. The movement causes the dissolution of the EFG-GTP complex and the hydrolysis of GTP. The tRNA$^{fMet-Ala}$ has now moved to the P site, with the uncharged tRNA in the E site. The E site tRNA is now expelled so that the ribosome is now ready to commence another cycle of aminoacyl-tRNA entry, peptide bond formation, translocation and tRNA release.

Termination

In comparison to the complex nucleotide sequence termination signals of the replicative DNA polymerase and RNA polymerase, the ribosome ceases to form peptide bonds when it encounters a simple three nucleotide stop codon on the mRNA. There is some evidence that the identity of the few nucleotides to the 5′ of the stop codon governs the efficiency of termination, but the primary signal is a stop codon of sequence UAG, UGA or UAA in Bacteria. The stop codon is sometimes incorrectly called a nonsense codon, since no complementary tRNA exists. Instead, the stop codon binds a protein, called **release factor**. The release factor causes the cleavage of the completed polypeptide C-terminal from the tRNA in the P site. The departure of the polypeptide, followed by the now uncharged tRNA and the release factor, causes the ribosome to release the mRNA and dissociate into its 30S and 50S subunits.

The polysome

It is tempting to think of a single ribosome binding to the RBS of a single mRNA, manufacturing its polypeptide until a stop codon appears, and then dissociating. However, in Bacteria and presumably Archaea, transcription and translation is **tightly coupled**. As soon as an RBS emerges from an RNA polymerase during mRNA synthesis, a ribosome binds and begins translation. As more mRNA is synthesized the ribosome moves down the codons until there is enough room for another ribosome to bind at the RBS. By the time the RNA polymerase has completed synthesis of mRNA, many ribosomes will be attached, each with a partially completed polypeptide. This multi-ribosome-, mRNA- and RNA polymerase-containing structure is sometimes called the polysome. The proximity of the RNA polymerase and ribosome plays an important role in the regulation of some genes (see Section F6). In an organism growing quickly on a readily available carbon and energy source, the polysome may also be spatially very close to the replisome, and the interplay between the two is the source of a lot of speculation.

F9 SIGNAL TRANSDUCTION AND ENVIRONMENTAL SENSING

Key Notes

Cellular response to environmental stimuli

Free living, single cells must be able to adapt quickly to relatively large changes in their immediate environmental conditions. Signal transduction mechanisms allow these external changes to be reflected in differential gene expression. Some molecules, such as lactose, diffuse directly into the cell and act on an inducer or repressor, but other more toxic compounds are prevented from doing so. In common with physical changes (pH, temperature, etc.) these toxic molecules can be detected via sensors in the cell membrane, and then a signal sent to the genome via phosphorylation of cytoplasmic proteins. The process of translocation via this phosphorelay pathway can be branched to include many genes or operons.

Two component sensors

The simplest sensor consists of a sensor kinase in the cell membrane and a response regulator in the cytoplasm. Binding of an external signal results in the hydrolysis of ATP at the cytoplasmic side of the sensor kinase, and activation of the response regulator by transfer of a phosphate group.

Oxygen sensing

Facultative anaerobes are able to grow in both atmospheric oxygen concentration as well as less aerobic conditions. Accordingly they must have mechanisms to sense and respond to oxygen concentrations to employ different biochemical pathways appropriate to aerobiosis. An example of an oxygen sensing system is FNR (after *f*umarate and *n*itrate *r*eduction). The group of genes and operons controlled by FNR (the regulon) allow *E. coli* to use fumarate and nitrate as alternative electron acceptors to oxygen. There is some evidence that there is crosstalk between sensors with similar but not identical function.

Chemotaxis

A control mechanism to efficiently move towards food or away from toxins is essential to motile microorganisms. Chemotaxis towards an attractant (inducer) is initiated by the stimulation of a cell-membrane located transducer. The signal is translocated to the flagellar motor by means of the Che proteins, and the balance between attraction and repulsion maintained by methylation of the transducer.

Related topics

Electron transport, oxidative phosphorylation and β-oxidation of fatty acids (E3)
Transcription (F6)

Cellular response to environmental stimuli

Single-celled organisms, such as bacteria, can face enormous changes in their external environment over very short periods of time. If we consider *Escherichia coli* and its life cycle, we can see how significant these changes are. In its normal habitat of the colon, the *E. coli* cell is kept at a more or less constant temperature of 37°C, and is surrounded by other microbes, nutrients (in the form of partly or completely digested food) and is in a low-oxygen environment. After excretion, and before the cell re-enters the digestive system of the same or a different host, *E. coli* is suddenly thrust into a colder, well oxygenated, more aqueous and nutrient-poor existence. To be able to cope with all these changes in life style, the cell must quickly turn off some genes and turn on others. At the gene level the switches are repressor and activator proteins (see Section F6 on the *lac* and *trp* operons), while at the protein level enzymes can be switched on or off by the presence or absence of metabolites (see Section F6). These cytoplasmic responses have come into effect due to changes outside the cell, and the way in which an indicator that the outside world has changed is carried from the cell wall and to the genes or proteins that might be involved in a response is called **signal transduction**.

Some molecules that have a significant effect on the cell are small and can diffuse or be actively transported. Once in the cell they can have a direct effect on their target, for example sugars, such as lactose, and alcohols, such as methanol. However, some useful growth components must be isolated from the cytoplasmic contents (toxic substances such as cyanide, formaldehyde, or even oxygen) or are unable to pass easily through the cell wall and membrane (large molecules such as poly dextrans). Other changes on the outside of the cell must be transformed into a form that enzymes can recognize. Change in pH, u.v. or heat are pertinent examples. Even the enzymes of extremophiles (see Section C2) can only tolerate relatively small ranges of heat and pH, yet the cell as a whole can survive a greater range of temperature or acidity. Rather than let environmental changes into the cell in their external concentrations the exterior of the cell must have systems to sense change, then relay the change in the sensor to a target. Frequently this relay involves the transfer of phosphate groups from a sensor protein on to one or more relay proteins and finally to the effector protein. For this reason it is sometimes called a **phospho-relay pathway**. When we examine cell response pathways it is convenient to think of them as linear pathways in which a sensor relays a series of phosphate molecules via a defined set of proteins to an effector. In the cytoplasm, the **translocation** of a signal from sensor to effector is certainly much more branched, with overlapping sets of relay proteins accepting phosphates from many different sensors. For this reason our understanding of the entire process of cellular sensing has become intertwined with computational techniques such as a neural networking. To gain access to the entire branched cellular response, we must first understand simple examples such as two component regulatory systems, and then develop it to look at a more global response such as the effect of oxygen on the cell or chemotaxis.

Two component sensors

Taken in isolation, the simplest bacterial sensing system allows the cell to adapt to an external factor by means of two protein components: a **sensor kinase** and a **response regulator**. The sensor kinase is buried so that there is an environmental signal binding site outside the cell, and an ATP binding cassette on the cytoplasmic side of the cell membrane (*Fig. 1*).

Oxygen sensing

One of the more frequent changes all microorganisms face is a change in the external oxygen tension. Some organisms can adapt between oxygen concentrations equivalent to atmospheric levels and almost completely anoxic conditions. These **facultative anaerobes,** such as *Escherichia coli*, achieve this by changing the terminal electron acceptor (See Section E3) used in oxidative phosphorylation. Under aerobic conditions they use water as a terminal electron acceptor and

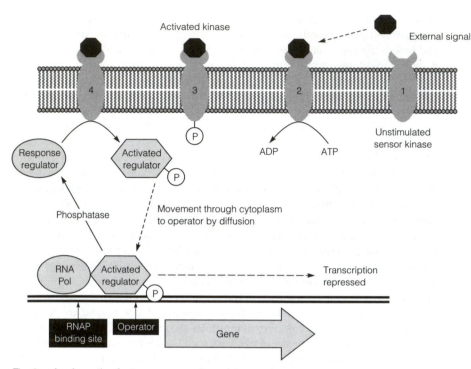

Fig. 1.   A schematic of a two component regulatory system.

cytochrome aa_3 oxidase. In the absence of oxygen and the presence of fumarate, cytochrome aa_3 oxidase is switched off and the fumarate/nitrate system is switched on. The primary sensor for this oxygen-dependent switch is a regulatory protein called **FNR** (after fumarate and nitrate reduction). This is a global regulator in that it alters the transcriptional properties of hundreds of *E. coli* genes, acting as an inhibitor of some while elevating the transcription of others. The group of promoters and genes that are regulated by FNR are known collectively as the **FNR regulon**, and are all characterized by a distinct sequence in the relevant operator, an FNR binding site. Where FNR acts as an activator, the FNR binding site replaces the −35 sequence of the standard σ^{70} promoter, so that transcription cannot take place unless active FNR is present. The FNR binding site is found in a similar place to the LacI binding site (see Section F6) when FNR is acting as a repressor.

The way in which FNR senses the presence of molecular oxygen is based more on chemistry than molecular biology. In the absence of oxygen, a cluster of iron and sulfur molecules forms at the N-terminal end of the protein, which allows FNR to dimerize and interact with DNA at its specific binding sites. When oxygen is present, the $[2Fe-2S]^{2+}$ cluster is destroyed and the protein can no longer function in DNA binding. This oxygen-sensing capability is carried out in concert with a number of other regulons, such as the Arc cluster. CRP (see Section F6) participates in another regulon as well, and it is interesting to note that the binding sites for FNR and CRP differ by only one base pair, raising the possibility that a considerable amount of **cross-talk** occurs between regulatory systems, allowing the cell to simultaneously adapt to many conditions at once.

Chemotaxis   Microorganisms, by definition, exist on a microscopic scale, and this means that, in relative terms, separation from a food source by only a few millimeters presents an enormous challenge. Most bacteria have some form of motility, either using flagella (see Section C10), gliding, or merely by extending colony size. In order to conserve energy, movement must not be random, depending on a chance encounter with a food source, but directed. The sensing process provides this direction. If movement is towards an organic compound, this is named **chemotaxis**, but other forms of taxis exist. These include geotaxis (movement towards or away from gravity) and phototaxis (movement towards or away from light). The best understood chemotactic system is formed by the Che proteins, present in many Gram negative Bacteria.

In response to the presence of a suitable carbon source (the **inducer**), a **transducer** (MCP) forms a complex with a sensor kinase CheA. The transducer functions as a medium to pass a signal across the Bacterial cell wall, periplasm, and cytoplasmic membrane. Unlike the more simple two component sensors, the transducer and sensor kinase interaction is mediated by a coupling protein, CheW. However, the result is still the phosphorylation of CheA. The signal is carried through the cytoplasm by transfer of the phosphate group to CheB and CheY. Phosphorylated CheY-P has a direct interaction with the flagellar motor switch. Once CheY-P has switched the motor on, it is dephosphorylated by ChyZ (*Fig. 2*).

As the microbe approaches the food source, and the concentration of the inducer changes, the cell must adapt its chemotactic response so as not to overshoot. This adaptation is provided by the second phosphorelay protein, CheB. A methylation protein, ChyR, is constantly methylating the transducer, but in its phosphorylated form CheB-P can either signal to the flagellar motor just as CheY-P does, or it can demethylate the transducer. The number of methyl groups attached to the transducers in the cell membrane influences the efficiency of sensor kinase phosphorylation. The competition to methylate (ChyR) or demethylate (CheB-P) the transducer thus adapts the cell's response to a stimulant.

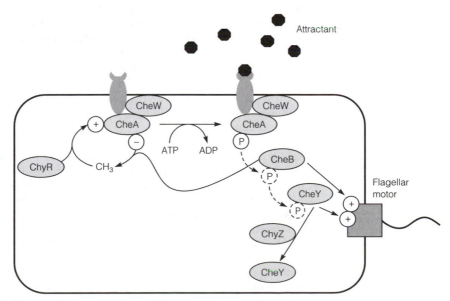

Fig. 2. Chemotaxis is controlled by the Che proteins.

F10 MUTATION

Key Notes

Mutation and adaptation	Stable inheritance of genetic information is maintained by systems that check and remediate DNA. Microorganisms need these systems to cope with factors, such as ultraviolet light and mutagenic chemicals to prevent induced mutagensis. Errors in the DNA replication machinery can also cause spontaneous mutagenesis, though in some genomes (at mutational hot spots) this type of mutagenesis seems to be allowed. Any strain derived from this with any change in its genomic make up compared to this wild type is known as a mutant. If the difference in genotype results in an observable change to the properties of the organism, this is a change in phenotype. The mutation of the wild type with a nutritional requirement results in an auxotroph.
Types of mutation	Single base pair mutations are called point mutations and can be of several types depending on the change made. A transition is the change of purine base to another purine, or pyrimidine to pyrimidine, whereas changes from purine to pyrimidine are transversions. Changes in gene expression from mutation can result in up regulation or down regulation of transcripts or proteins, whereas mutations with no observable effect are said to be silent. When a point mutation does cause a change in amino acid sequence that is phenotypically detectable, the mutation is classed as a missense mutation. A special case of missense is the nonsense mutation, when the result of a point mutation is the creation of a stop codon. Reversion to wild type can happen because a back mutation has occurred at exactly the same nucleotide(s) changed in the original mutant, a same site mutation. However, changes in another region of the genome (a second site mutation) can compensate for the mutation at the first site. Insertion of one or many base pairs of DNA can occur, as can the deletion of nucleotides. Large sections of genomic DNA can be moved to a completely different place in the genome, a phenomenon known as translocation.
Chemical mutagens	Base analogues mimic the structure of A, C, G, or T and are incorporated into DNA during replication. Base modifiers change the subgroups of existing bases in a DNA double helix. Intercalating agents insert into the helix itself, physically disrupting the shape and causing replication to miss out one or more bases. Common examples of chemical mutagens are the base analogue 5-bromo uracil, the base modifier methylmethane sulphonate and the intercalator ethidium bromide.
Physical mutagens	Ionizing radiation damages DNA, from natural sunlight to X-rays and radio-isotopic emissions. Ultraviolet light will cause the formation of these thymine dimmers between adjacent Ts, forming a bulge in the DNA helix. This is detected and repaired by the nucleotide excision repair system.

Use of mutagenic compounds	Mutations can accumulate to the point of lethality, and this can be plotted as a kill curve to determine the optimum concentration (or does) for single or multiple mutagenic events. Once suitable mutagenic conditions have been established a process of replica plating is used to mark which colonies may be useful in genetic analysis.
Related topics	Transcription (F6) Recombination (F13) DNA repair (F11) Bacteriophage (F14)

Mutation and adaptation

As with all organisms, microorganisms have mechanisms to preserve their genome so that their genetic complement is stably passed from generation to generation. However, in contrast with higher organisms, Bacteria and Archaea have no method of sexual reproduction so have to rely on the same processes of genomic change to enable diversity in their populations. Microorganisms face many environmental factors that could possibly cause mutation in their genomic DNA. Permanent changes in the DNA caused by external factors are known as **induced mutagenesis**. Those changes caused by errors in the DNA replication machinery, or other mistakes made by the cellular DNA metabolizing enzymes are known as **spontaneous mutagenesis**. The rate of error in DNA replication is between 10^{-7} and 10^{-11} per base pair per round of replication, equivalent to around 10^{-4} to 10^{-8} errors per gene, per generation in a bacterium the size of *E. coli*. Spontaneous change can also include those genomic rearrangements that an organism may make to its own genome, as exhibited by *Streptomyces*. These mutation **hot spots** on genome are normally associated with an abundance of short inverted repeats. It is thought that the repetitive nature of the sequence causes polymerase to stutter and make more errors at these particular points.

A strain isolated from its environment and held in pure culture in the laboratory is defined as the **wild type** of a particular strain. Any strain derived from the wild type with any change in its genomic make up compared to this wild type is known as a **mutant**. This change can be referred to as a change in **genotype**. If the difference in genotype results in an observable change to the properties of the organism, such as a sudden inability to use a particular carbon source, this is said to be a change in **phenotype**. Changes in dependence on medium components are frequently used in the elucidation of biochemical pathways or in the deduction of regulatory systems, so the mutation of the wild type with a nutritional requirement is a common experiment. These strains are known as **auxotrophs**. For example, if *E. coli* is subjected to a chemical mutagen and as a result of mutation now requires the vitamin B12 to be present in its medium for the strain to grow, the strain is **auxotrophic** for B12, or can be called a B12 auxotroph.

Types of mutation

Changes to an organism's DNA can either involve a change to many adjacent base pairs simultaneously, or just to a single base pair. Single base pair mutations are celled **point mutations** and can be of several types depending on the change made. A **transition** is the change of purine base (adenine or guanine, see Section F3) to another purine. There are similar pyrimidine transitions involving cytosine and thymine (*Table 1*).

Point mutations outside of a coding region may have two effects if they are within a regulatory region such as a promoter or operator. If the mutation causes

Table 1.   Transitions and transversions

Point mutation	Wild type → Mutant	Point mutation	Wild type → Mutant
Purine transition	A → G	Pyrimidine transition	C → T
	G → A		T → C
Pu/Py Transversions	A → C	Py/Pu Transversions	C → A
	A → T		C → G
	G → C		T → A
	G → T		T → G

the gene to be transcribed more, the gene is **up regulated**. Conversely, **down regulation** is where the mutation reduces transcription. Given the high information content of the bacterial and Archaeal genomes (see Section F2), it is more likely that a random point mutation will fall inside a gene. Changes in the DNA sequence of a gene will not necessarily disrupt gene function. The high redundancy of codons (see Section F7) means that, for example, a point mutation in the triplet ATT to ATC will still result in isoleucine being inserted into the final protein. These changes in DNA sequence that have no effect on phenotype are called **silent mutations.** Silent point mutations are also positional: if the mutation occurs outside the active site in a loop region of a protein, there will be no effect on phenotype even if the amino acid sequence changes. Over a third of the amino acids of β-galactosidase can be changed without affecting the ability of the enzyme to convert lactose into galactose and glucose. However, when a point mutation does cause a change in amino acid sequence that is phenotypically detectable, the mutation is classed as a **missense** mutation. A special case of missense is the **nonsense** mutation, when the result of a point mutation is the creation of a stop codon. This will cause the premature termination of protein synthesis, for example GGA (Glycine) to TGA (stop).

Mutations can accumulate with time in microorganisms, and we can trace some of these mutations through long periods of evolution. Sometimes treatment of a mutant to a further round of mutagenesis restores the phenotype that was lost. This **reversion** to wild type can happen because the **back** mutation has occurred at exactly the same nucleotide(s) changed in the original mutant, a **same site mutation**. However, sometimes changes in another region of the genome (a **second site mutation**) can compensate for the mutation at the first site. If the first site of mutation was the region of protein-protein interaction in a c-type cytochrome, the first round of mutation could interrupt the flow of electrons to an acceptor such as pseudoazurin. If the second site mutation changed pseudoazurin slightly so the interaction with the cytochrome was restored, this would return the bacterium to the wild type strain despite being a **double mutant**.

Changes of single base pairs are not the only sorts of mutations that can occur. The **insertion** of one or many base pairs of DNA can occur, as can the **deletion** of nucleotides. Due to the action of bacteriophage (see Section F14) or insertion sequences (see Section F13), large sections of genomic DNA can be moved to a completely different place in the genome, a phenomenon known as **translocation**. Although the large changes in gene sequence are more likely to disrupt genes

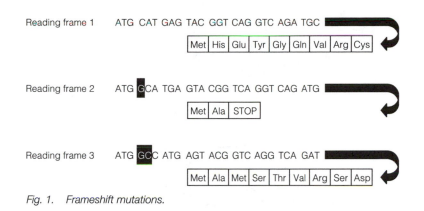

Fig. 1.   Frameshift mutations.

than point mutations, insertion, deletion or translocation of DNA sequences within genes may have no effect if the number of base pairs changed divides by three – this may fortuitously delete or insert codons that have no effect on enzyme activity. However, gene disruption is sure to occur if the number of base pairs deleted or inserted does not divide by three. This is certain to result in a frameshift mutation (*Fig. 1*).

Mutational events in individual genes are often tolerated by Bacteria in the laboratory. The loss of a gene such as *lac*Z only means that the strain can no longer be grown on minimal medium with lactose as sole carbon source, it will still grow on a general medium such as nutrient agar. However, the loss of some gene functions can result in a **lethal mutation**. These include changes in the genomic DNA sequence that disrupts the function of enzymes such as the replicative DNA polymerase and other cell division proteins. Occasionally a change in DNA sequence results in the translation of a protein that can function at the optimal growth temperature of the cell, yet cannot function only a few degrees higher. If the wild type strain can grow without apparent change in phenotype at both temperatures, such a temperature sensitive (**Ts**) mutation can be useful in studying the function of the mutated gene. The **conditional mutant** created might be subjected to further detailed biochemical study, at the **permissive** temperature where the enzyme can function and the **non-permissive** temperature when it cannot. Although conditional mutants are frequently temperature sensitive, other mutations have been made that result in permissive conditions of pH or salinity.

Chemical mutagens

Many chemical compounds can damage the base pairs in DNA. When microbiologists want to cause mutations for experiments in molecular biology and genetics, three main classes of compounds are available to them. These are the **base analogues,** chemicals that mimic the structure of A, C, G, or T and are incorporated into DNA during replication; **base modifiers** which change the subgroups of existing bases in a DNA double helix; and **intercalating agents** which insert into the helix itself, physically disrupting the shape and causing replication to miss out one or more bases.

A commonly used base analogue is **5-bromouracil** (5-BU). It is incorporated instead of thymine, as it has mostly the same composition as this base except for bromine in the place of a methyl group (*Fig. 2*). However, 5-BU is a tautomer, in that the compound is in equilibrium between two chemical structures (*Fig. 2*), the rarer

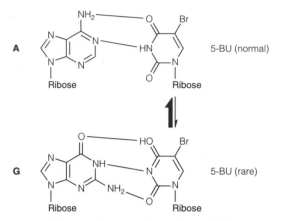

Fig. 2.   Tautomerism of 5-bromouracil changes base pairing.

of which now pairs with guanine. The switch between the rare and normal state is random, and if it occurs during replication a base pair change from AT to GC can result.

The base modifiers such as **methylmethane sulphonate (MMS)** also act on specific base pairs. MMS is an **alkylating agent**, methylating guanine to produce O^6-methylguanine, reducing the number of hydrogen bonds that can be formed from three to two. This means that the methylated G can now only pair with thymine (*Fig. 3*). After resolution during DNA replication, this means that a GC base-pairing becomes AT.

Some fluorescent intercalating agents are used in molecular biology, where their ability to target the double helix is used to visualize populations of DNA. However, the property of intercalation also makes them mutagenic to both bacteria and humans, so compounds such as **ethidium bromide** (see Section F16) and the cyanine dyes should be used with caution. Intercalating agents mutate by stretching the DNA helix out to the extent that DNA polymerase puts in extra bases in a complementary strand in an attempt to compensate.

Physical mutagens   X-rays, other forms of ionizing radiation, and ultraviolet light can all cause damage to DNA and mutation as a consequence. Neither X-rays nor radioisotopes are used routinely to cause mutation in microorganisms, but use of u.v. light is

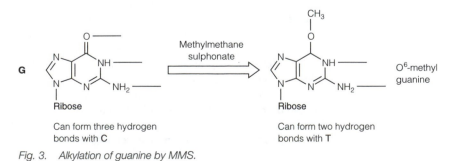

Fig. 3.   Alkylation of guanine by MMS.

(a)

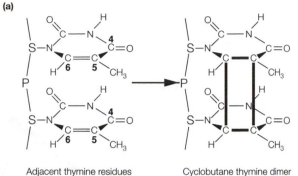

Adjacent thymine residues   Cyclobutane thymine dimer

Fig. 4. Formation of cylcobutane thymine dimers from adjacent thymine residues. S = sugar, P = phosphate.

more common. Ultraviolet damage is also a problem faced by microorganisms in their natural environment during their exposure to sunlight. Ultraviolet is a relatively low energy radiation, but both purines and pyrimidines absorb energy at the same wavelength. The low energy normally means that exposure to u.v. results in point mutations, but with long enough doses u.v. will eventually kill a cell. The most common effect of u.v. is to catalyze the covalent bonding of adjacent thymines, either next to each other on the same strand or between strands either side of a helix. The formation of these **thymine dimers** (*Fig. 4*) produces an abnormal bulge in the DNA helix, which induces **nucleotide excision repair** (see Section F11). However, the repair system has a higher error rate than replication, so a mutation can arise.

Use of mutagenic compounds

When using physical or chemical methods to generate mutant microorganisms, a balance must be struck between making useful changes to the DNA and modifying the entire genome so completely that the organism dies. All mutagens have a concentration-dependent effect, so a **kill curve** is prepared. This curve is a plot of amount of mutagen per number of unmodified cells (e.g. concentration of MMS or length of exposure to u.v.) versus the number of colony forming units recovered after exposure. When the number of CFU before and after exposure is about the same, it is likely that the mutagen is only having a mild effect and mutants with single locus changes will be recovered.

Once suitable mutagenic conditions have been established for a Bacterial, Archaeal or yeast culture, a process of **replica plating** is used to mark which colonies may be useful in genetic analysis. The mutated culture is first spread on an agar plate which has been chosen so that both the wild type and the desired phenotype can grow. Normally the primary culture is diluted so that a manageable number of colonies will appear, i.e. less than 200. The plate is incubated until the colonies are visible, and then blotted with a sterile velvet pad (*Fig. 5*). The velvet pad is then used to inoculate two further plates, one of which is grown under permissive conditions, the other under non-permissive. These conditions might be the presence or absence of a particular growth substrate, or at two different temperatures in the search for a Ts mutant. These replica plates are then incubated again, and it is then possible to deduce which colonies contain the desired mutation.

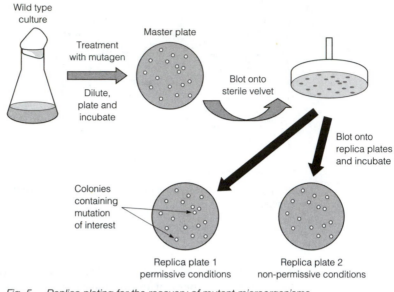

Fig. 5.   Replica plating for the recovery of mutant microorganisms.

F11 DNA REPAIR

Key Notes

DNA damage limitation	The damage caused to the structure and sequence of DNA helix must be repaired by the cellular machinery before efficient replication and transcription can begin again.
Repair by polymerase proof reading	The DNA polymerases themselves have the ability to proofread nucleotides during replication. However, this system cannot cope with bases that have been extensively modified and sometimes proofreading can cause mutation in the regions of DNA modification.
Mismatch repair	Newly replicated DNA spends some time in a hemimethylated form. The methyl-directed mismatch repair mechanism can correctly identify the non-mutated strand by means of this methylation. A series of proteins can then rectify the mutation using the parental strand as a template.
Nucleotide excision repair	The UvrABC system scans the genome and initiates the process in regions of distortion of the helix, such as those caused by thymine dimmers. After the incorrect bases have been removed, the system recruits DNA polymerase I and DNA ligase to mend the break in the strand.
Base excision repair	Chemical base modification of intact DNA can be excised by specific DNA glycosylases, which break the ribose-base glycosyl bond. Apurinic or apyrimidinic sites are then removed by specific endonucleases. Together this is known as the base excision repair (BER) system.
Direct repair	The DNA photolyases are used for photoreactivation or light repair of pyrimidine dimmers and proteins such O^6-methylguanine-DNA methyltransferase can repair the effects of alkylating agents such as methylmethane sulphonate.
The SOS response	Should a mutation still persist despite the action of the other systems, the cell can create a drastic system to resolve mismatches without reference to mother and daughter strands. The induction of the system depends on the balance between the proteins LexA and RecA. The cell can thus continue replication, but this response is extremely error prone.
Related topics	Transcription (F6) Recombination (F13) Mutation (F10) Bacteriophage (F14)

DNA damage limitation

The damage caused to the structure and sequence of DNA helix must be repaired by the cellular machinery before efficient replication and transcription can begin again. If the damage is not repaired, then the **mutation** will become inherited by subsequent generations. The existence of modified bases signals the induction of systems designed to remove specific types of damage. Some types of damage can be corrected by an enzyme-catalyzed reversal of the nucleotide modification, but more lasting damage means that stretches of one strand of DNA must be removed and replaced with new nucleotides. There are limitations to the capacity of the cell to repair its own DNA in that in most cases the repair machinery relies on a correct template against which to make corrections. If both sides of the helix, i.e. both base pairs are damaged, then the cell is less likely to correct the error. Thus events that cause deletion of regions of the DNA are more likely to be inherited than those that only damage one strand, or a few bases of one strand, of the helix.

Although DNA repair systems are extremely efficient in organisms such as the Bacterium *Deinococcus radiodurans* (see Section C6) and the Archaeon *Pyrodictium abysii* (see Section C6), there is a greater error in DNA repair mechanisms than in DNA replication.

Repair by polymerase proofreading

The DNA polymerases themselves have the ability to proofread nucleotides during replication (see Section F5). As such this provides a first line of defence in order to preserve the integrity of the genome. However, this system cannot cope with bases that have been extensively modified and sometimes proofreading can cause mutation in the regions of DNA modification.

Mismatch repair

Despite the high fidelity of DNA replication, and the backup proofreading system, it is still possible for the incorrect base to be opposite a template base. To fix such potential mutations, the repair mechanism must have some way of distinguishing between the template strand and the daughter strand synthesized against it. In many bacteria a mature chromosomal DNA has methylated bases at the sequence GATC. Daughter DNA strands experience some delay in methylation, so newly replicated DNA spends some time in a **hemimethylated** form (template strand methylated, daughter strand not). The **methyl-directed mismatch repair** mechanism can correctly identify the non-mutated strand by means of methylation (*Fig. 1*). The system can only resolve single base-pair mismatches and needs the participation of Dam methylase, MutH, L and S, DNA helicase II, single stranded binding protein, DNA polymerase III, exonuclease I, exonuclease VII, RecJ nuclease, exonuclease X and DNA ligase.

Nucleotide excision repair

The removal of many nucleotides (by UvrABC exonuclease) and their replacement (by DNA polymerase I and DNA ligase) is ubiquitous in the Bacteria. In *E. coli* this mechanism has been shown to be largely error-free. It enables the removal of large areas of mutation, e.g. as a result of the presence of pyrimidine dimers and is thus known as nucleotide excision repair (**NER**). The UvrABS system scans the genome and initiates the process in regions of distortion of the helix (*Fig. 2*). Nucleotide excision repair differs from base excision repair in that the entire nucleotide is removed, causing a complete break in one strand of the helix.

Base excision repair

Some mutagenic events lead to the modification of the base part of the nucleotide. The action of the base modifiers can sometimes be corrected by very specific DNA glycosylases, which break the ribose-base glycosyl bond. This leaves the backbone of the DNA strand intact, with some sites de-purinated or de-pyrimidinated. These **apurinic** or **apyrimidinic** sites (AP sites) are then removed by AP endonucleases. In contrast to NER, only one or two nucleotides are removed. The DNA strand is made entire again by the action of DNA polymerase I and DNA ligase, by the same mechanism as outlined in NER. The abnormal bases repaired by **BER** include

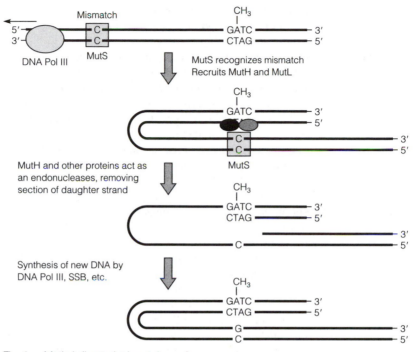

Fig. 1.   Methyl-directed mismatch repair.

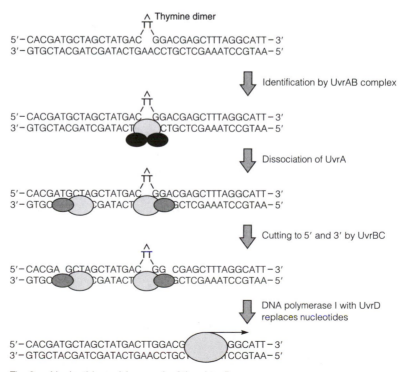

Fig. 2.   Nucleotide excision repair of thymine dimers.

uracil, hypoxanthine, xanthine, and alkylated bases. In some organisms it is also the mechanism used to resolve pyrimidine dimers.

Direct repair

In addition to proofreading and the excision repair mechanisms, the Bacterial cell has a range of mechanisms for correcting chemical changes in DNA *in situ*. The **DNA photolyases** are used for **photoreactivation** or **light repair** of pyrimidine dimers and proteins such O^6-methylguanine-DNA methyltransferase (the Ada protein) can repair the effects of alkylating agents such as MMS. This transferase is unusual in that it is not an enzyme in the strictest sense of the word. The methyl group is directly transferred to the Ada protein, which renders it completely inactive. Thus it cannot participate in another reaction, so does not play a catalytic role.

The SOS response

Should the base pairing between strands still remain incorrect, the Bacterial cell has a final mechanism to excise corrupted DNA. This mechanism may actually result in the persistence of mutations in the next generation, but prevents the accumulation of mismatches to the point where they become lethal. Normally a protein called **LexA** represses an overall response, but as mismatches accumulate, the concentration of another protein (**RecA**) also rises. A high concentration of RecA causes the cleavage of LexA. In the absence of LexA, 17 genes are turned on which constitute the **SOS response**. This system quickly resolves all DNA mismatches, but does not distinguish between parental and daughter strands. The cell can continue replication, but this response is extremely error prone.

F12 TRANSFER OF DNA BETWEEN CELLS

Key Notes

DNA passes between microorganism

Vertical transfer between members of the same species is well character-ized, but more recently it has become apparent that microorganisms are capable of horizontal transfer between species, genera, phyla or even kingdoms. The best studied systems of transfer between donor and recipient include conjugation, transduction, and transformation.

Conjugation

A plasmid known as the F′ factor encodes genes that enable the plasmid to be passed from one individual to another. It has also shown that F′ can integrate into the genome to form Hfr strains. When an Hfr strain conjugates with an F-recipient, parts of the donor's chromosome may be copied and transferred as well, but normally only sections of the genome close to the site of integration. If the acquisition of the donor's DNA causes the recipient to change its phenotype after recombination, the recipient is said to be a transconjugant.

Transformation

Transformation is the most widely used technique to introduce DNA into *Escherichia coli*, but the least well understood. Washing in ice-cold magnesium chloride solution renders the cells competent, a state in which they can take up DNA when heat shocked. After incubation in a recovery medium, a small proportion of the total cells maintain the plasmid and are known as transformants. A similar protocol allows *Bacillus subtilis* cells to take up linear DNA.

Generalized and specialized transduction

DNA transfer can occur between cells by the action of bacteriophage. If the phage is lysogenic and excises from the host's genome imperfectly, some host genes can be transferred to a new recipient on reinfection. This is the process of specialized transduction. In contrast, generalized transduction arises when parts of the host genome are packaged instead of phage DNA. In both cases the resulting recipient cells are known as transductants.

Related topics

Recombination (F13)   Plasmids (F15)
Bacteriophage (F14)

DNA passes between microorganisms

While the Bacteria lack the sophisticated systems of sexual reproduction found in eukaryotes, they do not rely on mutation alone to generate diversity within their populations. Spontaneous mutagenesis alone is insufficient to respond to the varying environments the majority of Bacterial species encounter. Bacteria can acquire variation in genes, complete new genes or even complete biosynthetic pathways via the acquisition of plasmids (see Section F15), infection by bacterio-phage (see Section F14) or transposons (see Section F13) and by direct, non-sexual interchange of genetic information (see below).

The phenomenon of Bacterial transfer of genetic information has long been studied in the context of same-species or **vertical gene transfer**. However, it has become increasingly apparent that transfer occurs between species too. **Horizontal gene transfer** can account for the rapid spread of antibiotic resistance in a variety of hospital-acquired infections, but also can be used to our advantage in the generation of consortia for the biodegradation of xenobiotics (see Section G2). In a mixed population, such as that found in a hospital or bioreactor, it is hard to pinpoint the exact mechanisms used to acquire DNA, but looking in the genomes of isolated organisms it can be easy to pinpoint the genomic regions which originated in other organisms by their unusual codon usage or other sequence signatures.

The best understood mechanism of gene transfer is **conjugation**, a method that gave the first genetic map of *Escherichia coli*. This is a plasmid-mediated effect, in contrast to **transduction** in which DNA moves from the **donor** to the **recipient** strains by means of a Bacterial virus or **bacteriophage**. The method used most commonly to introduce foreign genes into laboratory strains of Bacteria is **transformation**, a poorly understood process relying on a shock response to take up plasmids or other fragments of DNA.

Conjugation   Some strains of *Escherichia coli* possess a plasmid (the **F′ factor**) which allows them to transfer DNA to a recipient of the same species. The transfer of DNA is

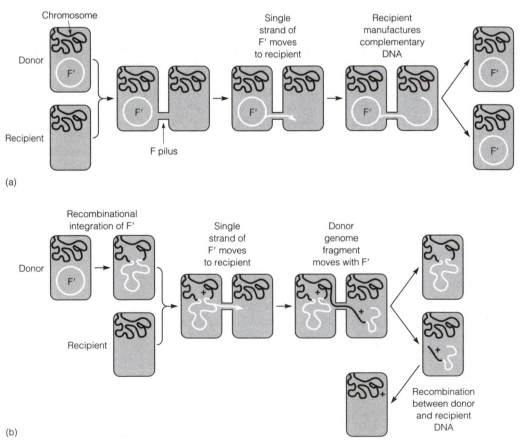

Fig. 1.   Mating of donor and recipient a) between an F⁺ strain and recipient b) between an Hfr strain and recipient.

unidirectional from the donor carrying the F′ factor (the strain designated F⁺) to a recipient (F⁻) that does not have the plasmid (*Fig. 1a*). The copying of the F′ factor plasmid is accompanied by a physical joining of donor and recipient via a plasmid-encoded structure, the F-pilus. After conjugation is complete the recipient becomes F⁺.

Conjugation was first described in 1946 by Lederburg and Tatum, who noted that two Bacterial auxotrophs could be mixed together to make a new non-auxotrophic strain. They noted that the transfer of information could be easily stopped by shaking the cells, even though the shaking process did not kill the cells outright. Others separated the donor and recipient by a fine filter and demonstrated that physical contact between the cells was an absolute requirement for conjugation. The process was not fully understood until it became apparent that the strains that could act as donors (**high-frequency recombination** or **Hfr** strains) all possessed the large F′ plasmid. It was also shown that F′ could **integrate** into the genome of the Hfr strains. During the procedure, parts of the donor's chromosome may be copied and transferred as well, though rarely the entire complement (*Fig. 1b*). If the acquisition of the donor's DNA causes the recipient to change its phenotype after recombination, the recipient is said to be a **transconjugant**.

Transformation

Although conjugation and transduction provide simple ways to introduce genes into microorganisms, both are time-consuming processes. The most commonly used way of introducing DNA into Bacteria in today's laboratory is via transformation. Most routine molecular biology experiments result in the generation of plasmids (see Section F16), in the form of DNA libraries, cloning experiments, or preparation of DNA constructs for other experiments.

The technique of transformation is well-defined, but the mechanism is poorly understood. In *E. coli*, in order to make the cells available for manipulation, plasmid-free cells are grown to mid log phase, and then washed in a cold magnesium chloride solution to remove traces of growth medium. A high concentration of washed cells is suspended in cold calcium chloride. At this stage the cells are **competent**, a state mimicking a physiological state in which the Bacteria are able to take up DNA from their environment. It is thought that this is a starvation response and perhaps the influx of external DNA may provide an extra gene that might help the cell survive. In the laboratory, the plasmid to be **transformed** is added to the competent cells. The plasmid remains in solution until the cold cells are suddenly heat shocked by raising the temperature to 42°C. Only a small percentage of the cells take up plasmid DNA as a result of the heat shock, and all the cells are in a weakened state. A rich recovery medium is added, and then the cells are plated out on a selective medium after incubation. The cells that have successfully taken up the plasmid are known as **transformants**. Not all Bacteria can be transformed in the lab, but a similar protocol is used to generate *Bacillus subtilis* transformants, though they can only take up linear DNA and must maintain the acquired gene(s) by recombination into their genome. However, the way in which either plasmid DNA in *E. coli* or linear DNA in *B. subtilis* crosses the cell membrane intact, or in what form it does so, is currently the subject of speculation.

Generalized and specialized transduction

DNA transfer can occur between cells by the action of **bacteriophage** (see Section F14). Occasionally, the excision of a lysogenic phage's genome from the host chromosome is imperfect. Instead of just phage genes, the excision includes one or two host genes from locations adjacent to the site of phage recombination.

When the altered phage infects another cell, it is possible that the extra genes will complement mutated genes in the second host's genome. This process is known as **specialized transduction**. In contrast **generalized transduction** can lead to the transfer of any gene in a host genome, not just those adjacent to a site of lysogeny.

An example of a generalized **transducing phage** is the bacteriophage P1 of *Escherichia coli* (*Fig. 2*). Using this phage to transfer DNA between *E. coli* strains is a relatively quick and easy way of making chromosomal mutations, and is some-times abbreviated to P1 transduction. The principle is outlined in *Fig. 2*, and for the most part follows the normal lifecycle of a lysogenic bacteriophage (see Section F14). The difference occurs just as the prophage leaves the host genome on the induction of the lytic cycle. The phage genome leaves completely and induces the degradation of the host chromosomes. The phage replicates inside the cell, but during this process some random Bacterial DNA is packaged as if it were phage DNA.

When the donor cell is lysed, the P1 phage released can be used to infect a recipient. If the donor was a wild type cell and the recipient was, for example, an auxotroph for histidine, some of the P1 phage may carry the DNA to correct the mutation. A small number of infected recipients will receive a fragment of wild-type donor DNA instead of phage P1 DNA and this can recombine with recipient chromosome. While a useful technique in *E. coli*, where P1 phage has been engi-neered to increase the frequency of generalized transduction, few similar systems exist. The number of genes that can be transferred is limited to the maximum amount of DNA that can be packaged into a P1 phage head.

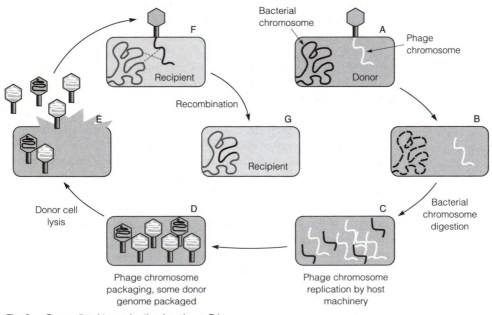

Fig. 2.   Generalized transduction by phage P1.

F13 RECOMBINATION

Key Notes

Homologous recombination

This is sometimes called generalized recombination and involves the exchange of DNA between two regions of homology (i.e. similar but not necessarily identical DNA sequence) on two different DNA molecules. RecBCD digests dsDNA until it reaches a chi sequence, where the activity of RecBCD then changes to a single strand-specific 5′–3′ exonuclease, so that a single DNA strand remains as a long 3′ overhang. The free single stranded DNA becomes coated with RecA protein, and can migrate into the other DNA molecule in a process of strand invasion. Once the nicks in the DNA strands have been joined, the two DNA molecules adopt a cross-shaped formation known as the Holliday junction. The binding of RuvA to the Holliday junction leads to the binding of RuvB to adjacent double stranded DNA. RuvB is a translocase, and promotes the movement of the Holliday junction along the DNA molecules (branch migration). The single-stranded parts of the junction are cleaved (strand exchange) and the two molecules resolve when RuvC binds to the RuvAB DNA complex.

Site specific recombination

Recombination between regions of DNA homology of only a few base pairs can occur via site specific recombination mediated by XerC and XerD. A similar method is used to integrate bacteriophage λ DNA into its host's genome. λ integrase makes cuts on both strands about 15 bp apart at a specific site. Integration host factor (IHF) catalyzes recombination between the homologous sites on the phage and host genome, resulting in the λ genome insertion into the *E. coli* DNA. The reverse of the process is catalyzed by the phage-encoded excisionase, stimulated by host IHF and the protein Xis.

Transposition

Transposition is the insertion of short DNA fragments into any position in the genome. This is sometimes called illegitimate recombination as it appears to require no homology between sequences. Insertion sequences use this mechanism to move between sites and can be detected in many chromosomes and plasmids. Transposable elements could be looked on as very basic viruses without the means to encode a protein coat for protection outside the cell. The IS encoded transposase makes cuts at another site in the host chromosome and is excised to be integrated into the new site. The IS is flanked by inverted terminal repeats, which aid in the transposition process. As the transposase cuts are staggered, the bases either side of the repeats are duplicated after integration.
The transposable viruses do not excise themselves during transposition, but place a copy elsewhere on the genome instead (replicative transposition). Transposon mutagenesis can offer a convenient method of generating mutants via Tn5 derivatives.

Related topics

Transcription (F6)
Bacteriophage (F14)

Homologous
recombination

This process is sometimes called **generalized recombination** and involves the exchange of DNA between two regions of homology (i.e. similar but not necessarily identical DNA sequence) on two different DNA molecules. One of these regions might be on a plasmid, the other on the chromosome. In rarer instances these two regions could be on the same chromosome but many hundreds of base pairs apart or more. After recognition and alignment of the regions of homology by the enzyme RecA (*Fig. 1a*), a protein complex, **RecBCD** digests dsDNA until it reaches a **chi** sequence (*c*rossover *h*ot *i*nstigator, sequence GCTGGTGG). The activity of RecBCD then changes to a single strand-specific 5′ -3′ exonuclease, so that a single DNA strand remains as a long 3′ overhang. This is represented by the arrowheads in *Fig. 1b*. The free single-stranded DNA becomes coated with RecA protein, and can migrate into the other DNA molecule in a process of **strand invasion** (*Fig. 1c*). Once the nicks in the DNA strands have been joined, the two DNA molecules adopt a cross-shaped formation known as the **Holliday junction**. This is shown in *Fig. 1d*, and is named after the person to first propose it.

The binding of RuvA to the Holliday junction leads to the binding of RuvB to adjacent double stranded DNA. RuvB is a translocase, and promotes the movement of the Holliday junction along the DNA molecules (*Fig. 1e*). This is known as **branch migration**. The single-stranded parts of the junction are cleaved (**strand exchange**) and the two molecules become separate molecules again (**resolve**) when RuvC binds to the RuvAB DNA complex. How the resolution proceeds is still the subject of some debate, but it has been postulated that an equilibrium is

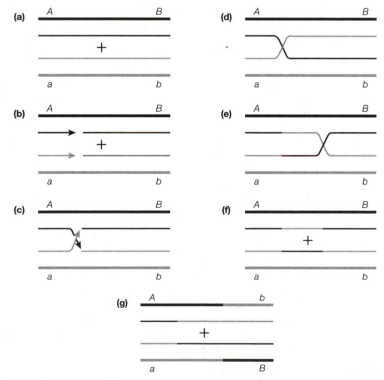

Fig. 1.   The Holliday model for homologous recombination. From Instant Notes in Molecular Biology, *2nd Edn, F4, Fig. 1, p. 100.*

established between a RuvA planar Holliday junction and tetrahedral junctions stabilized by another protein, RusA. Depending on the direction of a final cleavage, there are two possible outcomes (*Fig. 1f* and *Fig. 1g*).

Site specific recombination

Recombination can occur between regions of DNA with apparently no homology, where only a few base pairs are the same. This is an important mechanism in the XerC/XerD mediated resolution of the two circular molecules of DNA in *E. coli* chromosomal replication (see Section F5), and does not require RecA. A similar method is used to integrate bacteriophage λ DNA into its host's genome (see Section F14). The integration site is a 15 bp sequence present both in the host and in λ DNA. The **λ integrase** makes cuts on both strands about 15 bp apart. **Integration host factor** (IHF) catalyzes recombination between the homologous sites and the λ genome is inserted into that of *E. coli*. The reverse of the process is catalyzed by the phage-encoded **excisionase**, stimulated by host IHF and Xis.

Transposition

Transposition is the insertion of short DNA fragments into any position in the genome, a process carried out by **transposons**, **insertion sequences** (IS) and some specialized phage (*Table 1*). This is sometimes called **illegitimate recombination** as it appears to require no homology between sequences. However, analysis of transposon positions after transposition suggests it is a pseudo-random event, and that the recombination is site-specific. However, the recognition sites are extremely degenerate and occur so frequently in most genomes that almost all genes can be interrupted by the entry of a transposon.

The presence of insertion sequences can be detected throughout the genomes of many Bacteria and Archaea as well as in their plasmids, and are essential to some functions. The integration of the F′ plasmid into the *E. coli* chromosome is mediated by homologous recombination between identical insertion sequences (see Section F12). The ecology and microbiology of transposons is extremely complex, given that some of these elements can move around the genome in the absence of chromosomal replication and initiate a form of conjugation to transfer themselves from one cell to another. In some senses, transposable elements could be looked on as very basic viruses especially given that transposition is the method of replication for bacteriophage such as Mu.

A simple transposition event is illustrated by the conservative transposition of an IS element. The IS encoded **transposase** makes cuts at another site in the host chromosome (in a pseudo-random manner) and is excised and integrated into the new site in much the same way as λ achieves lysogeny. The IS is flanked by **inverted terminal repeats**, which aid in the transposition process. As the transposase cuts are staggered, the bases either side of the repeats are duplicated after integration (*Fig. 2*).

Table 1.   Transposable elements in E. coli

Type of element	Example	Notes
Insertion sequence	IS21	Carries only the information necessary for transposition
Transposon	Tn5	Carries transposition genes plus other genes e.g. for antibiotic resistance
Transposable bacteriophage	Mu	Carries many genes and also has a protein coated form for survival outside the host.

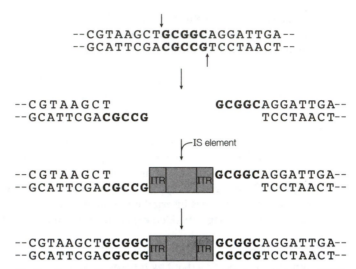

Fig. 2.   *Transposition of an insertion sequence (is) element into a host DNA with duplication of the target site (shown in bold). From* Instant Notes in Molecular Biology, *2nd Edn, F4, Fig. 2, p. 101.*

The transposable viruses do not excise themselves during transposition, but place a copy elsewhere on the genome instead. This process of **replicative transposition** allows Mu to replicate its genome in the host both in terms of copy number per chromosome as well as during cell division.

Transposons offer a convenient method of generating mutants, particularly when gene interruption is required. **Transposon mutagenesis** is most developed for Tn5 derivatives, in which additional genes such as those for kanamycin resistance or β-galactosidase activity have been added. A bacterial strain can be transformed with a plasmid bearing the modified transposons. The transposons can then jump from the plasmid to the chromosome and cause random interruption of genes. Auxotrophs or other mutants can then be identified as normal (see Section F10).

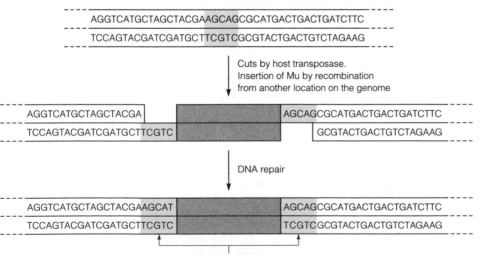

Fig. 3.   *Insertion of bacteriophage Mu by transposition into a bacterial genome.*

F14 BACTERIOPHAGE

Key Notes

Viruses of bacteria

The viruses that infect the prokaryotes are the bacteriophage, sometimes abbreviated to phage. The replicatative form (the virion) can be a variety of shapes including filamentous and icosahedral, or they may have a distinct head and tail section. Most possible variations of nucleic acid content (dsDNA, ssDNA (single-stranded DNA), dsRNA, ssRNA (single-stranded RNA)) have been observed in bacteriophage. In common with other viruses, after entry to the cell they may either replicate and cause lysis of the cell, or enter into a dormant form by integration of viral DNA into the host genome.

Modes of replication

Virulent phage have a lytic mode of replication, where the bacteriophage genome is injected into a cell, replicates, and then kills its host on the formation of many new virions. The alternative used by lysogenic phage is where (post-infection) rather than triggering more virion production, the viral genome recombines with the host genome (known as lysogeny) and does no further damage. In response to an environmental stimulus, the viral DNA excises from the host genome and a lytic cycle ensues.

The lytic cycle

The lytic or virulent bacteriophage attack and lyse their hosts during the replication cycle. After injection of the viral genome, bacteriophage T4 first transcribes nucleases to cut up the Bacterial genome. T4 genomes are synthesized as a concatamer, with each copy linked end to end with another. T4-specific nucleases cut the concatameric DNA into pieces. Phage head, tail, sheath, and other proteins necessary for the construction of the protein coat are then synthesized, assemble around the T4 chromosome, and are released on cell lysis.

The lysogenic cycle

The lysogenic phages are exemplified by bacteriophage λ. On entry to the host cytoplasm, the phage genomic *cos* sites at either end of the λ genome allow the molecule to circularize into a plasmid-like form. At this point the bacteriophage can replicate and lyse the cell much as T4 does, or enter into a lysogenic form where the entire λ genome integrates into the host genome by recombination. The phage genome (in this state known as a prophage) is replicated as a single copy during each round of host cell division. Induction of a further recombinational event excises the phage DNA from the host. The replication cycle then follows a lytic path. The replication of the λ genome is not accomplished using a bi-directional replication fork but by rolling circle replication, resulting in a long double-stranded, multi-copy molecule. This then has to be cleaved to yield individual genomes ready for packaging.

Related topics   Transcription (F6)   Recombination (F13)

Viruses of bacteria

In common with most living organisms, microorganisms can be infected by viruses. The virology of higher organisms is discussed further in Section L. The viruses that infect the Bacteria and Archaea are the **bacteriophage**, sometimes abbreviated to **phage**. Bacteriophage are not living entities and are incapable of reproduction outside of their host cell. In essence they are genetic elements that have acquired the ability to maintain themselves outside the host cell by coating their nucleic acids with proteins, but are no more alive than the protein-free DNA of plasmids, transposons or insertion sequences (see Sections F13 and F15). As with the viruses of eukaryotes, the form and reproductive cycle of bacteriophage varies considerably. The extracellular **virion** form can be filamentous (for example bacteriophage M13), icosahedral (for example, bacteriophage M13, but see Section L1 for a further discussion of icosahedral symmetry) or have a distinct head and tail section (*Fig. 1*). Some bacteriophage are surrounded by a membrane, which is often derived from the host. The diversity of shapes of the protein part of the virus is accompanied by a diversity of nucleic acid complements. Most possible variations of nucleic acid content – dsDNA, ssDNA, dsRNA, ssRNA – have been observed in bacteriophage. Almost all species of Bacteria and Archaea have been found to have one or more bacteriophage that infects them, and some phage have an extremely broad host range.

All bacteriophage have common methods of replication. Attachment to the host cell is followed by entry to the cytoplasm by penetration or injection. Phage gene expression commences and the phage genome is copied. Genomes and phage proteins are then assembled into virions ready for release and to begin the infection cycle again. An archetypal bacteriophage has a head containing the genome, and a tail to attach to the host. The tail fibres target specific receptors on the surface of the host cell membrane. A sheath surrounds a core tube, through which the viral genome is injected into the host cytoplasm. Contraction of the sheath is ATP dependent, and in a manner analogous to muscle contraction. Transcription of the bacteriophage genome is tightly controlled, but very quickly takes over the host cell's functions, shutting down all activities and dedicating all transcription and translation towards the manufacture of new virions.

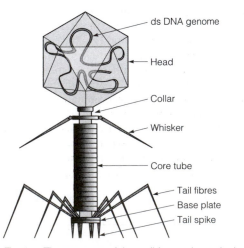

ds DNA genome

Head

Collar

Whisker

Core tube

Tail fibres
Base plate
Tail spike

Fig. 1.   The structure of the well-known bacteriophageT4.

Modes of replication

Bacteriophages can have one of two reproductive strategies. In the first **lytic** mode of replication, the bacteriophage enters a suitable cell, replicates and then kills its host. The alternative is a **lysogenic** mode. A whole virion infects the cell, but rather than triggering more virion production, the viral genome recombines with the host genome and does no further damage. The lysogenized virus's only function is to code for proteins that prevent re-infection with bacteriophage of the same type. The viral genome is passed on to daughter cells during normal host cell division. In response to an environmental stimulus the viral DNA excises from the host genome and a lytic cycle ensues. **Virulent bacteriophage** such as T4, which only have a lytic replication cycle, run the risk of exterminating the host population. In the lab, these bacteriophages are characterized by the complete lysis of all the host cells in a liquid culture. Those with the ability to perform **lysogeny** (such as bacteriophage λ) can maintain a pool of host cells capable of releasing virions gradually. In liquid culture, Bacteria infected with lysogenic phage fail to reach the expected optical density, but still appear to be dividing normally.

The lytic cycle

The lytic or **virulent** bacteriophage attack and lyse their hosts during the replication cycle (*Fig. 2*). This is exemplified by bacteriophage T4, whose relatively large genome contains about 200 genes. T4 attaches to the cell membrane and injects its genome into the cytoplasm. The first genes transcribed code for nucleases to cut up the bacterial genome, effectively stopping any further host cell activity. The remainder of the phage replication cycle relies on existing host ribosomes RNA polymerases and enzymes in the cell, in addition to the phage-encoded products. Phage genes are now expressed that enable the T4 DNA to be copied. The T4 genomes are synthesized as a **concatamer**, with each copy linked end to end

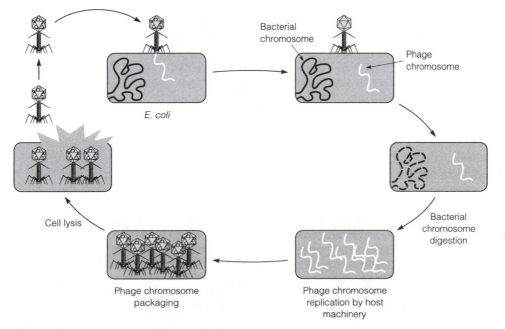

Fig. 2.   Replication cycle of a lytic bacteriophage.

with another. T4-specific nucleases cut the concatameric DNA into pieces, each approximately 105% of the genome. Phage head, tail, sheath and other proteins necessary for the construction of the protein coat are now synthesized, and coalesce around the T4 chromosome. Each viral head receives the same overall complement of genes, but as the DNA molecule is slightly larger than the minimum number of genes required, the genes at each end repeated. This **terminal redundancy** allows a simple but reliable method of ensuring that each phage head always receives at least one copy of each gene.

Finally, phage genes that cause the autolysis of the infected cell are transcribed, with the release of around 300 virions. Each of these has the potential to infect more cells, but if left in a population without phage resistance lytic phages will eventually wipe out all their local hosts.

The lysogenic cycle

The problem of killing the host that allows replication is overcome by the **lysogenic phages,** exemplified by bacteriophage λ. Free virions attach and inject their genome into the cytoplasm in the same way as the virulent phages, but then have the option of either establishing a carrier population amongst the host cells, or replicating, lysing and reinfecting (*Fig. 3*). On entry to the host cytoplasm, the *cos* sites at either end of the λ genome allow the molecule to circularize into a plasmid-like form. At this point the bacteriophage can replicate and lyse the cell much as T4 does, or enter into a form that perpetuates as the host divides.

The carrier or lysogenic state results from the integration of the entire λ genome into the host genome by recombination (see Section F13). The regions of homology allowing the recombination are found in the centre of the λ genome, and are known as *att* sites. The phage genome (in this state known as a **prophage**) is replicated as a single copy during each round of host cell division so that λ is

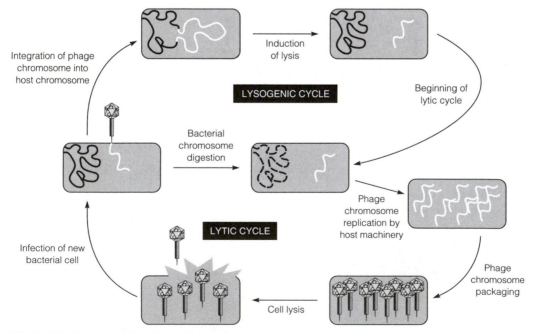

Fig. 3. The lysogenic cycle of bacteriophage λ.

passed vertically through the generations. To ensure that new colonies of hosts can be infected, a stimuli related to DNA damage (ultraviolet light, chemical mutagens, see Section F10) induce a further recombinational event to excise the phage DNA from the host. The replication cycle then follows a lytic path, with digestion of the host genome, phage genome replication, phage head protein formation and packaging until the host cell is lysed and phage progeny released. However, the mechanism by which the λ genome is copied differs considerably from bacterial replication.

The replication of the λ genome is not accomplished using a bi-directional replication fork. Instead, the plasmid-like cytoplasmic form is copied by **rolling circle replication**. This allows copies of the genome to be made in one direction only, but results in a long double-stranded, multi-copy molecule. This then has to be cleaved to yield individual genomes ready for packaging, but has the advantage of being very rapid.

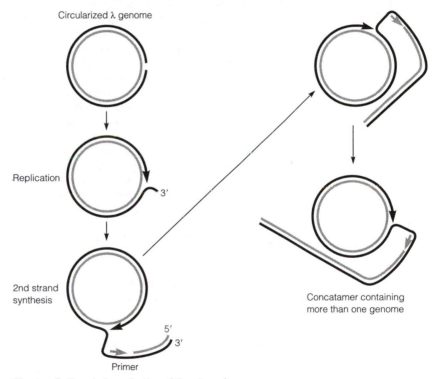

Fig. 4.   Rolling circle replication of the phage λ genome.

F15 PLASMIDS

Key Notes

Structure of plasmids

A plasmid is defined as an entire molecule of DNA that is replicated independently of the host chromosome. When over 100 kbp or more, they are known as megaplasmids. Removal of a plasmid is known as curing. There are more than 300 different types of naturally occurring plasmids for *Escherichia coli* alone, some of which have been adapted for applications in recombinant DNA technology.

Replication of plasmids

The plasmids of *Escherichia coli* are complete, closed, circular (ccc) double-stranded DNA molecules. Plasmids are replicated by the host cell machinery. Episomal plasmids integrate themselves into the chromosome and are replicated along with it. The number of plasmids per cell is known as the copy number, and this can vary between very low (stringent plasmids) with only one or two copies per cell, to very high (relaxed plasmids) with hundreds of copies per cell. As maintenance of a plasmid in a cell is energetically expensive, the cell will try to delete the plasmid. Thus to maintain a plasmid in a strain, some form of selective pressure is applied to ensure that as many cells as possible keep the plasmid.

Plasmid incompatibility

In common with the bacteriophage and other viruses, some plasmids have evolved mechanisms to repel reinfection by different types of plasmid. When this occurs, the two plasmids are said to be incompatible. Testing which plasmids are compatible with which gives a table of plasmid incompatibility groups.

Function of plasmids

The best known of the *E. coli* plasmids is the F' plasmid. The genes that govern transfer are known as *tra* genes and code for proteins with a variety of functions including enablement of the movement of DNA and the rolling circle replication associated with it. Other plasmids carry genes that confer heavy metal resistance, entire metabolic pathways, toxins, virulence factors or antibiotics. There is transfer of plasmids between members of the same species, but promiscuous plasmids may be transferred between cells of different species.

The plasmids used in molecular biology have been engineered to remove most of the unfavourable aspects of the wild type. Most are small (<5 kbp) so that they are easy to manipulate in vitro and are less likely to recombine with the host chromosome. The *tra* genes are absent so that the recombinant DNA will be contained in the designated host cell.

Related topics

Transfer of DNA between cells (F12)
Manipulation of cellular DNA and RNA (F16)

Structure of plasmids

A plasmid is defined as an entire molecule of DNA that is replicated independently of the host chromosome. This alone does not differentiate them from viruses, so in addition a virus is defined as having extracellular structures for protection outside the host. A plasmid should carry no genes that are essential for the existence of the cell. This helps us to distinguish between very large plasmids (over 100 kbp or more, known as **megaplasmids**) and the smaller chromosomes. Plasmids often carry genes that help the cell to survive environmental challenges, such as a sudden rise in the concentration of mercury, but do not carry any of the genes for the central metabolic pathways. This can be tested experimentally by **curing** the cell, which is removing the plasmid. If the cell can still grow when cured, the DNA is a plasmid and not a chromosome. As there are more than 300 known types of naturally-occurring plasmid for *Escherichia coli* alone, this section will be restricted to a discussion of Bacterial plasmids, specifically those from *E. coli*. These have also been adapted to be the most useful, with applications in **recombinant DNA technology**.

Replication of plasmids

The plasmids of *Escherichia coli* are complete, closed, circular (**ccc**) double-stranded DNA molecules. This means that they are circular, joined at both ends and are supercoiled within the cytoplasm. Typically the plasmids are about 5% of the genome size, and the minimum size is only around 2000 bp. The plasmid is replicated by the host cell machinery, using DNA polymerase III and all the other apparatus associated with it. Some plasmids (the **episomes**) integrate themselves into the chromosome and are replicated along with it, much like the lysogenic bacteriophage (see Section F14). The number of plasmids per cell is known as the **copy number**, and this can vary between very low (**stringent plasmids**) with only one or two copies per cell, to very high (**relaxed plasmids**) with hundreds of copies per cell. The relaxed plasmids in particular place a considerable energetic burden on the cell, since replicating each base pair requires one molecule of ATP. The cell will try and relieve itself of the burden of carrying the plasmid if there is no reason to keep it, so to maintain a plasmid in a strain, some form of **selection** must be applied. In order to be replicated by the host machinery, a plasmid requires an origin of replication and a region site-specific recombination much like a chromosome. However, some plasmids are replicated uni-directionally.

Plasmid incompatibility

There can be more than one different type of plasmid in a cell, as exemplified by even the wild type of *Borrelia burgdorferi* which carries 17 circular and linear plasmids. In common with the bacteriophage and other viruses, some plasmids have evolved mechanisms to repel re-infection by different types of plasmid. When this occurs, the two plasmids are said to be incompatible. Testing which plasmids are compatible with which gives a table of **plasmid incompatibility groups.** This can be useful to the biotechnologist who may wish to introduce recombinant genes in successive experiments.

Function of plasmids

The best known of the *E. coli* plasmids is the F′ plasmid (see Section F12), which confers the ability of a cell to transfer this plasmid by conjugation. The genes that govern transfer are known as *tra* genes, and take up about one third of the F′ plasmid's 99 159 bp. These genes code for proteins with a variety of functions including enablement of the movement of DNA and the rolling circle replication associated with it. The F plasmid confers a very specific property on *E. coli*, but some plasmids carry genes which confer heavy metal resistance, while others can carry entire metabolic pathways to deal with unusual carbon sources. The plasmids of most concern in medical microbiology encode toxins, virulence factors or antibiotics. Enteropathogenic *E. coli* (see Section H1) colonize the small intestine and produce toxins that causes diarrhea and haemolysis in humans, and both the protein that enables attachment to the intestine wall and the toxins themselves are plasmid encoded.

The current model for plasmid ecology is that a great diversity of genes is held extrachromosomally in any indigenous bacterial population. There is constant exchange of genetic material, but the majority of cells do not retain the plasmids acquired after a few cell divisions. When the population is stressed, only those bacteria carrying the plasmid-borne genes essential to deal with that stress survive and quickly divide to outnumber plasmid-free organisms. There is some transfer between members of the same species, but **promiscuous** plasmids may be transferred between cells of different species. This horizontal gene transfer has been noted in the spread of antibiotic and mercury resistance.

Vectors for recombinant DNA technology

The plasmids used in molecular biology have been engineered to remove most of the unfavorable aspects of the wild type. Most are small (<5 kbp) so that they are easy to manipulate in vitro and are less likely to recombine with the host chromosome. The *tra* genes are absent so that the recombinant DNA will be contained in the designated host cell.

F16 MANIPULATION OF CELLULAR DNA AND RNA

Key Notes

Recombinant DNA

Recombinant strains of *Escherichia coli* have been developed that can harbor foreign DNA but cannot survive outside the laboratory. Adapted *E. coli* plasmid cloning vectors maintain foreign DNA in an easily accessible form, while expression vectors are plasmids used to generate large amounts of recombinant protein. Shuttle vectors allow the easy creation of specific arrangements of genes or gene fragments in *E. coli* before the construct is transferred to another organism. Part of the attraction of *E. coli* is the ease with which genomic and plasmid DNA can be isolated. Plasmids can be isolated from lysed *E. coli* by techniques such as caesium chloride density centrifugation or by the use of the many commercially available spin columns.

Gel electrophoresis

Two common methods of DNA separation by size are polyacrylamide gel electrophoresis (PAGE) and horizontal submerged agarose gel electrophoresis. Both methods rely on the fact that DNA is slightly negatively charged, so if a current is applied through a matrix or solution, DNA molecules will move towards the positive electrode (the cathode). PAGE is useful for the fine separation of DNA fragments for applications such as sequencing but gels are difficult to handle after running. Agarose gel electrophoresis still makes relatively fragile gels, but they can be more easily manipulated. Molten agarose at is poured into a gel former along with a comb and allowed to cool. The comb is removed to leave wells in the surface of the agarose slab which is then submerged under TAE (tris-glacial acetic acid-DTA) or TBE buffer. After sample loading, electrophoresis begins. The gel can then be stained with ethidium bromide and the DNA bands can be visualiszed by placing the stained gel on a u.v. lightbox or transilluminator.

DNA sequencing

Most sequencing performed today is done enzymatically by the Sanger and Coulson method. The method uses extension by DNA polymerase from a primer with a mixture of dNTPs (deoxynucleoside triphosphate) and ddNTPs (dideoxydinucleoside triphosphates). The incorporation of a ddNTP leads to premature chain termination. If four reactions are performed in parallel, each with either ddATP, ddCTP, ddGTPor ddTTP, the reactions will be randomly terminated at either A, C, G or T respectively. Running the terminated products with PAGE gives a ladder of bands which corresponds directly to sequence.

PCR

DNA can be amplified in a test tube using the polymerase chain reaction (PCR). A thermostable type I DNA polymerase from the bacterium *Thermus aquaticus,* known as Taq forms the basis of PCR along with a block capable of rapidly heating and cooling the reaction mixture. PCR is performed in three stages in small thin-walled test tubes: denaturation, annealing, and extension. Each set of three stages is known as a cycle. The two template strands are first denatured by raising the temperature, followed by rapid cooling to between 45 and 65°C. The oligonucleotide primers to anneal to their specific targets and when the PCR block has been raised to the activity optimum of the polymerase (72–74°C), the primers are extended. A normal PCR reaction proceeds over 30 cycles. The power of PCR lies not just in its ability to produce lots of DNA, but more in the specificity of the annealing of the forward and reverse primers.

Site directed mutagenesis

PCR can be used to change the sequence of DNA in a very precise way, by lengthening the primers slightly and introducing mutations during the amplification cycle.

Real time and quantitative PCR

The endpoint of the PCR reaction should be constant for a standard buffer, the number of cycles at which fluorescence from dsDNA can be first detected (the threshold or Ct value) is proportional to the initial template concentration. Methods developed for the quantitation of DNA in this way have become known as quantitative PCR (QPCR). The addition of a reverse-transcriptase step (QRT-PCR) allows the quantification of mRNA, so the level of transcription of individual genes can be monitored in isolation or in comparison to others.

Related topics

Inference of phylogeny from rRNA (B3)
Enumeration of microorganisms (C4)
Prokaryotes and their environments (C11)
Structure of nucleic acids (F3)

Hybridization (F4)
DNA replication (F5)
Plasmids (F15)

Recombinant DNA   The Bacterium *Escherichia coli* plays a pivotal role in what is popularly known as genetic engineering. Our detailed knowledge of its microbiology, physiology, and biochemistry have allowed us to develop mutants which can be used to clone pieces of DNA from most known organisms, safe in the knowledge that the strains harboring foreign DNA (**recombinants**) cannot survive outside the laboratory. Adapted *E. coli* plasmids (see Section F15) are used to maintain foreign DNA in an easily accessible form (plasmids known as **cloning vectors**). **Expression vectors** are plasmids used to generate large amounts of recombinant protein (sometimes up to 10% of the total cell protein) so that enzymes can be purified for research, biotechnological, food, and medical uses. **Shuttle vectors** can also be constructed which allow the easy creation of specific arrangements of genes or gene fragments in *E. coli* before the **construct** is transferred to another organism via transformation (see Section F12), transduction (see Section F12), electroporation or transfection (the latter two techniques are not discussed in this text). As the demand for very specific proteins subject to post-translational modification (see Section F8) becomes more sophisticated, other organisms have been used for cloning and expression, but *E. coli* is still the most widely used.

Part of the attraction of *E. coli* is the ease with which its genomic and plasmid DNA can be isolated. *E. coli*, in common with many other bacteria, will lyse on the addition of a sodium hydroxide solution. After neutralization, the genomic DNA can be purified by techniques such as **caesium chloride density centrifugation** or by the use of the many commercially available **spin columns**. The many plasmids used in biotechnology can also be separated from the rest of the cytoplasmic contents after alkaline lysis. If the lysate is treated with sodium dodecyl sulphate (SDS) and neutralized with sodium acetate, larger macromolecules (proteins, membrane debris, and genomic DNA) become entangled (both physically and chemically) with the white precipitate formed by SDS. The plasmids remain in solution and can be purified further with a few simple steps. This means that plasmid DNA can be isolated from *E. coli* in less than 15 minutes, rendering the organism suitable for building up libraries of DNA fragments in plasmids for screening projects or archiving.

In a text such as this it is not appropriate to discuss the full range of DNA and RNA techniques available to the microbiologist. However, a few of the most commonly used have their origins in the understanding of bacterial DNA metabolism and are summarized in this section.

Gel electrophoresis

Many molecular biology experiments require that DNA fragments are separated by molecular weight – for example to see if a restriction enzyme has cut a plasmid, or to see if a PCR product is the right size. Two common methods are used widely, vertical **polyacrylamide gel electrophoresis (PAGE)** and horizontal submerged **agarose gel electrophoresis**. Both methods rely on the fact that DNA is slightly negatively charged, so if a current is applied through a matrix or solution, DNA molecules will move towards the positive electrode (the cathode).

PAGE can be used to separate DNA fragments that differ in size by as little as one base pair, but is not useful if DNA fragments bigger than about 2000 bp are loaded. A very thin gel is allowed to polymerize between two glass plates, with

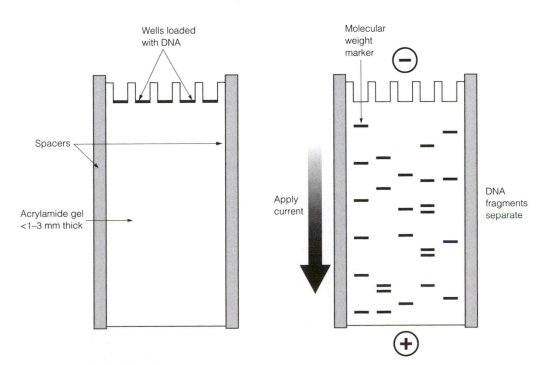

Fig. 1.   Vertical polyacrylamide gel electrophoresis.

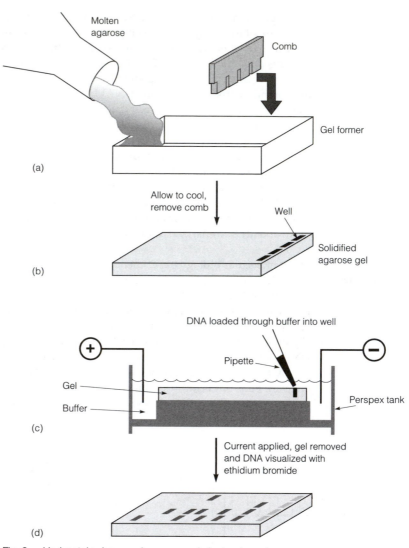

Molten
agarose

Comb

Gel former

(a)

Allow to cool,
remove comb

Well

Solidified
agarose gel

(b)

DNA loaded through buffer into well

Pipette

Gel

Buffer

Perspex tank

(c)

Current applied, gel removed
and DNA visualized with
ethidium bromide

(d)

Fig. 2.   Horizontal submerged agarose gel electrophoresis.

a comb inserted so that wells form ready to accept the DNA samples. The glass
plates and polyacrylamide are attached to tanks containing Tris-borate-EDTA
(TBE) or Tris-acetate-EDTA (TAE) buffer. The tanks have electrodes in them,
which are connected to a power supply so that the current runs through the gel
(see *Fig.* 2). The DNA is visualized by labeling the DNA before electrophoresis
most commonly with a radiolabel such as ^{32}P (see Section F4). However the recent
use of fluorescently primers has superseded radioactivity.

While PAGE is useful for the fine separation of DNA fragments for applications such as sequencing (see below), it take at least five hours to run. The gels
are difficult to handle after running, and it is also hard to excise the individual
bands from the acrylamide matrix, should they need to be cloned. **Agarose gel
electrophoresis** (*Fig.* 2) overcomes many of these problems, as while the gels are

still fragile, it is possible to handle them and also to recover a high percentage of the DNA from the gel after electrophoresis. Agarose is a sugar purified from seaweed, and a 0.7 to 2.0% w/v solution in TAE or TBE is generally used. Molten agarose at is poured into a **gel former** to a depth of about 8 mm along with a **comb** and is allowed to cool (*Fig. 2a*). The comb is removed to leave wells in the surface of the agarose slab (*Fig. 2b*), which is then submerged under TAE or TBE buffer. The samples are loaded into the wells through the surface of the buffer (*Fig. 2c*) and electrophoresis begins. After between one and three hours (depending on the size of the gel, the voltage used and the percentage of agarose in the gel), the gel is taken out of the buffer and stained with ethidium bromide (EtBr), a DNA inter-calating dye (see Section F10). When bound to DNA, EtBr fluoresces under ultraviolet light, so the DNA bands can be visualized by placing the stained gel on a u.v. lightbox or **transilluminator**.

DNA sequencing Although many new technologies are emerging, most sequencing performed today is done **enzymatically** by the **Sanger and Coulson method**. However, the first method developed (Maxam–Gilbert sequencing) relied on the chemical digestion of DNA fragments using toxic compounds such as hydrazine and dimethyl sulphide. While accurate and requires no previous knowledge of the DNA sequence, the method is difficult to perform and hard to interpret, so has fallen out of favor except in the sequencing of very small fragments of DNA (e.g. checking the sequence of short oligonucleotides).

Due to the complementary nature of DNA (see Section F4), the problem of sequence determination can be reduced to solving one strand – the sequence of the other strand can then be inferred from this. The sequencing reaction must generate a mixture of labeled single-stranded DNA molecules with one 'fixed' end and one base specific end, and all possible base-specific ends should be present in the mixture. The mixtures specific for each of the four DNA bases can then be sepa-rated according to size and compared with each other. The order in which the fragments occur from smallest to largest will then reflect the original sequence. The way in which the single stranded molecules are gene rated is to perform a polymerase-mediated extension reaction from a primer near to the region to be sequenced. This is an isothermal (single temperature) equivalent of the extension stage of PCR (see below) and has the same components as a PCR reaction, except with the addition of a low concentration of dideoxynucleotides (see Section F3).

The dideoxynucleotide (ddNTP) lacks an OH group at the 3' position, so although it can be added to a polynucleotide chain, there is no accepting group for the next dNTP. The incorporation of ddNTP leads to **premature chain termination**. If four reactions are performed in parallel, each with either ddATP, ddCTP,

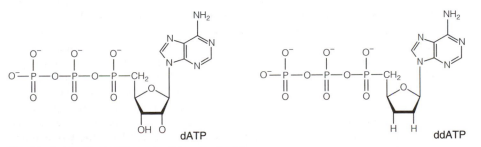

Fig. 3. A deoxynucleotide and a dideoxynucleotide.

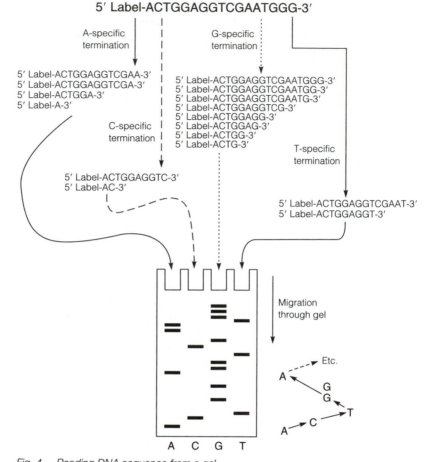

Fig. 4.   Reading DNA sequence from a gel.

ddGTP or ddTTP, the reactions will be randomly terminated at either A, C, G or T respectively. Running the terminated products with PAGE gives a ladder of bands which corresponds directly to sequence.

PCR

Although cloning of DNA in plasmids and its subsequent transformation allows a single molecule to be replicated many times by the bacterial cellular machinery (**amplification**), the DNA produced can be subject to the effects of the cell itself – some foreign DNA and even DNA from *E. coli* itself is toxic when cloned into high-copy number plasmids. The **polymerase chain reaction (PCR)** is a cell-free method of DNA amplification. It had already been noted in the early 1970s that DNA polymerase I protein (isolated from *Escherichia coli* or the bacteriophage T7), whose normal function is in DNA repair (see Section F11) and the joining of Okasaki pieces during DNA replication, could add nucleotides to single-stranded DNA in a complementary manner as long as a primer was provided to initiate the synthesis of this new strand. However, this conversion of ssDNA to dsDNA could only be performed once before laborious purification of the new DNA had to be performed, since *E. coli* and T7 DNA polymerase I have limited stability. Kary Mullis postulated that many such replications could be linked together to amplify very small amounts of DNA into much larger quantities if a suitable thermostable polymerase could be found. Around that time, the bacterium *Thermus aquaticus* had been isolated from Octopus Springs in Yellowstone by Thomas Brock, one of the first pure cultures of thermophiles capable of growing at 75°C. Kary Mullis reasoned that thermophiles such as *T. aquaticus* would have a thermostable DNA replication machinery too, and thus a thermostable DNA polymerase I. This proved to be the case, and the enzyme from *T. aquaticus*, known as **Taq** forms the basis of PCR along with a block capable of rapidly heating and cooling the reaction mixture.

PCR is performed in three stages in small thin-walled test tubes: denaturation, annealing and extension (*Fig. 5*). The PCR solution in the tubes has specific components listed in *Table 1*. Each set of three stages is known as a **cycle**. In the first stage, the two template strands are **denatured** by raising the temperature of the PCR block to 94°C or more (*Fig. 5b*). After around 20 seconds, the block is quickly cooled to between 45 and 65°C, which allows the oligonucleotide **primers** to **anneal** to their specific targets (*Fig. 5c*). The *Taq* in the solution now has a primer from which to begin synthesis, and once the temperature of the block has been raised to the activity optimum of the polymerase (72–74°C), the primers are **extended** (*Fig. 5d*). Now one cycle has been completed there is one extra copy (**amplicon**) of the target spanning the region between the forward and the reverse primers.

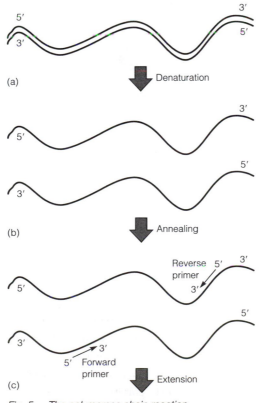

Fig. 5. The polymerase chain reaction.

Continued

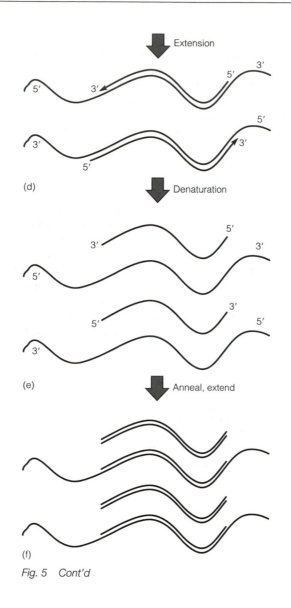

Fig. 5   Cont'd

Table 1.   Components of the PCR reaction

PCR solution component	Approximate concentration	Function
Buffer	25–100 mM	Maintains pH at optimum for *Taq* activity
Magnesium ions	1–5 mM	Essential for *Taq* activity, binds to template, oligonucleotides and dNTPs. Governs stringency of PCR (see Section F4)
Oligonucleotides (primers)	In excess	Prime the polymerization. Sequence governs specificity of the PCR
dNTPs	In excess	Substrate for polymerase
Taq polymerase	1 unit per 100 μL	Enzymatic addition of dNTPs complementary to template
Template	As little as one copy	DNA to be amplified
Water	n/a	Added to bring other components to the correct concentration

Once the cycle of denaturation, annealing and extension is repeated again, (*Fig. 5e, f*) there are now considerably more new strands synthesized by *Taq* than template strands. Within four cycles, the concentration of amplicons outnumbers the template by eight to one, and within 12 cycles the ratio is over 4000:1. A normal PCR reaction proceeds over 30 cycles, and even if there is only one copy of the target, this will result in 536 870 912 amplicons (theoretically). In many PCR reactions the number of initial template DNA molecules is around 50, which after 30 cycles of PCR would result in nearly 30 thousand million amplicons.

The power of PCR lies not just in its ability to produce lots of DNA and the simplicity of the reaction components (*Table 1*). The specificity of the annealing of the forward and reverse primers can be changed by changing the sequence. PCR primers are generally 18 to 21 nt long, and if their sequence occurs only once each in the template, this will allow the amplification of only one PCR product. The amplicon can then be used for cloning or sequencing. If a sequence such as the inverted repeat of a common insertion sequence (see Section F13) is chosen then more than one product will be amplified. The same insertion sequence can occur many times in a genome, and the position of the IS in the genome is frequently stable for a particular strain, but can differ between strains of the same species. By using PCR to amplify products between ISs and running the results on an agarose gel, we can make a fingerprint characteristic of a particular strain (*Fig. 6*). In many bacteria insertion sequences do not occur frequently enough to enable efficient PCR between them, but there are many other sequences that do give good fingerprints. The REP (repetitive extragenic palindromic), BOX and ERIC (enterobacterial repetitive intergenic consensus) sequences can be used for PCR typing of *Rhizobium*, Gram positive and enterobacterial sub-groups of the bacteria with considerable reliability.

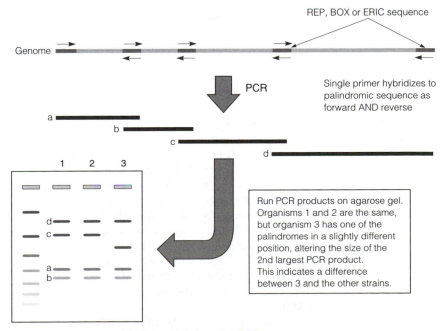

Fig. 6.   The principle of DNA fingerprinting using PCR.

Site directed
mutagenesis

The principles illustrated by PCR are also used to change the sequence of DNA in a very precise way. The techniques of chemical mutagenesis (see Section F10) are indiscriminate. For example, if a mutant is required of the following sequence (*Fig. 7*):

```
5′ – NNNNNNNNNNN GGG ATC GAT TCG ATG GGT ACG CAT CGA NNNNNNNNNNN – 3′
3′ – NNNNNNNNNNN CCC TAG CTA AGC TAC CCA TGC GTA GCT NNNNNNNNNNN – 5′
                 Gly  Ile Asp Ser Met Gly Thr His Arg
                  1    2   3   4   5   6   7   8   9
```

Fig. 7.   An example of a sequence that is a target for mutagenesis. The sequences 'NNNNN' represent the flanking genomic DNA.

The 'N' represents the rest of the genome. The fifth codon in *Fig. 7* is to be changed to isoleucine from methionine. The use of MMS will either alter every single guanine in the fragment, or if the concentration is low, will randomly alter one or more of the guanines. This means that after every round of mutagenesis, a long process of screening and sequencing has to take place to ensure that the desired mutation has taken place. Alternatively an oligonucleotide could be designed in *Fig. 8*:

```
5′ –    GGG ATC GAT TCG ATA GGT ACG CAT CGA    – 3′
        Gly  Ile Asp Ser Ile Gly Thr His Arg
```

Fig. 8.   An oligonucleotide for PCR mutagenesis of the sequence in Figure 7.

The use of this oligo in PCR will generate products that all have a very specific single base pair change. The primer will still function (along with a suitable reverse primer designed to the 3′) as the bases around the mutation still have sufficient homology to promote efficient annealing. If the PCR product can then be cloned into the correct position in the genome, this change will be adopted in the recombinant. The PCR method has been developed from an older one using a whole plasmid as a template, a mutagenic oligo and T7 DNA polymerase to manufacture a complete copy of one strand of the plasmid.

Real time and
quantitative PCR

Quantitation using PCR is complicated by the PCR reaction itself. Over a large dynamic range, whatever the starting template concentration the final concentration of PCR will be the same. This means that the results of a standard PCR reaction of 30–40 cycles followed by agarose gel electrophoresis will have no relevance to how much template DNA was added at the beginning of the reaction. The advent of fluorescent DNA probes has allowed microbiologists to track the accumulation of dsDNA in a PCR reaction in real time and determine a relationship between DNA template concentration and the number of PCR cycles. Although the **endpoint** (*Fig. 8*) of the PCR reaction should be constant for a standard buffer, the number of cycles at which fluorescence from dsDNA can be first detected (the **threshold** or **Ct** value) is proportional to the initial template concentration. Methods developed for the quantitation of DNA in this way have become known as **QPCR**.

If a standard curve of the PCR amplicon is made with known amounts of template, it is possible to deduce the template concentration in an unknown

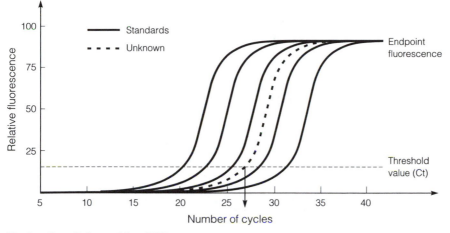

Fig. 9.   Quantitative real time PCR.

sample. This technology is used in the determination of viral load of DNA viruses in clinical samples, but is most powerful in combination with a reverse transcription (RT) step. **QRT-PCR** allows the quantification of mRNA, so the level of transcription of individual genes can be monitored in isolation or in comparison to others. Applications include validation of microarray data (QRT-PCR is more sensitive and more accurate) as well as clinical applications such as RNA virus quantitation (e.g. HIV-AIDS) and oncogene quantitation (e.g. monitoring of the onset of mammarian cancer via detection of BRACA1/BRACA2 transcripts in blood).

G1 PROKARYOTES IN INDUSTRY

Key Notes

Biotechnology

Biotechnology is the use of living organisms in technology and industry. Prokaryotes have been exploited for many years in the manufacture of food and other useful products. For large-scale applications, prokaryotes are grown in fermenters using either batch, fed-batch or continuous-culture processes.

Gene technology and biotechnology

The efficient production of some proteins and chemicals can only be carried out by mutant or recombinant microorganisms. Gene technology provides a means of providing virus-free human proteins, as well as for the overexpression of useful enzymes from extremeophiles.

Fermented foods

Prokaryotes are used in the manufacture of a surprising diversity of processed food, e.g. cheese, yoghurt and soy sauce. The role of Bacteria is predominantly to convert carbohydrates and alcohols into organic acids, lowering the pH and increasing the storage life and palatability of the food. Starter cultures are used to reduce the duration of the fermentation process.

Biocatalysis

Microorganisms are a convenient source of enzymes. The regiospecific and stereospecific properties of enzymes can be exploited to produce enantiomerically pure preparations of some chemicals. These bacterial enzymes may either be used directly, or held in immobilized cells for ease of use. The use of prokaryote enzymes can significantly reduce the cost and increase the yield of reactions an organic chemist would find difficult to perform.

Bioremediation

The biodeterioration of man-made compounds by microbes can be a problem (e.g. in the case of paints and plastics) or a benefit when it comes to cleaning up the environment. Bioremediation is the use of microbes to treat waste water and soil where unacceptable levels of pollutants have accumulated. Even highly toxic compounds such as polychlorinated biphenyls (PCBs) can be metabolized by some Bacteria.

Biotechnology

Prokaryotes have been used to make a wide variety of products for many thousands of years, bur lately this has been called biotechnology. In order to produce products or intermediates, microbes are grown in large vats (fermenters) protected from contamination and changes in pH, temperature and dissolved oxygen concentration. Industrial fermenters vary in size from a few litres to several thousand. Modern biotechnology relies heavily on developments in chemical engineering to regulate and monitor the processes occurring in these fermenters, and this regulation makes the growth of prokaryotes under these conditions some-what different from that found in ordinary shake flasks. The optimization of fermenters requires a thorough understanding of the kinetics of prokaryotic growth.

Industrial fermenters allow the growth of prokaryotes by one of three processes: batch, fed-batch and continuous fermentation. A batch fermenter is sterilized, inoculated and allowed to grow before the entire culture volume is harvested. A fed batch system follows a similar regimen, but instead only some of the culture is harvested after growth, and fresh medium is added to restart the process. Continuous culture uses a constant in-flow of medium offset by the exit of spent medium, product and biomass and is not used so much in the production of products for human consumption because of perceived problems with culture sterility. Most developments in fermentation systems focus on the provision or exclusion of oxygen from the vessel, depending on the process required. The amount of dissolved oxygen in a culture can be increased by a variety of methods, from the simple mechanical stirred vessel to the air-lift fermenter.

Biotechnology now plays an important role in many industries, including the food-processing industry, providing both enzymes (e.g. invertase for the manufacture of glucose syrup) and biomolecules (e.g. sodium glutamate).

Gene technology and biotechnology

The exploitation of biological potential for industrial processes currently relies on metabolic diversity of microorganisms to provide biological solutions to common problems. Improvement of biotechnological processes would seem an attractive start for the modern molecular biologist, but public opinion and government regulation of the use of genetically modified organisms has restricted the widespread use of gene technology.

However, the production of many drugs and fine chemical products relies on **recombinant** or **mutant microorganisms** to produce enzymes (see below) and many human proteins have been cloned in *Escherichia coli* (*Table 1*) so that significant quantities can be produced for therapeutic use without the danger of viral contamination. As more enzymes are studied with a view to industrial application, those from the extremeophiles seem more attractive. Although bacteria growing at high temperature or extremes of pH are often easier to keep sterile on a large scale, protein yields are often low. Cloning into well studied prokaryotes such as *E. coli* or *Bacillus subtilis* allows much more controlled and efficient **heterologous overexpression** of useful enzymes such as amylases and lipases from extreme halophiles for washing powder.

Fermented foods

Today's industrial fermented foods have their origins in much older processes. Foods that use bacteria in their production include soy sauce (*Pediococcus* species),

Table 1. A selection of human proteins cloned in E. coli *and their therapeutic use*

Protein	Function	Therapeutic use
Urokinase	Plasminogen activator	Anticoagulant
Serum albumin	Major blood protein	Synthetic plasma constituent
Factor VIII, Factor X	Blood-clotting	Prevention of bleeding in hemophiliacs
Interferons	Can cause cells to become resistant to some viruses	Antiviral therapy
Growth hormone releasing factor (HGH)	Permits the action of growth hormone in the body	Growth promotion, recovery from physical stress
Erythropoietin (EPO)	Stimulates production of red blood cells	Replacement of cells after chemotherapy; treatment of anemia

Table 2.   What are microbes used for in the food industry?

Foods	Flavorings
Fermented meat	Vinegar
Soy sauce	Nucleotides
Cheeses/milk/yoghurt	Amino acids
Mushrooms and edible fungi	Vitamins
Baker's yeast	
Coffee	
Pickles, sauerkraut, olives	

Organic acids	Starter cultures
Citric acid	Dairy industry (cheese etc., silage)
Itaconic acid	Leguminous crops

cheese, yoghurt (*Lactobacillus, Enterococcus, Bifidiobacterium* species as well as many others), sauerkraut (*Lactobacillus, Streptococcus, Leuconostoc* species) and the traditional manufacture of vinegar (*Acinetobacter* species). In general, prokaryotes are used to acidify foodstuff to allow storage, and a table of foods made by microorganisms is shown in *Table 2*.

Bacterial food fermentations are generally complex and involve a succession of different organisms to render the final product palatable. The process rarely involves bacteria alone, but the production of fermented cabbage (sauerkraut) uses bacteria with progressively higher tolerance to acid. *Streptococcus faecalis* and *Leuconostoc mesenteroides* initiate the conversion of plant sugars into organic acids, and are succeeded by *Lactobacillus brevis* and finally *Lactobacillus plantarum*. A similar process is used to preserve the human foods cucumbers (as gerkhins) and olives, while in agriculture *Enterococcus, Pediococcus* and *Lactobacillus* species successively acidify cut grass to produce silage for winter feeding of cattle. Many of these processes will occur because of endemic prokaryotes without the aid of man, but the use of **starter cultures** grown up in the laboratory increases the speed of the fermentation.

Biocatalysis

Microorganisms can be used as a convenient source of enzymes, since these same proteins provide them with the means to grow on a variety of complex organic compounds. The enzymes can be used in their purified form (either immobilized or in solution) but are sometimes more stable when whole cells are used. Bacterial cells may be rendered non-viable by **immobilization** or **permeabilization** but still retain catalytic activity. This can alleviate the need to provide cofactors such as ATP, which can add significantly to costs. The reactions the enzymes catalyze are often **regiospecific** (attacking a single group on a molecule but leaving others of the same chemical composition) and **stereospecific** (attacking one enantiomer such as D-glucose, but not the corresponding stereoisomer). The specificity of enzymes has allowed industrial chemists to perform reactions that would be impossible by normal synthetic routes, but mostly **biotransformations** are cheaper to perform and have a higher yield.

Table 3.   Examples of industrial production of organic compounds by prokaryotes

Prokaryote source	Chemical	Major application
Acetobacter	Acetic acid	Solvent, starting compound for many synthetic reactions
Clostridium	Isopropanol	Solvent, antifreeze
Clostridium	Acetone	Solvent, starting compound for many synthetic reactions
Bacillus	Acrylic acid	Precursor for acrylonitrile and other polymers
Bacillus	Propylene glycol	Solvent, antifreeze, antifungal compound

Examples of biotransformations are provided in *Table 3*, but one of the most economically significant biotransformations is the production of acrylamide (a polymer used in many chemical processes as well as in the cosmetics industry). It is possible to use a copper catalyst to convert acrylonitrile into acrylamide, but this reaction must be performed at 100°C, after which the catalyst must be regenerated and the unreacted highly toxic acrylonitrile must also be rigorously separated from the product. However, these problems are avoided by the use of immobilized *Pseudomonas chlorophis* performing the biotransformation in a bioreactor. The reaction can be run at 10°C, so heating costs are reduced and the bacterial enzyme responsible (nitrile hydratase) converts over 99.9% of the acrylonitrile into acrylamide. Around half a million tonnes of acrylamide are produced annually by this process.

Although biotransformations have been used in the bulk chemical industry, the main application is in the production of **fine chemicals** such as antibiotic derivatives. The cost savings can be dramatic: cortisone was first synthesized as a 31-step organic synthesis starting from 615 kg of deoxycholic acid. This yielded 1 kg of cortisone, which was sold as an anti-inflammatory drug in the 1940s at around $200 per gram. Use of enzymes from the fungus *Aspergillus niger* in some of the steps reduced the cost to $6 per gram in 1952. The use of mycobacterial enzymes allowed plant sterols to be used as much cheaper-starting compounds so that by 1980 the price of cortisone had dropped to 46 cents in the United States, about a quarter of one percent of the original cost.

Bioremediation   The metabolic diversity of microbes has allowed their commercial exploitation in the field of waste water and soil clean-up. The **biodeterioration** of paints, plastics and other man-made compounds by microbes is often seen as a problem, but where unacceptable levels of these compounds accumulate, this property can be a benefit. Although the technology is still in its infancy, compounds previously considered **recalcitrant** are now being subjected to treatment by a variety of microbes. Pilot studies have shown the efficacy of the use of bacteria against compounds such as trichloroethylene (**TCE**) and polychlorinated biphenyls (**PCBs**). The use of microbes is not without problems: the **biosurfactants** of *Pseudomonas* species have been used to allow endemic soil bacteria **bioavailability** to emulsified crude oil after tanker spillages. However, the emulsified oil proved much more mobile than the crude slick, and a bigger problem was created as the hydrocarbons moved deep into gravel and towards potable water sources. Experimental waste water treatment plants, where polluted medium is much more contained, have been far more successful than *in situ* approaches.

H1 HUMAN BACTERIAL INFECTIONS

Key Notes

Types of human diseases	These include infectious diseases, inherited diseases, neoplastic diseases, immunity-related diseases, degenerative diseases, nutritionally deficiency diseases, endocrine diseases, Iatrogenic diseases, environmental diseases and idiopathic diseases.
Koch's postulates	1. The microbe must be present in all people with the disease and should be associated with the lesions of the disease. 2. The microbe must be isolated in pure culture from a person who has the disease. 3. The isolated microbe, when administered to susceptible individuals, must produce the disease. 4. The microbe must be re-isolated in pure culture from the intentionally-infected host.
Major bacterial pathogens of humans	These are grouped into: i) spirochetes (*Borrelia, Leptospira, Treponema*), ii) aerobic Gram negative bacteria (*Helicobacter, Bordetella, Brucella, Legionella, Neisseria, Pseudomonas*), iii) facultative anaerobic Gram negative bacteria (*Calymmatobacterium, Escherichia, Gardnerella, Haemophilus, Pasteurella, Proteus, Salmonella, Shigella, Streptobacillus, Vibrio, Yersinia*), iv) anaerobic Gram negative bacteria (*Prevotella*), v) *Rickettsia* and *Chlamydia* (*Rickettsia, Coxiella, Ehrlichia, Bartonella, Chlamydia*) vi) Mycoplasma, vii) Gram positive bacteria (*Enterococcus, Staphylococcus, Streptococcus, Bacillus, Clostridium, Listeria, Actinomyces, Corynebacterium*), and viii) *Mycobacterium*.

Types of human diseases

There are different kinds of human diseases (*Table 1*). Among them, infectious diseases are caused by microbial agents (also known as infectious agents), and include viruses, Bacteria, protozoa, fungi and multicellular organisms such as worms. Although the focus here is Bacterial diseases, much of the terminology is similar to all infectious agents. For example, the term 'communicable' is used for infectious diseases that can spread from person to person and microbes that can spread more easily are termed 'contagious'.

Koch's postulates

Bacteria were first associated with human diseases in 1870s by Robert Koch. In a series of elegant experiments, he identified *Bacillus anthracis* as the causative agent of anthrax and created a number of postulates. Koch's postulates provided a basis for advances in the Bacterial diseases and prove that a specific microbe is responsible for a particular disease. These postulates are

1. The microbe must be present in every case of the disease and should be associated with the lesions of the disease.

Table 1.   Types of diseases

Types of diseases	Cause
Infectious diseases	Microbial agents, transmitted to susceptible hosts via exposure to contaminated environment or infected organisms
Inherited diseases	Abnormal genes, passed from one generation to the next
Neoplastic diseases	Abnormal cell growth leading to formation of benign or malignant tumors. Causes include genetic, environmental factors, chemicals, radiation and viruses
Immunity related diseases	These develop when immune system fails so body is unable to defend itself or becomes abnormal so immune system begins attacking normal tissues, autoimmune diseases
Degenerative diseases	Associated with aging. For example, there are significant reductions in cardiac efficiency, kidney filtration, etc.
Nutritional deficiency diseases	Caused by lack of appropriate nutrition, vitamins, proteins, carbohydrates, etc.
Endocrine diseases	Abnormal production of hormones
Latrogenic/Nosocomial diseases	Hospital acquired, results from activity or treatment of physicians, e.g. post-surgery (also known as nosocomial infections)
Environmental diseases	Results from the exposure to environmental poisons
Idiopathic diseases	Causes are not yet known

2. The microbe must be isolated in pure culture from an infected host with the disease.
3. The isolated microbe, when administered to susceptible host must produce the disease.
4. The microbe must be re-isolated in pure culture from the intentionally-infected host.

However, in practice, it is difficult to fulfill all of these criteria for all microbial pathogens and there are exceptions. Pure cultures of Bacteria in the laboratory have been a problem with some strains. For example, *Mycobacterium leprae* (the causative agent of leprosy) can only be cultured in animal models and can not be grown on laboratory media.

Major bacterial pathogens of humans   For simplicity, the major Bacterial pathogens are divided into several groups as indicated below.

Spirochetes   These are spiral-shaped, motile, Gram-negative organisms. Their size varies from 0.1–3 μm but are very thin and can best be observed by dark-field microscopy or special staining. Spirochetes may be anaerobic, facultative anaerobic, or aerobic and include causative agents of human and zoonotic diseases. A zoonosis is an infectious disease transmitted from animals to humans.

Borrelia burgdorferi   Causative agent of lyme disease. It is initially (within days) characterized by fever, headache, enlarged local lymph nodes, muscle pain, joint pain and within months may lead to long-lasting arthritis or involve the central nervous system.

Biology and diagnosis	Size is 0.2–0.5 μm wide and 20–30 μm long, spiral-shaped, motile organisms. They are Gram negative but are easily observed using Giemsa staining. Lyme disease is diagnosed based on signs and symptoms such as erythema migrans, facial palsy, or arthritis and a history of possible exposure to infected ticks. Laboratory tests involve immunofluorescence staining or by direct demonstration of Bacteria in blood specimens by dark field microscopy.
Transmission	The disease is transmitted by ticks (*Ixodes* spp.). Other mammals such as mice are hosts for ticks and are important for maintaining *B. burgdorferi* in nature.
Borrelia recurrentis	Causative agent of epidemic relapsing fever. It is characterized by fever, chills, headache, muscle pain but is normally self-limiting.
Biology and diagnosis	Size is 0.3 μm wide and 10–30 μm long, spiral-shaped, motile organisms. They are Gram negative but are easily observed using Giemsa staining. As the appearance of fever is not a specific diagnostic method, immunofluorescence staining or by direct demonstration of *Borrelia* in blood specimens are recommended.
Transmission	Epidemic relapsing fever (etiologic agent- *B. recurrentis*) is transmitted by body lice (*Pediculus* spp.). Endemic relapsing fever (etiologic agents-*B. duttoni, hispanica, persica, hermsii,* and *turicatae*) is transmitted by ticks (*Ornithodoros* spp.).
Leptospira interrogans	Causative agent of leptospirosis. It is characterized by fever, chills, headache, muscle pain, and involves liver, kidney or blood vessels damage but is normally self-limiting.
Biology and diagnosis	Size is 0.15 μm wide and 5–15 μm long, spiral-shaped, motile organisms that are best observed using dark-field microscopy or immunofluorescence staining.
Transmission	Transmission of infections from animals to humans occurs through contact with contaminated soil or moist soils where rodent urine, contaminated tissues or blood are present. *Leptospira* enter humans via skin abrasions or mucous membranes.
Treponema pallidum	Causative agent of syphilis. Initially (within 2–3 weeks), it is characterized by a swollen painless lesion (termed a chancre) at the site of infection and may lead to blisters on the body. If untreated, the disease can progress to secondary and tertiary stages which can occur within 5–20 years, leading to extensive damage of tissues such as the cardiovascular system and the central nervous system (CNS).
Biology and diagnosis	Size is 0.15 μm wide and 5–15 μm long, spiral-shaped, motile organisms that are best observed using dark-field microscopy or immunofluorescence staining.
Transmission	Sexually transmitted disease (STD) but can also spread through blood transfusions, mother to fetus, or direct contact with a chancre that may be present on the external genitals, lips and in the mouth of the infected patient.
Aerobic Gram negative bacteria	
Helicobacter pylori	Causative agent of stomach ulcers. Although it co-exists with its host, 1% of infected individuals develop gastric cancer and 20% develop duodenal ulcers.

Biology and diagnosis	Size is 0.5–1 µm wide and 2.5–5 µm long, motile, spiral, curved or rod-shaped. Diagnosis can be made by the examination of tissue biopsies, Gram staining or using serological tests including enzyme immunoassay.
Transmission	Occurs throughout the world with most likely transmission via fecal-oral or oral-oral route.
Bordetella pertussis	Causative agent of pertussis (whooping cough). The organism grows in the mucosal membranes of the human respiratory tract. It is characterized by violent coughing for 20–50 times a day for up to a month.
Biology and diagnosis	Size is 0.2–0.5 µm wide and 0.5–2 µm long, non-motile, cocco-bacillus, cells are present singly or in pairs. Diagnosis can be made by the examination of specimens (nasopharyngeal tissues) together with enzyme immunoassay (EIA).
Transmission	Human to human transmission via airborne droplets and highly communicable.
Brucella spp.	Causative agent of brucellosis. It is characterized by fever, chills lasting for years and may associate weight loss, body aches, enlarged lymph nodes, spleen and liver.
Biology and diagnosis	Size is 0.5 µm wide and 0.3–0.9 µm long, non-motile, cocco-bacillus, cells are present singly or in pairs. Four species of *Brucella* are pathogenic to humans, *B. aborus*, *B. melitensis*, *B. suis*, *B. canis*) and their natural hosts are cat, goat/sheep, swine and dog, respectively. Diagnosis can be made by the examination of biopsies or isolation of bacterium from the blood, and enzyme immunoassay.
Transmission	Through skin abrasions, through conjunctival sac of the eyes, inhalation of bacteria containing droplets or via contaminated food.
Legionella pneumophila	Causative agent of Legionnaires' disease (pneumonia-like infection) and Pontiac fever (influenza-like infection). Pontiac fever is self-limiting. Legionnaires' is characterized by fever, chills, muscle pain, coughing, breathing difficulties, and may lead to vomiting, diarrhea, or involving the CNS, resulting in fatal consequences.
Biology and diagnosis	Size is 0.3–0.9 µm wide and 2–20 µm long, motile, rod-shaped. Diagnosis can be made by isolation of the organisms or by direct immunofluorescence testing or enzyme immunoassay.
Transmission	Transmission is via airborne droplets or waterborne.
Neisseria gonorrhoeae	Causative agent of gonorrhoeae. In males, it is characterized by pus-like discharge, painful urination and may even result in complete closure of the urethra (canal that connects bladder to the outside). In females, there is rarely urethral discharge but rather dysuria, painful urination, and inflammation of the uterine tubes, and are major causes of infertility in women. Endocervicitis may present as slight vaginal pus discharge. Pelvic inflammatory disease may occur and disseminate throughout the reproductive tract. In rare cases, cells may disseminate via the bloodstream to produce rash and arthritis. Babies born to infected mothers may develop eye infections (gonococcal ophthalmia/ophthalmia neonaturum), which can result in blindness if eye drops/ointment is not administered at birth.

Biology and diagnosis	Size is 0.6–1 µm in diameter, non-motile, diplococci. Diagnosis can be made by isolation of the organisms or by immunofluorescence testing or enzyme immunoassay.
Transmission	Sexually transmitted disease (STD) but can also be spread by mother to fetus when it passes through the infected birth canal.
Neisseria meningitidis	A major causative agent of bacterial meningitis (also known as meningococcal meningitis). The organism spreads from the nasopharynx to the bloodstream and finally invades the CNS to produce disease. While in the bloodstream, the bacterium may grow in large number to produce bacteremia, causing severe damage to the blood vessels, resulting in hemorrhages (noticeable rash on the skin). Disease is characterized by fever, stiff neck, vomiting, severe headache, coma, and death within a few hours.
Biology and diagnosis	Size is 0.6–1 µm in diameter, non-motile, diplococci. Diagnosis can be made by direct examination of the organisms and serological test using latex agglutination for the detection of capsular polysaccharides or enzyme immunoassay.
Transmission	Human to human via respiratory droplets.
Pseudomonas aeruginosa	Causative agent of lung infection, eye infection, and skin infection. It is an opportunistic pathogen and colonizes the injured epithelial surfaces, resulting in serious consequences.
Biology and diagnosis	Size is 0.5–1 µm wide and 1.5–5 µm long, motile, rod-shaped, cells are present singly or in pairs. Diagnosis can be made by direct examination of the organisms or immunoassay.
Transmission	Exposure of injured epithelial surface to the contaminated surfaces, water is the most common form of transmission.
	Other aerobic Gram negative Bacteria which may cause human infections are *Moraxella* spp. *Francisella* spp. and *Alcaligenes* sp.
Facultative anaerobic Gram negative bacteria	
Calymmatobacterium granulomatis	Causative agent of granuloma inguinale with the skin lesions and subcutaneous tissue of genital and anal regions.
Biology and diagnosis	Size is 0.5–1.5 µm wide and 1–2 µm long, rod-shaped. Diagnosis can be made by direct examination of the organisms, Giemsa but not Gram staining is effective.
Transmission	May be sexually transmitted (although this claim is controversial).
Escherichia coli	Causative agent of the gastrointestinal and CNS infections. The gastrointestinal infections are associated with fever, diarrhea which may contain blood, vomiting within days and may result in severe dehydration, shock and death. *E. coli* meningitis is observed in neonates and is characterized by fever, stiff neck, vomiting, severe headache, coma, and death.

Biology and diagnosis	Size is 1–1.5 µm wide and 2–6 µm long, motile, rod-shaped, cells are present singly or in pairs. Diagnosis can be made by isolation of the organisms, Gram staining or by immunofluorescence testing.
Transmission	Through contaminated food/water.
Gardnerella vaginali	Causative agent of vaginosis (foul-smelling vaginal discharge).
Biology and diagnosis	Size is 0.5 µm wide and 1.5–2.5 µm long, non-motile, rod-shaped. Diagnosis can be made by examination of organisms in the vaginal secretions and growing on selective agar.
Transmission	Sexually transmitted disease.
Haemophilus influenzae	Causative agent of pneumonia, bronchitis, and meningitis. It is characterized by sore throat, fever, breathing problems, invasion of the bloodstream, and death.
Biology and diagnosis	Size is less than 0.3–0.8 µm in diameter, non-motile, rod-shaped. Diagnosis can be made by isolation and direct examination of the organisms in the clinical specimens or by serological testing.
Transmission	Human to human via respiratory droplets.
Pasteurella multocida	Causative agent of cellulitis (infection of the subcutaneous) tissue.
Biology and diagnosis	Size is 0.3–1 µm wide and 1–2 µm long, non-motile, ovoid or rod-shaped, cells are present singly or in pairs. Diagnosis can be made by culture of pus taken from the lesions.
Transmission	Normally present in the respiratory tract of dogs and cat. Transmitted to humans via bite of domestic animal.
Proteus vulgaris/ P. mirabilis	Causative agent of gastrointestinal, urinary tract, and extraintestinal infections.
Biology and diagnosis	Size is 0.4–0.8 µm wide and 1–3 µm long, motile, rod-shaped, cells are present in singly or in pairs. Diagnosis can be made by culture and biochemical tests.
Transmission	Through contaminated food/water.
Salmonella spp.	Some species can cause gastroenteritis (inflammation of stomach and intestines) and may result in enteric fever. *Salmonella typhi* is the causative agent of typhoid fever and enteric fever. Gastroenteritis is associated with nausea, vomiting, and diarrhea. In typhoid fever, the *Salmonella* cells penetrates the intestine, enters local lymph nodes and gains access to the bloodstream and disseminates to various organs. Typhoid is characterized by fever, chills, aches, weakness, and constipation rather than diarrhea, leading to bacteremia and infection of other organs such as CNS. Of interest, enteric fever typically involves the liver and/or spleen with prolonged fever.

Biology and diagnosis	Size is 0.5–1.5 µm wide and 2–5 µm long, motile, rod-shaped. Diagnosis for gastroenteritis is made by culture from stools, blood for enteric fever and stool and blood for typhoid fever.
Transmission	Through contaminated, under cooked food (meat and dairy products are typical) or untreated water.
Shigella spp.	Causative agent of shigellosis, i.e., bacterial dysentery (diarrhea with blood and mucus). It is characterized by severe abdominal pain and cramps but is self-limiting within days.
Biology and diagnosis	Size is 0.5–0.8 µm wide and 2–3 µm long, non-motile, rod-shaped, cells are present singly or in pairs. Diagnosis can be made by culture from stools and biochemical tests.
Transmission	Through contaminated food/water, fecal-oral route.
Streptobacillus moniliformis	Causes fever, chills, vomiting, headache, rash, and arthritis.
Biology and diagnosis	Size is 0.1–0.8 µm wide and 1–5 µm long, non-motile, rod-shaped, present in long chains. Diagnosis can be made by culturing from clinical specimens and biochemical tests.
Transmission	Through bites of infected mice, rats, cats or by ingesting food/water contaminated with rat faeces.
Vibrio cholerae	Causative agent of cholera. It is characterized by vomiting, abdominal pain, and excessive amounts of watery dehydrating diarrhea.
Biology and diagnosis	Size is 0.5–0.8 µm wide and 1.3–2.5 µm long, motile, rod or curved-shaped. Diagnosis can be made by culturing from stool specimens and serological tests.
Transmission	Through contaminated water/food.
Yersinia pestis	Causative agent of plague. There are three major clinical forms: bubonic plague, septicemic plague, and pneumonic plague. The former is characterized by fever and appearance of bubo (swollen local region in the groin or armpit), and if not treated leads to bacteremia (septicemic plague), septic shock, and can spread to the lungs leading to pneumonic plague. Pneumonic plague is characterized by fever, breathing difficulties, and death in some patients.
Biology and diagnosis	Size is 0.5–0.8 µm wide and 1–3 µm long, non-motile, rod-shaped. Diagnosis can be made by culturing from clinical specimens and Gram staining.
Transmission	Bubonic plague is transmitted by the fleas of urban and sylvatic rodents mainly (bites or ingestion) to humans and pneumonics plague is airborne person-to-person.
Yersinia enterocolitica	Causative agent of enterocolitis (usually in children). It is characterized by fever, abdominal pain, and diarrhea but is self-limiting.

Biology and diagnosis	Size is 0.5–0.8 µm wide and 1–3 µm long, non-motile, rod-shaped. Diagnosis can be made by examining stool specimens.
Transmission	Through ingestion of contaminated food/water.
	Other facultative anaerobic Gram negative Bacteria which may cause human infections are *Serratia* spp. *Klebsiella* spp. *Enterobacter* spp. *Morganella* spp. and *Citrobacter* spp.

Anaerobic Gram negative bacteria

Prevotella spp.	Found in infected oral cavities but their clinical significance is not clearly understood.
Biology and diagnosis	Size is 0.5 – 0.8 µm wide and 1 – 3 µm long, non-motile, rod-shaped. Diagnosis can be made by culturing from the clinical specimens.
Transmission	Unknown but likely to be through human to human contact.
	Other anaerobic Gram negative Bacteria which may cause human infections are *Porphyromonas* spp. (*P. gingivalis*, *P. endodontalis*).

Rickettsia and Chlamydia

	Obligate, intracellular bacteria that live and multiply within their eukaryotic host cells and cause zoonotic diseases.
Rickettsia spp.	*Rickettsia rickettsii* is the causative agent of Rocky Mountain spotted fever and *R. prowazekii* is the causative agent of epidemic typhus fever with life-threatening consequences.
Biology and diagnosis	Size is 0.2 µm wide and 0.5–1.3 µm long, rod-shaped, obligate intracellular organisms. Diagnosis can be made by serological tests or by culturing in host cell cultures.
Transmission	Through bite of infected tick/mite or through faeces of infected lice/flea.
Coxiella burnetii	Causative agent of Q-fever. It is characterized by flu-like symptoms, fever, headaches, pneumonia, and hepatitis.
Biology and diagnosis	Size is 0.2 µm wide and 0.5–1.3 µm long, cocco-bacillus, obligate intracellular organisms. Diagnosis can be made by serological tests or by culturing in host cell cultures.
Transmission	Through inhalation of contaminated droplets and associated with domestic animals.
Ehrlichia spp.	Causative agent of Ehrlichiosis. Several species of *Ehrlichia* (such as *Ehrlichia chaffeensis*) can cause this disease with symptoms such as fever, headache, fatigue, muscle aches, nausea, vomiting, diarrhea, cough, and joint pains. Symptoms normally appear after an incubation period of 5–10 days following the tick bite.
Biology and diagnosis	Size is 0.5–1.6 µm long, variable shapes, obligate intracellular organisms. Diagnosis can be made by Giemsa staining and serological tests.

Transmission	Transmitted by tick bites.
Bartonella spp.	Causes fever, malaise, hepatitis, bacteremia, mostly limited to immunocompromised patients.
Biology and diagnosis	Size is 0.5 µm wide and 1–2 µm long, non-motile, rod-shaped. These organisms can be cultured, thus they are not obligate intracellular organisms. Diagnosis can be made by culturing on selective agar and serological tests.
Transmission	Transmitted by domesticated cats, cat fleas, body louse, and ticks (*Ixodes* spp. and *Dermacentor* spp.).
Chlamydia trachomatis	Causes trachoma (inner eye lid infection and may include the cornea), conjunctivitis, and genital infections (major cause of non-gonococcal urethritis in males). May also cause corneal infections.
Biology and diagnosis	Size is 0.3–0.9 µm in diameter, cocci-shaped, obligate intracellular organisms. They have two stages in their life cycle, an infectious, rigid, metabolically inactive stage (elementary body) and a delicate, metabolically active stage (reticulate body). Diagnosis can be made by serological, immunfluorescence tests and by using nucleic acid probes.
Transmission	Through exposure to contaminated fingers, sexual contact, inanimate objects and flies.
Chlamydia pneumonia	Causes sore throat, bronchitis, and pneumonia, characterized by prolonged coughing and asthma by damaging epithelial lining of the respiratory tract.
Biology and diagnosis	Size is 0.3–0.9 µm in diameter, cocci-shaped, obligate intracellular organisms. They have two stages in their life cycle, an infectious, rigid, metabolically inactive stage (elementary body) and a delicate, metabolically active stage (reticulate body). Diagnosis can be made by serological, immunofluorescence tests and by using nucleic acid probes.
Transmission	Through inhalation of contaminated droplets.
Mycoplasmas Mycoplasma pneumonia	Causes infections of the respiratory tract (normally in 5–9 year old children). It is characterized by prolonged coughing, fever and headaches.
Biology and diagnosis	Size is 0.1 µm – 0.3 µm in diameter. Diagnosis can be made by culturing from specimens and serological tests.
Transmission	Through inhalation of contaminated droplets, close personal contact.
	Other *Mycoplasma* which may cause human infections are *Ureaplasma* spp. that causes genital infections in humans.
Gram positive bacteria Enterococcus faecalis	Causes urinary tract infections.

Biology and diagnosis	Size is 0.6–2.5 µm long, non-motile, cocci-shaped, cells are present in pairs or in short chains. Diagnosis can be made by isolation and biochemical tests.
Transmission	Through inhalation of contaminated droplets, or through contaminated food/water.
Staphylococcus aureus	Causes various infections including skin lesions (boils and carbuncles), pneumonia, meningitis, endocarditis (heart), toxic shock syndrome. The exotoxin but not the cells themselves cause food poisoning.
Biology and diagnosis	Size is 0.5–1.5 µm in diameter, non-motile, cocci-shaped, cells are present singly, in pairs or in grape-like clusters. They are non-motile, anaerobic organisms. Diagnosis can be made by biochemical tests and serological tests.
Transmission	Through inhalation of contaminated droplets, or through contaminated food/water and insects.
Streptococcus	Bacteria in this genus are non-motile, facultative anaerobes and consist of an enormous number of species, responsible for a variety of infections involving various tissues and organs. *Streptococcus* that are major human pathogens are divided into two groups.
Group A streptococci **(GAS)**	Most commonly associated with human diseases and include *Streptococcus pyogenes* and *S. pneumoniae*. The former is responsible for pharyngitis (strep throat), fever, acute rheumatic fever, sepsis, and shock. The latter (pneumococci) are responsible for pneumonia, sinusitis, bacteremia and meningitis.
Biology and diagnosis	Size is 0.5–2 µm in diameter, non-motile, cocci-shaped, cells are present in pairs or in short chains. Diagnosis can be made by isolation, Gram staining, biochemical and serological tests.
Transmission	Through inhalation of contaminated droplets, or through contaminated food/water.
Group B streptococci **(GBS)**	Cause neonatal pneumonia, sepsis and meningitis.
Biology and diagnosis	Size is 0.5–2 µm in diameter, non-motile, cocci-shaped, cells are present in pairs or in short chains. Diagnosis can be made by isolation, Gram staining, biochemical, and serological tests.
Transmission	Mother to fetus during birth.
Bacillus anthracis	Causative agent of cutaneous and inhalational anthrax.
Biology and diagnosis	Size is 1–1.5 µm wide and 3–10 µm long, rod/ovoid-shaped and form spores. Diagnosis can be made by demonstrating organisms in the specimen, Gram staining and serological tests.
Transmission	Through inhalation of contaminated droplets, or by contact with infected animals.

Clostridium botulinum

Causative agent of botulism (paralysis). Botulism is not an infectious disease but intoxification caused by bacterial toxins, i.e. botulinum toxin and bacteria may be completely absent. The species is divided into 8 types (A, B, C alpha, C beta, D, E, F, G) on the basis of specific exotoxins produced, all of which form spores.

Biology and diagnosis

Size is 0.3–2 µm wide and 1–2 µm long, rod-shaped, cells are present in pairs or in short chains. Diagnosis is by symptomatology and treatment is with specific antitoxins.

Transmission

The spores are present everywhere in nature and usually germinate when they are in an anaerobic environment, such as home canned beans, corn or other low acid food. The growing cells release a neurotoxin (the most powerful known to man). It is the toxin which if ingested before it is heat inactivated by cooking that is the killer. This toxin is notorious and may be used as a bioterror agent (also see section I).

Other species of *Clostridium* are *C. difficile*, which causes colitis (colon tissue destruction), *C. perfringens*, which causes gas gangrene and *C. tetani*, which causes tetanus or lockjaw.

Listeria monocytogenes

Causative agent of listeriosis. Usually limited to immunocompromised patients or pregnant women and can lead to bacteremia, abortion, meningitis, and meningo-encephalitis.

Biology and diagnosis

Size is 0.4 µm wide and 0.5–2 µm long, motile, rod-shaped, cells are present singly or in short chains. Diagnosis can be made by cultures of clinical specimens, Gram staining and serological tests.

Transmission

Through ingestion of contaminated food.

Erysipelothrix rhusiopathiae

Causes zoonotic infections (from animals to humans). Human infections are primarily limited to individuals who handle fish or pigs. Clinical categories include: i) a localized cutaneous form (most common) associated with a throbbing itching pain and swelling of the finger or part of hand, ii) a generalized cutaneous form, and iii) a septicemic form (associated with the heart disease endocarditis).

Biology and diagnosis

Size is 0.2–0.4 µm wide and 0.8–2.5 µm long, non-motile, Gram positive, non-sporulating rod and is a facultative anaerobe. Diagnosis is made by the isolation of the organisms from tissue biopsies or blood.

Transmission

The organism is likely to be found in faecally contaminated environments and enters into humans through scratches or lesions on the surface of the skin.

Actinomyces israelii

Causative agent of actinomycosis. Bacteria are normally present in the mouth but can cause infection following injury such as a broken jaw, tooth extractions, etc. which introduce them to the deeper tissues.

Biology and diagnosis

Size is 0.2–1 µm wide and 2–5 µm long, non-motile, rod-shaped, cells are present singly, in pairs or short chains forming distinct V or Y forms. Diagnosis can be made by cultures of clinical specimens, and staining.

Transmission

Injury to mucosal surfaces.

Corynebacterium diphtheriae	Causative agent of diphtheria. It is associated with respiratory tract infection by the bacterium and systemic effects of its exotoxin. Disease is characterized by sore throat, often with a thick adherent membrane composed of dead cells, bacteria, phagocytes, and fibrin; headache, nausea, and death can occur from heart tissue damage by the exotoxin.
Biology and diagnosis	Size is 0.3–0.8 µm wide and 1.5–8 µm long, non-motile, rod-shaped, cells are present singly, in pairs or short chains forming distinct V forms. Diagnosis can be made by clinical symptoms, isolation of the organism, and demonstration of its toxin.
Transmission	Through inhalation of contaminated airborne droplets, or by direct contact with skin infections.

Other Gram positive Bacteria which may cause human infections are *Aerococcus viridans*, *Micrococcus luteus*, and *Sarcina* spp. that may cause pulmonary, skin, gastrointestinal and subcutaneous infections.

Mycobacteria

Mycobacterium tuberculosis	Causative agent of tuberculosis. It is infection of the respiratory system characterized by cough, fever, and weight loss. Bacteria avoid host defenses by forming lesion (tubercle). Tissue injury results from cell-mediated hypersensitivity.
Biology and diagnosis	Size is 0.2–0.7 µm wide and 1–10 µm long, non-motile, rod-shaped. Diagnosis can be made by clinical symptoms (chest radiograph), and demonstration of the organism in sputum or other specimen (acid fast staining).
Transmission	Through inhalation of contaminated airborne droplets or by direct contact with infected patients.
Mycobacterium leprae	Causative agent of leprosy. It has two forms, tuberculoid leprosy and lepramatous leprosy. The former is associated with dry, pale patches on any surface of the body and may develop loss of nerve sensation due to bacterial invasion of peripheral sensory nerves. The latter is associated with extensive skin involvement and damage can result in loss of facial bones, fingers, and toes.
Biology and diagnosis	Size is 0.2–0.7 µm wide and 1–10 µm long, non-motile, rod-shaped. Diagnosis can be made by clinical symptoms and demonstration of the organism in the infected tissue (acid fast staining).
Transmission	Through inhalation of contaminated airborne droplets, or by direct contact with infected patients.

Other species of *Mycobacterium* are *M. avium*, which causes pulmonary diseases, infection of the skin, bones and joints.

H2 BACTERIAL INFECTIONS: TYPES, PORTALS OF ENTRY, MODES OF TRANSMISSION, CONTRIBUTING FACTORS IN THEIR EMERGENCE/SPREAD AND COMMON FEATURES IN BACTERIAL INFECTIONS

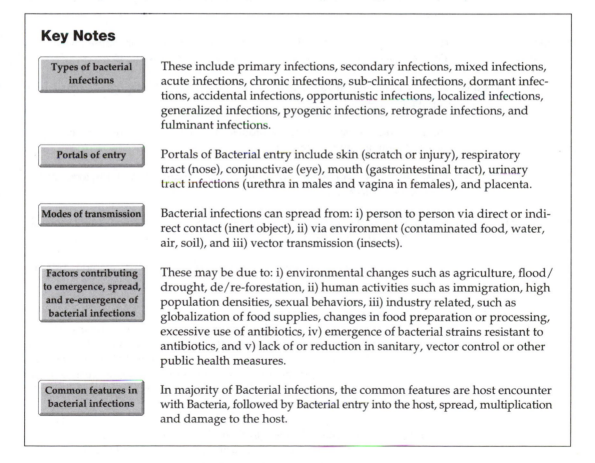

Key Notes

Types of bacterial infections	These include primary infections, secondary infections, mixed infections, acute infections, chronic infections, sub-clinical infections, dormant infections, accidental infections, opportunistic infections, localized infections, generalized infections, pyogenic infections, retrograde infections, and fulminant infections.
Portals of entry	Portals of Bacterial entry include skin (scratch or injury), respiratory tract (nose), conjunctivae (eye), mouth (gastrointestinal tract), urinary tract infections (urethra in males and vagina in females), and placenta.
Modes of transmission	Bacterial infections can spread from: i) person to person via direct or indirect contact (inert object), ii) via environment (contaminated food, water, air, soil), and iii) vector transmission (insects).
Factors contributing to emergence, spread, and re-emergence of bacterial infections	These may be due to: i) environmental changes such as agriculture, flood/drought, de/re-forestation, ii) human activities such as immigration, high population densities, sexual behaviors, iii) industry related, such as globalization of food supplies, changes in food preparation or processing, excessive use of antibiotics, iv) emergence of bacterial strains resistant to antibiotics, and v) lack of or reduction in sanitary, vector control or other public health measures.
Common features in bacterial infections	In majority of Bacterial infections, the common features are host encounter with Bacteria, followed by Bacterial entry into the host, spread, multiplication and damage to the host.

Primary infections   In these infections, Bacteria are the primary cause of infection with apparent clinical symptoms. Examples include *Shigella* dysentery.

Secondary infections Bacterial invasion subsequent to primary infection. Examples include Bacterial pneumonia following a viral pneumonia.

Mixed infections Two or more species of pathogen infecting the same tissue. Examples include anaerobic abscess due to *E. coli* and *Bacteroides fragilis*.

Acute infections Disease progresses rapidly, within hours or days. Examples include diphtheria or Bacterial meningitis.

Chronic infections Disease progresses slowly, takes months or years. Examples include mycobacterial diseases, e.g., TB.

Sub-clinical infections No detectable clinical symptoms. Examples include asymptomatic gonorrhea.

Dormant infections Bacteria uses host as a carrier (carrier state). Examples include typhoid carrier.

Accidental infections Environmental or accidental exposure to Bacteria. Examples include anthrax infection.

Opportunistic infections These are caused by normal flora when host defenses are compromised. Examples include *Pseudomonas aeruginosa* or urinary tract infections due to *E. coli*.

Localized infections Infections are limited to a small area or to a specific tissue. Examples include staphylococcal boil.

Generalized infections These are disseminated to many tissues or different body regions. Examples include Gram negative bacteremia.

Pyogenic infections These infections involve pus-formation and can be caused by staphylococcal and streptococcal infections.

Retrograde infections Bacteria ascend in a duct or tube against the flow of secretions. Examples include *E. coli* urinary tract infections.

Fulminant infections These infections occur suddenly and intensely. Examples include airborne *Yersinia pestis* (pneumonic plague).

Portals of entry Given the access and/or opportunity, microbes can attack nearly all tissues/organs in the human body. Some Bacteria produce infections at their portals of entry and may disseminate to other organs (*Fig. 1*) to produce multiple infections, while others only can cause tissue/organ-specific infections. Below are examples of major portals of entry for bacteria. It is noteworthy that many microbes may reside in their host as part of normal flora and produce infections under specific conditions such as weak immune system.

Skin – scratch or injury Many Bacteria can enter through broken skin such as *Staphylococcus aureus*, causing abscesses, and, *Yersinia pestis*, causing plague, or some bacteria can cause skin infections such as *Staphylococcus epidermidis*.

Respiratory tract – nose Examples include *Bordetella pertussis*, causing pertussis, *Streptococcus pneumoniae*, causing pneumococcal pneumonia and *Corynebacterium diphtheriae*, causing diphtheria.

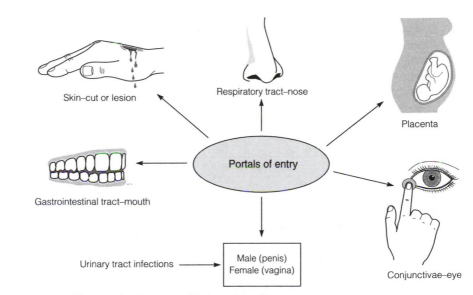

Fig. 1.   The portals of entry used by bacterial pathogens.

Conjunctivae (Eye)   Examples include *Chlamydia* spp., causing conjunctivitis, and *Neisseria gonorrhoeae*, causing conjunctivitis or ophthalmia neonatorum (eye infection of the newborn).

Gastrointestinal   For example, *Salmonella typhi*, causing typhoid, *Vibrio cholerae*, causing cholera,
tract (mouth)   and *Salmonella* spp., causing Salmonellosis and infections of gums caused by *Streptococcus* spp.

Urinary tract infections
Urethra   Examples include *Neisseria gonorrhoeae*, causing gonorrhea (males) and *Treponema pallidum*, causing syphilis (males).

Vagina   For example, *Gardneralla vaginalis*, causing vaginitis, *Neisseria gonorrhoeae*, causing gonorrhea (females) and *Treponema pallidum*, causing syphilis (females).

Placenta   Examples include *Treponema pallidum*, causing syphilis.

Modes of
transmission
Infectious diseases are transmitted to humans via exposure to contaminated environment or infected individuals (*Fig. 2*).

Person to person   This may involve direct contact with the infected host or indirect contact via inert
transmission   object (such as a towel) or droplets.

Environmental   This may involve exposure to contaminated food, water, air, or soil.
transmission

Vector transmission   This may be mechanical, i.e. fomites (inanimate objects) or insect acting as a carrier with no reproduction of the pathogen and transmit pathogen to a susceptible host; or biological, i.e. pathogen multiplies within insects before transmission to the susceptible host.

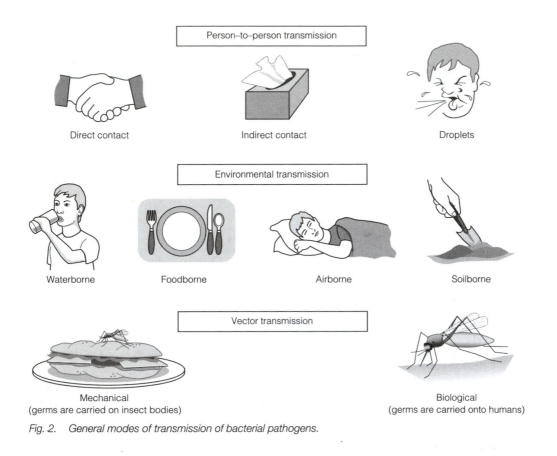

Fig. 2.   General modes of transmission of bacterial pathogens.

Factors contributing to emergence, spread and re-emergence of bacterial infections

Environmental changes	These may be due to agriculture, dams, de/re-forestation, flood / drought, climate changes and may contribute to emergence, spread or re-emergence of Bacterial infections.
Human activities	Population growth, immigration, use of high population densities (such as daycare centres, army, prisons), war or civil conflicts, sexual behaviors or use of intravenous drugs.
Industry	Globalization of food supplies, changes in food processing, transplantation, immunosuppressive drugs, widespread use of antibiotics.
Emergence of a resistant strain	Bacterial evolution in response to a given environment, i.e. antibiotics or harsh environmental conditions.
Public health measures	Lack or reduction in preventative measures, i.e. appropriate sanitary measures, vector control measures, vaccination. Other factors include poor nutrition and water supplies, lack of personal hygiene, limited access to hospitals/treatment.

Common features in bacterial infections

Encounter	Bacteria meet the host
Entry	Bacteria enter the host
Spread	Bacteria spread from the site of the entry
Multiplication	Bacteria multiplies in the host
Damage	Bacteria and/or an overwhelming host immune response produce damage
Outcome	i) excessive cellular destruction results in tissue damage/host death, ii) Bacteria are cleared from the host, or iii) they learn to coexist

H3 PATHOGENESIS AND VIRULENCE OF BACTERIAL INFECTIONS

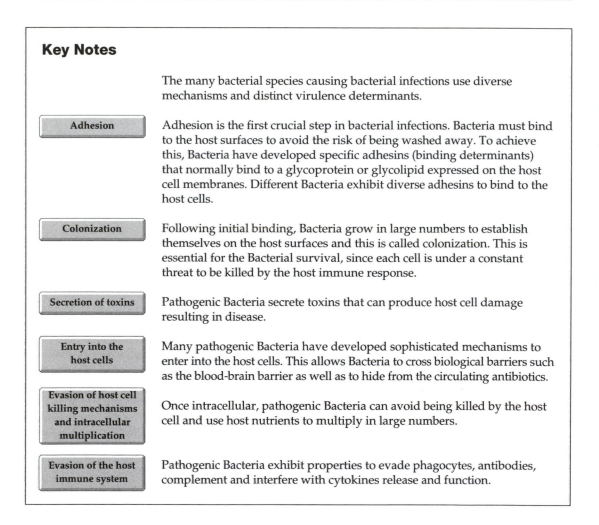

Key Notes

The many bacterial species causing bacterial infections use diverse mechanisms and distinct virulence determinants.

Adhesion

Adhesion is the first crucial step in bacterial infections. Bacteria must bind to the host surfaces to avoid the risk of being washed away. To achieve this, Bacteria have developed specific adhesins (binding determinants) that normally bind to a glycoprotein or glycolipid expressed on the host cell membranes. Different Bacteria exhibit diverse adhesins to bind to the host cells.

Colonization

Following initial binding, Bacteria grow in large numbers to establish themselves on the host surfaces and this is called colonization. This is essential for the Bacterial survival, since each cell is under a constant threat to be killed by the host immune response.

Secretion of toxins

Pathogenic Bacteria secrete toxins that can produce host cell damage resulting in disease.

Entry into the host cells

Many pathogenic Bacteria have developed sophisticated mechanisms to enter into the host cells. This allows Bacteria to cross biological barriers such as the blood-brain barrier as well as to hide from the circulating antibiotics.

Evasion of host cell killing mechanisms and intracellular multiplication

Once intracellular, pathogenic Bacteria can avoid being killed by the host cell and use host nutrients to multiply in large numbers.

Evasion of the host immune system

Pathogenic Bacteria exhibit properties to evade phagocytes, antibodies, complement and interfere with cytokines release and function.

There are hundreds of Bacterial species causing hundreds of diverse Bacterial infections. The basic pathogenic mechanisms associated with each bacterial infection involve diverse mechanisms and distinct virulence determinants. A complete description of each Bacterial pathogen and their virulence determinants is beyond the scope of this book. Here, a common scheme relevant to the majority of Bacterial infections is presented.

Adhesion

Adhesion is the primary step in Bacterial pathogenesis. Certain Bacteria may be limited in targeting specific tissues. For example, *Streptococcas mutans* is abundant in dental plaque but not on epithelial surfaces of tongue. In contrast, *S. salivarius* binds to the epithelial cells of tongue but is absent from dental plaque. This is due to the existence of precise interactions at the molecular level between specific cellular determinants (known as adhesins) and the specific receptors (usually glycoproteins or glycolipids) on the host cells. For example, uropathogenic *E. coli* binds to urinary tract epithelial cells using protein PapG on their fimbriae (see Section H5). The ability of Bacteria to bind to the host cells is crucial to remain in the host body. Nearly all surfaces in the body are protected by some form of flushing movements (e.g. intestinal tract and the bloodstream), which do not allow prokaryotes to remain unattached and constantly attempt to flush them out. Most pathogenic microbes express adhesins on their surface to allow the attachment of the pathogen to host cell surfaces. Many adhesins are proteins and are expressed on the surface of pili/fimbriae (see Section H5). The structures of the fimbriae are thought to help overcome the electrostatic repulsions resulting from the negative charge on both cells, i.e., Bacterium and the host cells (*Fig. 1*). The receptor(s) may be specific glycoprotein or glycolipid which may be limited to specific cell types, thus providing a species, tissue or cellular tropism. Thus Bacteria that infect various tissues or different organisms may require different adhesins.

Colonization

A single Bacterial cell is unlikely to produce an infection. The ability of bacteria to remain at a particular site and multiply is called colonization. It is essential for bacteria to grow in large numbers to persist, despite a competent host immune response. For some Bacteria, colonization is sufficient to produce infection. For example, enterotoxigenic *E. coli* (ETEC) bind to intestinal epithelial cells and colonize. This is sufficient to produce toxins and cause infection, i.e. secretory diarrhea. Similarly, enteropathogenic *E. coli* (EPEC) and enterohemorrhagic *E. coli* (EHEC) bind to the intestinal epithelial cells, colonize and secrete toxins to produce malabsorptive diarrhea and dysentery (bloody diarrhea). The ability of *E. coli* to produce distinct infections is due to their ability to produce distinct toxins (see Section H5). Thus binding and colonization is sufficient to produce infections in these microbes (see Section H5).

Secretion of toxins

Pathogenic bacteria produce toxins, which cause damage to the host cells (see Section H4). For example, following binding, ETEC produces toxins (heat-labile toxin, LT and heat-stable toxin, ST), which enter the intestinal epithelial cells and

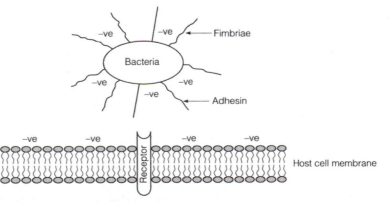

Fig. 1.   *The role of pili/fimbriae in binding to host cell membranes.*

disrupt water and ion flow, thus producing secretory diarrhea (see Section H4). In contrast, EPEC binds to the intestinal epithelial cells and injects toxins into the host cytoplasm (directly from the Bacterial cytoplasm into the host cytoplasm) producing malabsorptive diarrhea. While EHEC (O157H7, see section H4) injects toxins into the host cells to produce dysentery. In addition, EHEC produces notorious Shiga-like toxin, which enters the bloodstream and can cause endothelial cell damage as well as kidney failure (also see Section H5).

Entry into the host cells

The ability of some pathogenic Bacteria to produce infection is not mere binding and colonization but they exhibit complex pathogenic mechanisms to produce disease. This is particularly true for Bacterial pathogens that produce central nervous system (CNS) infections. For example, the majority of meningitis-causing Bacteria cross the gut, evade the immune responses, traverse the blood–brain barrier to gain entry into the CNS and produce disease (see Section H5). To achieve this, the majority of the meningitis-causing Bacteria have evolved mechanisms to invade the host cells. This ability allows Bacteria to cross a biological barrier as well as protects them from an overwhelming immune response and to hide from antimicrobials circulating into the bloodstream. Again, different Bacteria use diverse proteins/toxins to invade into the host cells. The Bacterial invasion of the host cells involves host cell cytoskeletal rearrangements (*Fig. 2*). Cytoskeleton is a network of filaments just beneath the plasma membrane that provides structural support to the cells. The ability of Bacteria to manipulate cytoskeletal proteins to gain entry into the host cells is crucial to remain viable and cross the barrier for the onset of the disease.

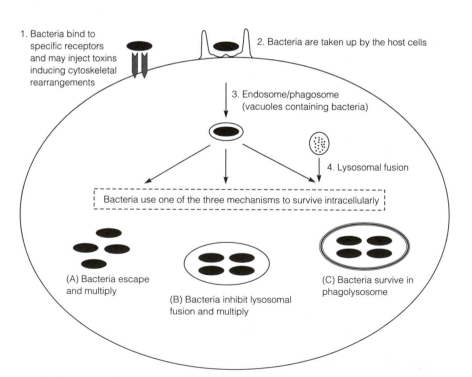

Fig. 2.   Bacteria induces cytoskeletal rearrangements resulting in its uptake. Once inside the host cell, bacteria use diverse mechanisms (indicated by A, B, C) to evade intracellular killing.

Evasion of host cell killing mechanisms and intracellular multiplication

Once taken up, bacteria use diverse strategies to avoid host cell killing (*Fig. 3*, also see Section H7). For example, *Listeria monocytogenes* invade the intestinal epithelial cells (also known as mucosa cells, M cells) escape phagosome, multiply, and move from one cell to another adjacent cell, eventually reaching the bloodstream. In contrast, *Salmonella* spp. invades intestinal epithelial cells and then infects the underlying macrophages. Once inside the macrophages, Bacteria inhibit fusion of lysosomes with phagosome (also see Section H7). Similarly, *Mycobacterium tuberculosis* inhibits fusion of lysosomes with phagosome in macrophages. Other Bacteria such as *Staphylococcus aureus* produce catalase and superoxide dismutase which detoxify toxic oxygen radicals and survive macrophage onslaught (*Fig. 3*, also see Section H7). Other mechanisms of avoiding host cell killing are to destroy the host and/or phagocytic cells. For example, *Streptococcus pyogenes* kill the phagocyte by secreting streptolysin that induces lysosomal discharge into the cell cytoplasm (see Section H7).

Evasion of the host immune response

A key component in bacterial infections is their ability to evade the host immune responses. This property of Bacteria is covered in detail in Section H7.

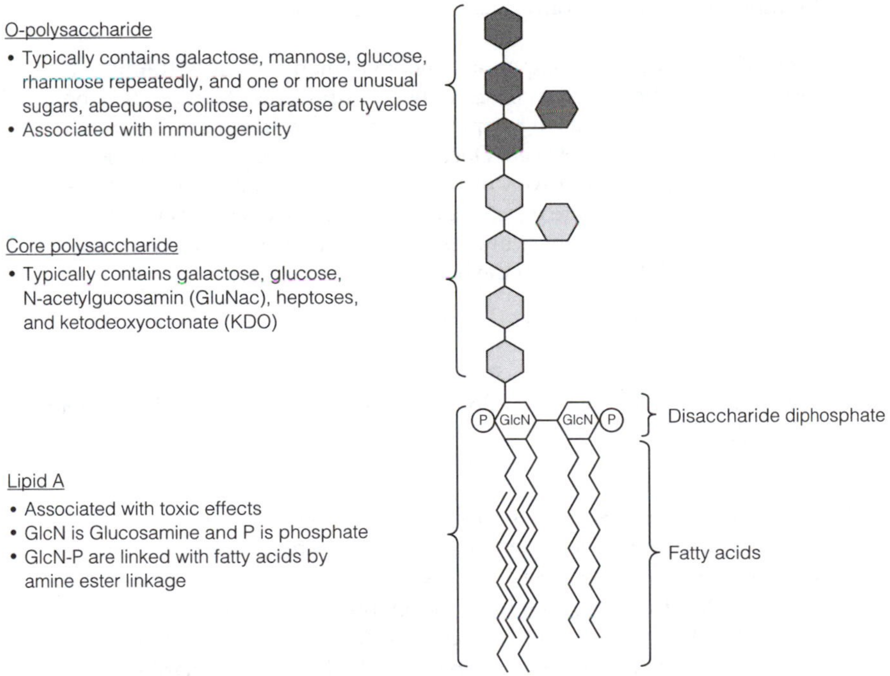

O-polysaccharide
• Typically contains galactose, mannose, glucose, rhamnose repeatedly, and one or more unusual sugars, abequose, colitose, paratose or tyvelose
• Associated with immunogenicity

Core polysaccharide
• Typically contains galactose, glucose, N-acetylgucosamin (GluNac), heptoses, and ketodeoxyoctonate (KDO)

Lipid A
• Associated with toxic effects
• GlcN is Glucosamine and P is phosphate
• GlcN-P are linked with fatty acids by amine ester linkage

Disaccharide diphosphate

Fatty acids

Fig. 3.   Structure of the endotoxin, lipopolysaccharide (LPS).

H4 BACTERIAL TOXINS

Key Notes

Endotoxin	Endotoxin is heat-stable, cell-associated, and a major component of the outer membrane of Gram negative Bacteria. If present in the bloodstream, it causes the release of cytokines by the host cells, resulting in septic shock and death.
Exotoxins	These are heat-labile, extracellular proteins of bacteria that induce damage to the host cells. Exotoxins can be divided into three groups based on the their sites of action which include: i) membrane-damaging toxins, ii) toxins that act on targets inside the host cell, and iii) superantigens, toxins that over-induce the host immune system.
Membrane-damaging toxins	These are toxins that may cause damage to the host cell membranes using enzymatic activitiy or by forming pores, ultimately resulting in cell lysis. Examples include shingomyelinases and phospholipases.
Intracellularly-acting toxins	Some of the most toxic compounds known to man such as botulinum toxin and diphtheria toxin, consist of two components, A and B. B is the binding part and it binds to receptors on the host cell and the A part (toxic part) enters the host cell. Once inside the cell, A part can act on variety of targets. Other toxins include toxins secreted by type III secretion system. These are are injected directly from the bacterial cytoplasm into the host cytoplasm, resulting in various outcomes such as host cell death or bacterial invasion of the host cell.

Traditionally, Bacterial toxins are divided into two broad categories, endotoxins and exotoxins.

Endotoxins

These are cell-associated toxins that are structural components of the outer membrane of Gram negative Bacteria, e.g. endotoxins are part of the outer membrane of Gram negative Bacteria such as *E. coli*, *Salmonella*, *Shigella*, *Pseudomonas*, *Neisseria*, and *Haemophilus*. In *E. coli*, endotoxin (lipopolysaccharide, LPS) is released from the cell in the bloodstream after lysis as a result of the action of the host immune system or by the action of some antibiotics. The LPS is heat-stable toxin and is composed of three components (*Fig. 1*):

1. O-polysaccharide, which is associated with immunogenicity of bacteria, attached to the core polysaccharide but much longer than the core polysaccharide. It maintains the hydrophilic domain of the LPS molecule and is a major antigenic determinant of the Gram negative bacteria (there are >170 different O-antigens).
2. Core polysaccharide – with some variations – the core antigen is common in Gram negative Bacteria.

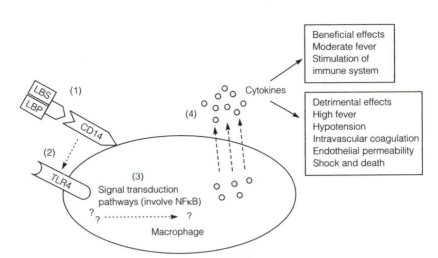

Fig. 1.   Mechanisms of LPS action and their role in disease. (1) LPS is released in serum and binds to a serum protein, LPS-binding protein (LBP), which binds to CD14 on phagocytes. (2) CD14 activates toll-like receptor-4, which (3) leads to signal transduction pathways inducing (4) cytokine release.

3.   Lipid A, which is associated with its toxic effects. Lipid A is hydrophobic, membrane-anchoring region of LPS and its structure is highly conserved among Gram negative bacteria.

Complete LPS is called smooth or S-LPS while LPS lacking (or modified) O-antigen is called rough LPS. O-antigens are targets for the action of the host complement system, but when the reaction takes place at the tips of the polysaccharide chains, complement fails to have its normal lytic effect on Bacteria (also see Section H6). Such Bacteria are virulent because of this resistance to immune forces of the host. If the projecting polysaccharide chains are shortened or removed, antibody reacts with antigens on the general Bacterial surface, or very close to it, and complement can lyse bacteria. Hence the Bacterial strains expressing 'rough' LPS are killed and are considered non-pathogenic.

Mechanisms of LPS action

Once released in the serum, LPS binds to a serum protein called LPS-binding protein (LBP). This LPS-LBP complex binds to the CD14 receptor, which is primarily expressed on the surface of phagocytes (macrophages and neutrophils), and endothelial cells. These interactions induce the activation of toll-like receptor-4 (TLR-4), leading to intracellular signaling pathways, thus stimulating the inflammatory response, i.e. the release of cytokines such as interferon-gamma (IFN-γ), tumour necrosis factor-alpha (TNF-α) and interleukin-1 (IL-1), as well as oxidative bursts (Fig. 2). The low levels of inflammatory mediators have beneficial effects such as moderate fever and stimulate the host immune system, which lead to Bacterial killing. However, overproduction of inflammatory mediators lead to the pathophysiological events such as high fever, hypotension, disseminated intravascular coagulation, increase in endothelial permeability leading to shock and could result in the host death. For example, injection of purified LPS into the experimental animals causes pathophysiological reactions such as fever, changes in white blood cell counts, disseminated intravascular coagulation, hypotension,

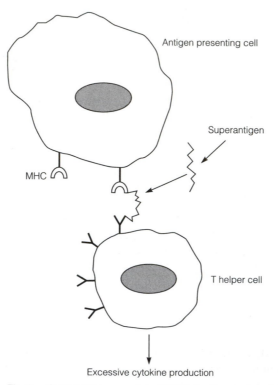

Fig. 2.   *Superantigens mediate MHC (of antigen presenting cell but without antigen) binding to T helper cell. This leads to hyper-activation of T cells, resulting in excessive cytokine release, which may result in shock.*

shock, and death. Although the Gram positive Bacteria lack endotoxin, they possess other components, i.e. lipoteichoic acids and peptidoglycans that can also induce septic shock and death.

Exotoxins

Exotoxins are proteins produced by both Gram negative and Gram positive Bacteria and are toxic to the host cells. They are highly immunogenic and are usually secreted by Bacterial cells but may be held on the surface of the Bacterial cell. Exotoxins are generally heat-labile (heat-sensitive) but there are some heat-stable toxins such as *E. coli* ST enterotoxin (also see Section H5). In some cases (such as *E. coli* strains causing intestinal infections), these toxins are sufficient to produce infection. Some toxins have broad specificities such as phospholipases that cleave phospholipids which are regular components of host cell membranes, resulting in the leakage of the cytoplasmic contents and cell death. Other toxins are highly specific in their actions. For example, *Clostridium botulinum* produces toxin that travels in the bloodstream to target neurons and produces botulism (flaccid paralysis). Similarly, toxin of *C. tetani* produces tetanus, i.e. spastic paralysis. In contrast, toxin of *Corynebacterium diphtheriae* targets a variety of cell types and organs such as the heart and causes its failure, i.e. diphtheria.

Membrane-damaging toxins

These toxins cause damage by destroying the integrity of the plasma membrane of the host cells. Usually they are active against a wide range of cell types and can be divided into two groups on the basis of their mechanisms of action as follows.

Toxins with enzymatic action

A large number of Gram positive Bacteria produce enzymes such as phospholipases and sphingomyelinases, which break down the lipid component on the plasma membrane. This destabilizes the membrane, eventually leading to the cell lysis.

Pore-forming toxins

This group of toxins includes the thiol-activated cytolysins, streptolysin O (produced by *Streptococcus pyogenes*) and listeriolysin O (produced by *Listeria monocytogenes*). These toxins form pores in the host cell membranes. This results in the leakage of nutrients and essential ions from the cell and eventually results in cell lysis.

Intracellular acting toxins

This is a potent group of toxins that act on the host intracellular targets. These include proteins, such as botulinum toxin which inhibits the release of neuro-transmitters in the peripheral nervous system, diphtheria toxin that inhibits protein synthesis resulting in cell death, and cholera toxin that acts in the gut to induce diarrhea (*Table 1*). Intracellular acting toxins can be divided into three broad categories.

A-B toxins

These are common type of toxins. One part of the toxin, B portion is responsible for binding to specific receptors, while the A portion carries the toxic enzymatic activity. Diphtheria toxin is one of the simplest of the A-B toxin consisting of just one A and one B portion. These are synthesized together as a single molecule, and then cleaved to give two peptides joined by a disulfide bridge. In contrast, cholera toxin consists of five B subunits surrounding an active A_1-A_2 molecule, which is synthesized separately. The B part is non-toxic and is responsible for cellular specificity. It will bind only to cells that have the correct receptor. Hence in the case of botulinum toxin, the receptor is only present on neurons. After binding to

Table 1. *Examples of bacterial exotoxins that act inside the cell*

Bacteria	Toxin	Site of action	Mode of action	Symptoms/role in disease
E. coli *Vibrio cholerae*	LT toxin Cholera toxin	Intestinal epithelial cells	ADP-ribosylation of G protein which regulate adenylate cyclase	Water secretion into the intestine resulting in watery diarrhea
Corynebacterium diphtheriae	Diphtheria toxin	Many cell types	ADP-ribosylation of EF-2 leads to inhibition	Cell death, general organ damage
Psudomonas aeruginosa	Exotoxin A		of protein synthesis	
Bordetella pertussis	Pertussis toxin	Many cell types	ADP-ribosylation of G protein which regulate adenylate cyclase	Cell damage, fluid secretion
Shigella dysenteriae	Shiga toxin		RNA glycosidase enzyme modifies 28S rRNA in	Cell death
E. coli	Vero (Shiga-like) toxin	Many cell types	60S ribosome subunit. Inhibits protein synthesis	
Clostridium tetani	Tetanus toxin	Neurons in CNS	Metallopeptidase inhibits release of neurotransmitters	Spastic paralysis

a specific receptor, the toxin-receptor complex is internalized either by endocytosis, followed by translocation of the A portion from the endocytic vacuole into the cytoplasm or by translocation of the A portion across the host plasma membrane directly into the cytoplasm. In many cases, the enzymatic activity of A-B toxin is ADP-ribosylation of a target site in the cell. The A portion catalyzes the removal of the ADP ribosyl group from NAD and the attachment of that group to a cellular protein. The target proteins and the consequences of the ADP-ribosylation are variable thus explaining various types of symptoms. Diphtheria toxin, for example, inhibits protein synthesis, with one molecule enough to kill the cell. In contrast, cholera toxin modifies G protein, a regulatory protein of adenylate cyclase causing it to be permanently switched to the production of cAMP. In the intestinal tract, where the toxin acts, the most significant effect of rise in cAMP is the production of an ion imbalance, leading to massive water loss from the cell, i.e. diarrhea.

Superantigens

These toxins attach to the T cell receptors of T helper cells and the major histocompatibility complexes (MHCs) of the antigen presenting cells (macrophages). This results in the hyper-activation of T helper cells (more than normal) without their recognition of an antigen (*Fig. 3*). The activation of so many T cells results in the release of large amounts of cytokines which can cause a symptom similar to septic shock. For example, *Streptococcus pyogenes* produce superantigens that produce extensive tissue damage around a wound (cellulitis).

Toxins secreted by type III secretion system

This is a recently discovered protein (toxin) secretion system in Gram negative Bacteria. It enables Gram negative Bacteria to inject proteins directly from their cytoplasm into the host cell cytoplasm. Once Bacteria make a close contact with

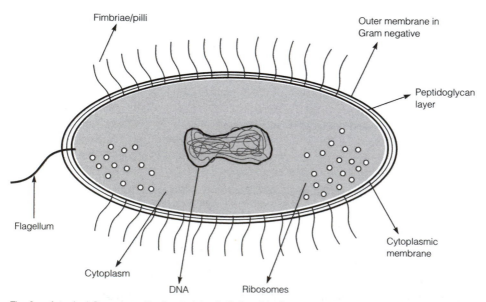

Fig. 3.   A typical Gram negative bacterial cell. Cell wall in Gram negative consists of an outer membrane and peptidoglycan layer. In contrast, Gram positive bacteria lack outer membrane and only contain a thick cell wall. The cell wall in Gram positive consists of severeral peptidoglycan layers forming a thick wall.

the host cell, it forms a tube between itself and the host cell. Many different proteins are required to build this organelle and consist of a thin tube with collars which insert into Bacterial membranes and the eukaryotic host cell membrane. Following this, toxins are directly injected from the bacterial cytoplasm into the host cell cytoplasm. These toxins (proteins) are not subjected to N-terminal processing during secretion, thus function differently from common secretory pathway. In addition, these toxins do not require a 20 amino acid signal sequence required for secretion but in most cases this process seems to be regulated by bacterial contact with the target cell. The injected proteins interfere with the host intracellular signaling pathways, resulting in distinct phenotypes. For example, when *Yersinia* encounters a macrophage, they inject toxins (YopH) into the host cell which interfere with signaling pathways and inhibit bacterial uptake. In contrast, *Salmonella* uses type III secretion system to inject toxins (SopE) in the intestinal epithelial cells which produce cytoskeletal changes and induce Bacterial uptake by the host cells.

H5 *ESCHERICHIA COLI* AS A MODEL BACTERIAL PATHOGEN

Key Notes

Escherichia coli	*E. coli* is a facultatively anaerobic, Gram negative Bacterium that is a normal microbial inhabitant of the gastrointestinal tract of humans and animals. Although the majority of *E. coli* strains are non-pathogenic, some can cause serious and fatal infections of the gut, bladder, and the central nervous system, i.e. meningitis. Meningitis due to *E. coli* mostly occurs due to *E. coli* strains possessing the K1 capsule.
Neonatal meningitis	*Escherichia coli* meningitis is largely limited to neonates (<30 days old) and results in up to 40% mortality.
Pathogenesis of *E. coli* K1 meningitis	The pathogenic mechanisms associated with *E. coli* K1 meningitis include: i) colonization of the intestinal tract, ii) crossing the intestinal lumen to gain entry into the bloodstream, iii) survival of the host defense mechanisms, iv) achieving a threshold-level of bacteremia, v) invasion of the blood–brain barrier, vi) crossing the blood–brain barrier as live bacteria and entry into the CNS, and, finally, vii) survival and replication in the subarachnoid space producing the inflammatory response, thus resulting in meningitis. So far, research has identified that:

1. *E. coli* K1 proteins required for binding to human brain microvascular endothelial cells (which form the blood-brain barrier) are FimH protein (expressed on the tip of fimbriae/pili) and OmpA protein (a major outer membrane protein).
2. *E. coli* K1 proteins required for the invasion of human brain microvascular endothelial cells are IbeA, IbeB, IbeC, AslA, TraJ, CNF1 proteins.
3. *E. coli* K1 factor required for its traversal of the human brain microvascular endothelial cells (crossing the blood-brain barrier as live bacteria) is its K1 capsule.

Escherichia coli

Escherichia coli was first described by Dr. Theodor Escherich in the late 19th century as a normal microbial inhabitant of healthy individuals. Now *E. coli* is recognized as a Gram negative short rod, facultatively anaerobe, which is a universal inhabitant of the intestinal tract of animals (Section 8, *Fig. 3*). In humans, this colonization starts to happen immediately after birth and they are commonly found in the bowels of neonates. Thus the human association with *E. coli* is life long and for the most part trouble free. However, some strains of *E. coli* are pathogenic and cause serious and fatal human infections (*Fig. 1*). *Escherichia coli* strains producing a diverse array of infections are classified by serotypes. Serotyping is based on three antigenic determinants:

1. lipopolysaccharide (O);
2. capsule (K); and
3. flagella (H).

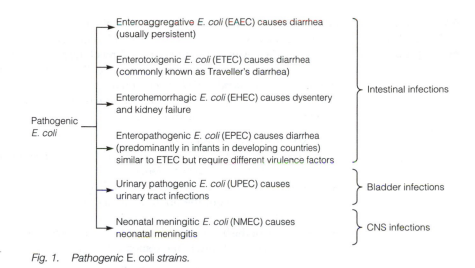

Fig. 1.   Pathogenic E. coli *strains.*

There are more than 700 serotypes but certain *E. coli* serotypes are more abundant in specific diseases. For example, the notorious EHEC O157:H7 commonly cause dysentery and NMEC O18:K1:H7 commonly cause neonatal meningitis. Here, NMEC is represented as a model organism and our current understanding of their pathogenesis and pathophysiology associated with meningitis is described.

Neonatal meningitis

Escherichia coli is a major cause of neonatal meningitis. The incidence rate of *E. coli* neonatal meningitis is estimated at 0.12 and 0.06 per 1000 live births with mortality rate ranging from 20 to 40%. It principally affects infants (<30 days old), with premature and special care babies particularly at risk, with a high risk of either death or long-term disability. A high level of mortality is partly a result of the rapidity of disease onset requiring immediate diagnosis and treatment. In addition, the common association with septicaemia (presence of bacteria in blood) itself has serious consequences, such as high fever, headache, nausea, and hypotension, i.e. low blood pressure due to inadequate blood flow to heart, brain and other organs (also known as shock), resulting in death. Septicaemia can develop quickly with the appearance of a rash under the skin (representative of bleeding under the skin). If untreated, spots/bruises get bigger and become multiple areas of bleeding under the skin and can appear anywhere on the body. In contrast, meningitis is inflammation of the meninges (membranes enclosing the brain). Normally clear cerebrospinal fluid (CSF) becomes cloudy with leukocytes (white blood cells) attracted to the site of inflammation. The swelling around the brain increases pressure inside the skull causing intense headache. Inflammation of spinal meninges affect nearby muscles causing a stiff neck. In addition, the brain function is affected as meninges inflammation can diminish flow of blood to the brain. If not diagnosed early and treated aggressively, meningitis results in death within hours to days. Even with treatment, up to 50% survivors may sustain developmental disability such as hearing loss, learning disability, loss of vision and other complications.

Pathogenesis of *E. coli* K1 meningitis

Among various serotypes of NMEC, *E. coli* strains expressing K1 capsule are significantly associated with meningitis. In fact more than 80% of NMEC cases are due to *E. coli* K1. Much of our understanding of the pathogenesis and pathophysiology of *E. coli* meningitis comes from the K1 strains. The basic pathogenic mechanisms associated with *E. coli* K1 meningitis are highly complex and remain unclear. However, based on our current understanding, they can most likely be divided into the following steps (*Fig. 2*):

1.   Colonization of the intestinal tract.
2.   Crossing the intestinal lumen to gain entry into the bloodstream.
3.   Survive host defense mechanisms.
4.   Magnitude of bacteremia (i.e. a threshold level).
5.   Bacterial invasion of the brain microvascular endothelial cells (blood-brain barrier).
6.   Traversal of the blood-brain barrier as live bacteria and entry into the CNS.
7.   Survival and replication in the subarachnoid space, producing disease.

The sequence of events initiates with *E. coli* acquisition from the mother's flora or from the environment. *E. coli* colonizes the infant's intestinal tract and translocates from the intestinal lumen to the bloodstream. This is the first critical step but the precise mechanisms remain unclear. It is shown that Bacterial translocation of the intestinal lumen requires an *E. coli* cell density of at least 10^8 per gram of feces. However, the urinary tract may also be a route of entry into the bloodstream (in up to 20% cases). This is followed by a sustained high-level bacteremia. This is due to the ability of Bacteria to survive intravascularly by evading and/or resisting host defense mechanisms and multiplication. Indeed, bacteremia is the second critical step in *E. coli* meningitis. For example, bacteremia below or equal to 10^3 colony forming units per milliliter leads to meningitis in 5% of cases, while bacteremia above 10^3 colony forming units per milliliter leads to neonatal

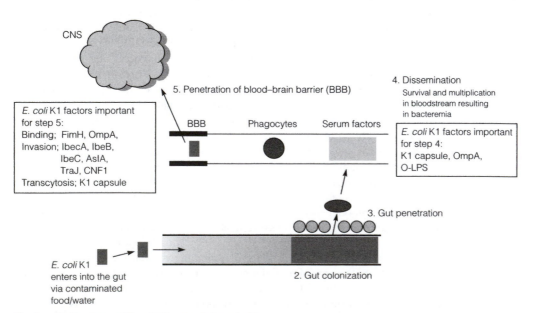

Fig. 2.   Mechanisms of E. coli *K1 neonatal meningitis.*

meningitis in more than 50% of cases. However, the mechanisms of *E. coli* K1 to survive and multiply in the bloodstream are highly complex. To this end, recent studies have shown that *E. coli* K1 can survive within macrophages and inhibit neutrophil NADPH-oxidase to produce reactive oxygen species intracellularly. The ability of *E. coli* K1 to survive macrophage onslaught and/or to evade other host immune responses is critical to achieve a high level of bacteremia. A limited number of findings suggest that an outer membrane protein A (OmpA) and K1 capsule are important *E. coli* determinants and are required to induce a high level of bacteremia. Outer membrane protein A is a major outer membrane protein (approximar molecular weight of 35 kilodaltons) in *E. coli*. *E. coli* K1 strains lacking OmpA show significantly reduced bacteremia in a neonatal rat model of *E. coli* meningitis. In addition, K1 capsule and *O*-lipopolysaccharide (*O*-LPS) are important Bacterial determinants in *E. coli* K1 for induction of a high level of bacteremia. This is due to their abilities to protect Bacteria from complement-mediated lysis as well as protection from opsonization-mediated phagocytic killing. Complement is activated via either the classical or the alternative pathway (see Section F). It is believed that the *O*-LPS is important in *E. coli* resistance to the classical pathway, while the K1 capsule may be involved in resistance to the alternative pathway but the precise mechanisms and additional responsible factors remain unclear.

Once a high level of bacteremia is achieved, *E. coli* K1 is then able to cross the blood-brain barrier, enter the brain, setting up an inflammatory response in the meningeal membranes, so causing meningitis. It is the mechanism by which the blood-brain barrier is breached that is the area of interest, not only in the search for novel treatments for meningitis, but to better understand the blood–brain barrier to create therapies that can target the CNS more efficiently while designing others that will not cross the blood-brain barrier, so limiting possible side effects.

The blood–brain barrier in humans is a monolayer of endothelial cells joined by tight junctions which form the blood vessel of the fine capillaries (microvessels) and supply the brain with nutrients. The blood-brain barrier regulates the exchange of molecules with the brain and protects it against pathogens as well as toxins. It is proposed that the process by which *E. coli* K1 crosses the blood–brain barrier is a stepwise sequence of events and includes adhesion to the host cell membrane, cell invasion, transfer through the cytoplasm and exiting the cell as live bacteria, so invading the CNS. The entire process requires the function of various bacterial determinants and host molecules and is termed bacterial translocation/crossing of the blood-brain barrier. Most of the work with regard to these considerations has been carried out over the last 10 years and is in its infancy.

Adhesion

To cross the blood–brain barrier, microbes have first to overcome the forces imposed by blood flow allowing them to move out of circulation. To do this they must adhere to the brain endothelial cells, which constitute the blood–brain barrier. It makes sense that external components of the Bacteria will come into contact with the host cell first, suggesting them as likely candidates for a role in the adhesion process. Fimbriae, the most external of all, are classified according to antigenic properties, type 1 fimbriae and S fimbriae. Type 1 but not S fimbriae are implicated in *E. coli* binding to the endothelial cells. *E. coli* typically expresses 100–300 fimbriae, each approximately 0.2–2.0 μm long and 0.7 nm wide. They are heteropolymers based around a 17 kilodaltons structural protein FimA and express a 29 kilodaltons protein FimH at their tips. FimH is a lectin-like protein with high affinity to mannose residues. It was additionally observed that mannose inhibited *E. coli* binding to the endothelial cells suggesting a role for type 1 fimbriae. A feature of interest, and with possible virulence implications, is the ability of *E. coli* to determine whether type 1 fimbriae are expressed. It is a phase dependent 'all or nothing' expression. This is a result of inversion or switching 'on' or 'off' of the promoter for the *fim*A gene coding for the fimbrial component FimA by upstream genes *fim*B and *fim*E. It is shown that *E. coli* K1 adherence to the endothelial cells is reduced in the non-fimbriated state and that increased levels of *fim* gene expression are associated with fimbriate *E. coli* K1 association with the brain microvascular endothelial cells (BMEC), tending to confirm the binding role of type 1 fimbriae. Highlighting the refinement of Bacterial host interaction, it has been shown, using guinea pig erythrocytes as target cells, that type 1 fimbriae expressing FimH are able to modulate the level of attachment to the receptor such that it is augmented by an increase in sheer force applied by a flowing liquid medium. This feature would certainly be of advantage to an invasive strain of *E. coli*, causing the initial host–pathogen interactions to be more effective and allowing other determinants to bind, resulting in more intimate contact for more effective interaction and subsequent invasion.

Among other determinants, Outer membrane protein (OmpA) has been shown to be important in *E. coli* K1 binding to human BMEC (HBMEC). OmpA is a 35 kilodalton molecule and is the most abundant protein in the outer membrane. Structurally it's N terminal forms an anti-parallel β -barrel with 8 transmembrane strands resulting in 4 extracellular hydrophilic loops. It was subsequently shown that OmpA is involved in the HBMEC binding through *N*-glucosamine (GlcNAc) epitopes of HBMEC glycoprotein gp96 (96 kilodaltons) expressed on the host membrane. This was confirmed with the finding that the chitooligomers (GlcNAc1, 4-GlcNAc oligomers) block the *E. coli* K1 binding and invasion of HBMEC *in vitro* and traversal of the blood–brain barrier *in vivo*.

Invasion

Following initial binding, *E. coli* must invade the host cell membrane to gain entry into the host cell. Both type 1 fimbriae and OmpA contribute to invasion of HBMEC as a result of their effects on HBMEC binding. In addition, several bacterial determinants responsible for *E. coli* K1 invasion of HBMEC have been identified and include Ibe proteins (IbeA, IbeB, IbeC), AslA, TraJ, and an A-B type toxin, cytotoxic necrotizing factor 1 (CNF1). The functional involvement of these *E. coli* determinants in *E. coli* K1 invasion of HBMEC *in vitro* and traversal of the blood–brain barrier *in vivo* have been shown with deletion and complementation experiments. Moreover, a 37 kilodaltons laminin receptor precursor (LRP) has been identified as HBMEC receptor for bacterial CNF1 protein (A-B toxin with approximate molecular weight of 110 kilodaltons). LRP is a ribosome-associated cytoplasmic protein shown to be a precursor of 67 kilodalton laminin receptor (LR). Upon maturation, LR is recruited to the cell membrane and acquires ability to bind to the beta-chain of laminin with high affinity. Recent studies have shown that incubation of HBMEC with *E. coli* K1 upregulates 67LR expression by HBMEC and recruits 67LR to the invading *E. coli* K1 in a CNF1-dependent manner. Overall, these findings suggest that *E. coli* K1 invasion of HBMEC is a complex process that requires the function of various Bacterial proteins as well as an A-B toxin.

Host cell cytoskeletal rearrangement

E. coli K1 invasion of HBMEC requires rearrangements of the actin cytoskeleton and blocking actin condensation with microfilament-disrupting agents such as cytochalasin D abolishes *E. coli* K1 invasion of HBMEC. Also, transmission and scanning electron microscopy show that *E. coli* K1 induces membrane protrusions at the entry site on the HBMEC surface confirming the involvement of the cytoskeleton. Thus much of the work has been focussed on *E. coli* and host factors involved in HBMEC cytoskeletal changes. The indication that invading *E. coli* K1 are in some way manipulating host cell cytoskeletal arrangements is shown with the finding that a protein tyrosine kinase (PTK) inhibitor, genistein, blocked invasion, while it did not reduce adhesion, demonstrating the role of PTK and a distinction between adhesion and invasion. Subsequently, the HBMEC signaling molecules, such as focal adhesion kinase (FAK), and its associated cytoskeletal proteins paxillin, phosphatidylinosital 3-kinase (PI3K), and cytosolic phospholipase A2 (cPLA$_2$), are involved in *E. coli* K1 invasion of HBMEC, most likely through their effects on actin cytoskeleton rearrangements. The relationship between FAK and PI3K in HBMEC was further confirmed with the finding that FAK operates upstream of PI3K in the signaling pathway. The tyrosine phosphorylation of FAK, paxillin and PI3K are mediated by *E. coli* K1 determinants such as OmpA.

Another *E. coli* K1 virulence factor, cytotoxic necrotizing factor 1 (CNF1), was evaluated for a role in invasion and penetration of HBMEC. CNF1 is an AB type bacteria toxin (see Section H4) that contributes to *E. coli* K1 invasion of HBMEC *in vitro* and traversal of the blood-brain barrier *in vivo* through its activation of RhoGTPases leading to cytoskeletal rearrangements. Rho GTPases are the key regulators of actin cytoskeleton in all eukaryotic cells and link external signals to the cytoskeleton by switching a GDP-bound inactive Rho to a GTP-bound active Rho, resulting in specific cytoskeletal rearrangements. These involve three major intracellular pathways including: i) RhoA pathway leading to stress fiber formation, ii) Rac1 activation triggers lamellipodia formation, and iii) Cdc42 activation promotes filopodia formation. The stress fiber formation plays an important role in *E. coli* K1 entry into the HBMEC suggesting the involvement of RhoA. In support, CNF1 is shown to activate RhoGTPases, such as RhoA, Cdc42 but not Rac1, by a site-specific deamidation of glutamine 63 of RhoA (or glutamine 61 of Cdc42). CNF1-mediated RhoA activation is correlated with levels of LRP expression in HBMEC further confirming LRP as a receptor for CNF1. RhoA activation leads to ezrin activation (i.e. phosphorylation). Ezrin is a member of ERM (ezrin, radixin and moesin) protein family, which connects F-actin filaments to plasma membranes and induces formation of membrane protrusions responsible for *E. coli* K1 internalization of HBMEC. Overall, CNF1-mediated cytoskeletal changes involve RhoGTPases and these pathways are distinct from OmpA-mediated cytoskeletal changes, which involve FAK, paxillin and PI3K (*Fig. 3*).

Traversal of the blood–brain barrier as live bacteria

Once invaded in the HBMEC, *E. coli* K1 remains in an enclosed vacuole, transmigrate through the HBMEC and exit from the other side as live bacteria to produce disease (*Fig. 4*). During the transport in an enclosed vacuole, *E. coli* does not multiply but remains viable, which in the normal course of events would be killed by lysosomal enzymes released into the vacuole by fusion with lysosomes (*Fig. 4*). It is not clear what Bacterial factors are important for these steps and their precise mechanisms remain unidentified. At least one of the factors seems to be K1 capsule. For example, *E. coli* K1$^+$ but not K1$^-$ strains can escape lysosomal fusion. This is shown with findings that vacuoles containing *E. coli* K1$^+$ recruit early endosomal markers such as early endosomal autoantigen 1 (EEA1), transferrin receptor as well as late endosomal markers such as Rab7 and Lamp-1.

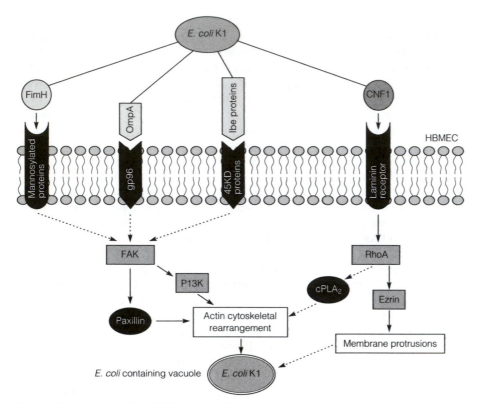

Fig. 3.   Mechanisms of E. coli *K1 invasion of human brain microvascular endothelial cell.*

But *E. coli* K1$^+$ containing vacuoles do not obtain cathepsin D, a lysosomal enzyme, thus avoid bacterial killing by lysosomal enzymes. After crossing the blood-brain barrier as live bacteria, *E. coli* K1 subsequently survives and multiplies in the subarachnoid space, and induces host inflammatory responses in the meninges. Again, *E. coli* K1$^+$ but not K1$^-$ strains can produce positive CSF cultures and cause *E. coli* meningitis, indicating that K1 capsule is a critical determinant

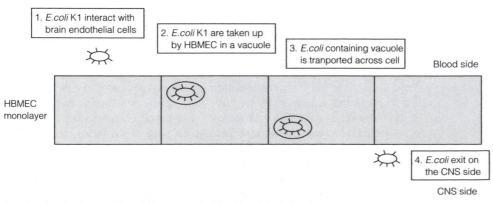

Fig. 4.   Mechanisms of E. coli *K1 traversal of the blood–brain barrier.*

in *E. coli* meningitis. Although there are several *E. coli* virulence determinants responsible for distinct functions, the involvement of some virulence factors such as OmpA and K1 capsule in more than 1 step is clear. However the complete elucidation of the pathogenic mechanisms associated with *E. coli* K1 meningitis is far from complete. The recent availability of *E. coli* K1 genome will undoubtedly help advance our understanding of this fatal infection.

H6 HUMAN DEFENSE MECHANISMS

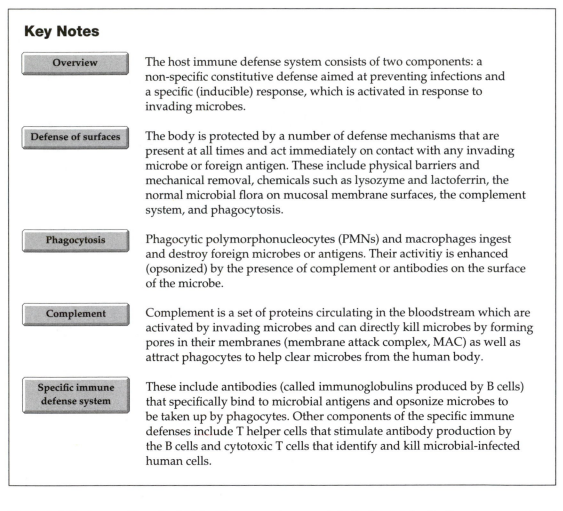

Key Notes

Overview	The host immune defense system consists of two components: a non-specific constitutive defense aimed at preventing infections and a specific (inducible) response, which is activated in response to invading microbes.
Defense of surfaces	The body is protected by a number of defense mechanisms that are present at all times and act immediately on contact with any invading microbe or foreign antigen. These include physical barriers and mechanical removal, chemicals such as lysozyme and lactoferrin, the normal microbial flora on mucosal membrane surfaces, the complement system, and phagocytosis.
Phagocytosis	Phagocytic polymorphonucleocytes (PMNs) and macrophages ingest and destroy foreign microbes or antigens. Their activitiy is enhanced (opsonized) by the presence of complement or antibodies on the surface of the microbe.
Complement	Complement is a set of proteins circulating in the bloodstream which are activated by invading microbes and can directly kill microbes by forming pores in their membranes (membrane attack complex, MAC) as well as attract phagocytes to help clear microbes from the human body.
Specific immune defense system	These include antibodies (called immunoglobulins produced by B cells) that specifically bind to microbial antigens and opsonize microbes to be taken up by phagocytes. Other components of the specific immune defenses include T helper cells that stimulate antibody production by the B cells and cytotoxic T cells that identify and kill microbial-infected human cells.

Human defense mechanisms

Once bacterial pathogens gain access into the human body, they are encountered by a highly professional defense system. Traditionally, the human defense system has been divided into two components:

1. Non-specific/constitutive/innate immune responses that include skin that acts as a physical barrier, neutrophils that police the bloodstream and attack any foreign invaders, complement and cytokines that direct the activities of neutrophils, and natural killer cells that kill virus-infected cells.
2. Specific/inducible/acquired immune responses that include antibodies (produced by B-lymphocytes) and T-lymphocytes.

These defense systems do not operate independently but communicate with each other to build an effective defense against the invading organism.

Non-specific immune responses

Defense of surfaces

The outer layers of the skin are dry, thick layers of keratinized cells that provide a physical barrier to bacterial pathogens. Natural openings in the skin such as pores, hair follicles, and sweat glands are protected by secetion of toxic chemicals such as fatty acids and lysozyme. Generally, microbes can only breach this barrier through wounds. However, underneath the skin is skin-associated lymphoid tissue (SALT). In SALT, the resident Langerhans cells (similar to macrophages) attack bacteria and initiate specific immune defenses.

Inside the body, the intestinal tract, respiratory tract, vaginal tract, and bladder are covered in mucosal membranes. The cells in these membranes are protected by thick layer of mucus/mucin. The mucin consists of proteins and polysaccharides that prevent microbes from reaching the epithelial cells. Mucin also contains:

1. Lactoferrin – iron binding protein that deprives microbes of iron.
2. Lysozyme – an enzyme that digests cell walls of Bacteria.
3. Defensins – small proteins that make holes in the Bacterial membranes.

Mucin is constantly shed and replaced, together with any trapped microbes. In addition, low pH (2) in the stomach, cilia movement over trachea, flushing of urinary tract and competition from the normal flora prevents colonization of these by pathogenic microorganisms. Similar to the skin, mucoal membranes are associated with the specific immune defense by mucosa-associated lymphoid tissue (MALT). For intestinal tract, it is called gastrointestinal-associated lymphoid tissue (GALT). In GALT, the resident macrophages (similar function to Langerhans cells of SALT) initiate specific defenses.

Violation of initial defenses may lead to inflammation. Inflammation is tissue reaction to infections or injury characterized by redness (increased blood flow), swelling (increased extravascular fluid and phagocyte infiltration), heat (increased blood flow and pyrogens, fever-inducing agents), pain (local tissue destruction and irritation of sensory nerve), and loss of function. Inflammation leads to recruitment of neutrophils to the infection site, which leads to ingestion and destruction of bacteria. Neutrophils are also known as polymorphonuclear leukocytes (PMNs) indicating that the nucleus is a multi-lobed structure that may appear to be multi-nuclei when viewed in cross-sections. The PMNs are primarily found in the bloodstream. The steps of ingestion and killing by phagocytic cells involve:

1. Attachment of the microorganism – formation of pseudopodia.
2. Uptake of the pathogen in a vacuole 'phagosome'.
3. Fusion of lysosome (containing antibacterial chemicals and enzymes) with phagosome. Lysosome contain myeloperoxidase enzyme that is normally inactive but, following fusion with phagosome, this enzyme becomes activated and produces superoxide and hypochlorite. Superoxide is a toxic form of oxygen that can oxidize and inactivate proteins and other molecules on microbial surface. When mixed with chlorine, it forms hypochlorous acid, highly toxic to invading organisms. In addition, toxic nitric oxide (NO) is produced in response to invading microbes.
4. Overall result is killing and digestion of the ingested bacterium. Often death of the PMN also occurs and if in large numbers, the dead PMNs forms pus.
5. Release of digested products to the outside.

Complement

In addition to neutrophils, blood/plasma contains potent complement system. Complement is a set of proteins produced in part by the liver that circulate in blood and are activated sequentially by proteolytic cleavage, a process called complement activation. Complement attaches to invading microbes and directly causes damage to the cell through a process known as membrane attack complex (MAC) or help phagocytes (macrophages and neutrophils) to eliminate microbes from the body. There are nine complement proteins, C1–C9. There are two pathways of complement activation:

1. The classical pathway which is initiated by antibody-antigen complexes; and
2. The alternative pathway which is triggered by microbial surface molecules such as LPS.

The activation causes complement proteins to be proteolytically cleaved. The cleaved products are indicated by small fragment 'a' or large fragment 'b'. For example, LPS binding to C3 causes its cleavage into C3a and C3b. C3b binds to bacteria and enhances its phagocytic uptake (opsonization-mediated uptake). The cleavage of other components, C5b initiate the deposition of C6, C7, C8, and C9 on Bacterial surface which forms pores in the bacterial membranes (MAC) leading to the Bacterial death. The ultimate effect of complement is damage of bacteria by creating pores (MAC) or their opsonization (help Bacterial uptake by phagocytes).

Cytokines

Cytokines are produced primarily by endothelial cells (which form blood vessels) and phagocytes (macrophages and neutrophils). Cytokine producing cells express CD14 receptor on their surface, which binds to the specific Bacterial components such as LPS (see Section H4). This leads to the release of cytokines. The function of cytokines is like smoke-signals, as they attract phagocytes to the site of infection and link non-specific and specific immune systems.

Specific immune response

This system involves cells that respond to the invading bacteria in a specific fashion. These include cytotoxic T-cells (CD8 cells), T helper cells (CD4 cells), B cells (produce antibodies), and active macrophages. Macrophages are phagocytes similar to neutrophils that are mostly found in the tissues. Macrophages are first produced by stem cells as monocytes which circulate in the bloodstream. On attraction to a site of action or entry into a tissue they differentiate into macrophages. The invading microorganisms are taken up by macrophages and killed (like in neutrophils) but instead of releasing the digested products to the outside, some bacterial products are loaded onto a protein complex called major histocompatibility complex class II proteins (MHC II) and presented on the cell surface. Hence, these cells are called antigen-presenting cells (APCs). Why certain microbial products are selected to be expressed on APCs (macrophages, dendritic, B cells) and not others is not known. Of interest, macrophages in the brain are called microglial cells, while macrophage in liver are called Kupffer cells. The B cells also act like APCs in uptaking and processing foreign antigens and loading them onto MHC II complex. Finally, the antigen-MHC II complex is recognized by T helper cells.

1. In the case of T helper-macrophage interactions, T helper cells become activated and produce cytokines such as interferon-gamma (IFN-γ), interleukin-2 (IL-2), which activate macrophages and stimulate B cells to produce antibodies.
2. In the case of T helper–B cell interactions, T helper cells become activated as above which in turn stimulate B cell to differentiate into antibody-producing cells, i.e. plasma cells.

This approach is used in response to extacellular microbial pathogens. In contrast, if microbial pathogens are intracellular their products are loaded onto MHC I complex. The MHC I display proteins on almost all cells. Finally, the antigen-MHC I complex is recognized by cytotoxic T-cells. Cytotoxic T-cells kill the host cells displaying antigen-MHC I complex, thus killing any intracellular microbe in it. This is particularly useful against viral infections.

Antibodies

Antibodies are proteins called immunoglobulins (Igs). They are made of two heavy and two light chains, which create two functionally important areas as shown in *Fig. 1*. The region that recognizes and bind antigen is called the Fab region, while Fc region binds to the phagocytes (phagocytes have Fc receptor). The result of antibody binding to specific antigen is:

- Enhanced phagocytosis by providing binding site (Fc) for phagocytes.
- Trap microbes in mucin.
- Complement activation.
- Neutralization of toxins.
- Inactivation of proteins.
- Inhibition of binding of toxins/bacteria to host cells.

This ultimately results in the neutralization and removal of the microbes and/or their toxins from the host. The antibodies present in the serum are IgG (70–75% of total antibodies) and IgM (10% of total antibodies), while IgA (15% of total antibodies) is present in secretions such as milk, tears, saliva, and mucosal secretions and is called secretory IgA (sIgA). In addition, trace amounts of IgE (mostly bound to deep layers of the skin and known as skin sensitizing antibody and very little in blood) and IgD are present in the serum and are involved in allergic reactions.

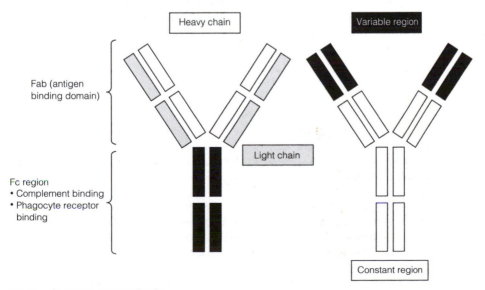

Fig. 1. Structure of an antibody.

H7 BACTERIAL EVASION OF IMMUNE DEFENSES

Key Notes

Evasion of antibodies	Pathogenic Bacteria use diverse mechanisms to inhibit antibody functions. These include: i) Bacterial secretion of glycosidases that remove carbohydrate chains of antibodies thus inhibiting their function, ii) bacterial secretion of proteases that cleave antibodies in the hinge region, and iii) by expressing antibody-binding proteins that bind to the Fc region of antibody (Fc region is normally required for phagocyte binding).
Evasion of cytokines	Cytokines are important molecules in building an effective immune response against the invading microbes and their antigens. Pathogenic bacteria inhibit cytokine synthesis, thus interfering with the immune response. In addition, Bacteria may bind to cytokines so these molecules can not reach their target sites to induce an effective immune response or may secrete proteases that destroy cytokines.
Evasion of complement	Pathogenic Bacteria can evade complement system by secreting enzymes that degrade complement proteins or by binding to protectins, thus interfering with complement system.
Evasion of phagocytic killing	Phagocytes provide an effective and efficient defense against microbial pathogens or their antigens. However, the pathogenic Bacteria can evade phagocytic killing by avoiding contact with phagocytes, inhibiting phagocyte chemotaxis, hiding their antigenic surfaces, inhibiting phagocytic engulfment, escaping phagosome, inhibition of fusion of lysosomes with phagosome or by killing phagocytes.

As indicated in Section H6, humans have developed a highly professional immune system to defend against bacterial or other microbial pathogens. As Bacteria evolve faster than humans they have developed a variety of strategies to overcome and/or evade these defense systems. An overview of Bacterial strategies to evade human immune responses is provided below.

Evasion of antibodies

At the mucosal surfaces, Bacteria have to cope with potent secretory immunoglobulins (sIgA) and then again in the bloodstream, IgG and other antibodies are highly effective in trapping microbes and neutralizing toxins. Bacteria exhibit the following evasion mechanisms to inactivate antibodies (*Fig. 1*)

Glycosidases

Glycosidases are enzymes that cleave carbohydrate chains. Antibodies such as sIgA are heavily glycosylated and glycosylation is essential for antibody function, conformation and net charge. Some pathogenic Bacteria secrete glycosidases removing carbohydrate chains from antibodies rendering to their inactivity.

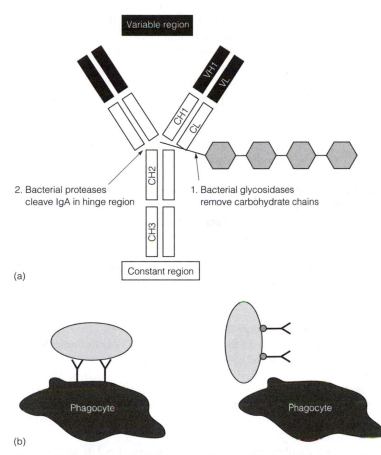

Fig. 1.   Bacterial evasion of antibodies. (a) Bacterial glycosidases cleave carbohydrate chains rendering them obsolete, (b) Bacterial proteases cleave antibody in hinge region, (c) Phagocytes normally bind to the Fc domain of antibody but some bacteria produce Fc-binding proteins thus preventing phagocyte binding.

Proteases

Some pathogenic Bacteria produce proteases (called Ig proteases) which cleave Ig in the hinge regions – between CH1 and CH2.

Antibody-binding proteins

Some pathogenic Bacteria produce Ig binding proteins. Such proteins bind to Fc region of Ig thus inhibiting phagocytosis (Fc regions are required for binding to phagocytes). For example, *Staphylococcus aureus* produces cell wall-anchored protein, Protein A, which binds to Fc domain of the IgG. Thus antibody can not react with Fc receptors on phagocytes and bacteria gets away. Protein G and Protein H are additional Streptococcal IgG binding proteins that result in the immune evasion.

Antigenic variation

Some pathogenic Bacteria shed old antigens and present new ones to the immune system making the available immune response obsolete. New antigens (fimbriae or other outer membrane proteins) do not bind to the existing antibodies. Thus host must produce new antibodies to challenge new antigens. The antibody production may take several days, providing the pathogen sufficient time to colonize and produce infection. Antigenic variation usually results from site-specific inversions or gene rearrangements in the DNA of microorganisms. *Neisseria gonorrhoeae* can change fimbrial antigens during the course of an infection.

Evasion of cytokines

Cytokines are smoke signals that warn human body of the invading microorganisms, help attract phagoctes to the site of infection, as well as link non-specific immune response with the specific immune response and thus help build an effective defense system against the invading microbes. There are several strategies employed by Bacteria to interfere with cytokine synthesis, release and function as indicated below.

Bacterial toxins inhibiting cytokine synthesis

Lymphostatin protein produced by enteropathogenic strains of *E. coli* inhibits the production of interleukin-2 (IL-2), IL-4 and interferon-gamma (IFN-γ). *Yersinia* can inhibit tumor necrosis factor-alpha (TNF-α) synthesis by injecting toxins (Yop proteins).

Binding to cytokines

Escherichia coli, *Shigella* spp. and *Mycobacterium tuberculosis* have receptors for cytokines such as IL-1, (TNF-α) and epidermal growth factor (EGF) thus interfering with their function.

Inactivating cytokines

Bacterial proteases can directly inactivate cytokines or cleave cytokine receptors on host cells so they can not respond or communicate with other immune cells to build an effective immune response.

Induction of anti-inflammatory cytokines response

Some Bacteria can induce an anti-inflammatory response by induction of IL-10 and transforming growth factor-beta (TGF-β) production, which have immunosuppressive effects as well as inhibit macrophage activation.

Evasion of complement

Complement is a set of proteins that circulate in blood and are activated through antibody binding to Bacterial surface or directly by Bacterial surface, leading to MAC attack (forming pores in Bacterial membranes), or opsonises Bacteria (help Bacterial uptake by phagocytes).

Inactivating complement proteins

Streptococcus agalactiae encode enzymes responsible for inactivating C5a. *Pseudomonas aeruginosa* produce elastase that inactivates C3b and C5a.

Avoiding complement by binding to protectins

To avoid host cells from unnecessary complement lysis, host produces proteins, called protectins, which bind to complement proteins and do not allow this process to proceed. They are also required to keep complement system in check. It is shown that *Streoptococcus pyogenes* can bind to protectins and can avoid complement lysis.

Evasion of phagocytic killing

As indicated in section H6, phagocytes (neutrophils and macrophages) play a crucial role in uptaking, killing and thus clearing Bacteria from the human body. However, the pathogenic Bacteria employ a diverse array of mechanisms to evade the uptake and onslaught by the phagocytes as described below.

Avoiding contact with phagocytes

Many pathogenic Bacteria invade or remain in regions inaccessible or less patroled by phagocytes, e.g. internal tissues such as the CNS and surface tissues (e.g. skin) are less guarded by phagocytes.

Avoid provoking an overwhelming inflammatory response

Some Bacteria inhibit phagocyte chemotaxis. *Mycobacterium tuberculosis* inhibit leukocyte migration. *Clostridium* toxin inhibits neutrophil chemotaxis.

Hide antigenic surface of bacterial cell

Bacteria cover their surface with components seen as 'self' by host phagocytes and other immune systems. For example, pathogenic *Staphylococcus aureus* produces cell-bound coagulase which clots fibrin on the Bacterial surface.

Inhibition of phagocytic engulfment

Many pathogenic Bacteria bear molecules on their surfaces that inhibit phagocytosis. For example, polysaccharide capsules of *Streptococcus pneumoniae*, *Haemophilus influenzae*, *Treponema pallidum* and *Klebsiella pneumoniae* may inhibit phagocytic engulfment.

Escape from the phagosome

If taken up, Bacteria are contained in a vacuole, i.e. phagosome. Early escape from the phagosome vacuole is used by some bacteria for their growth and virulence. For example, *Rickettsia* spp. produces phospholipase which lyses phagosomal membranes within 30 seconds of ingestion and thus evades phagocytic killing.

Inhibition of fusion of lysosomes with phagosome

Some pathogenic Bacteria can even survive inside phagosomes because they prevent discharge of lysosomal contents into phagosome environment. For example, phago-lysosome formation is inhibited in phagocyte by *Salmonella* and *M. tuberculosis*. With *M. tuberculosis*, Bacterial cell wall components are released from phagosome and modify lysosomal membranes to inhibit fusion. Many pathogenic Bacteria produce enzymes, i.e. superoxide dismutase that convert superoxide to water thus survive macrophage myeloperoxidase system. *Coxiella burnetii* is an intracellular bacterium that is phagocytozed by macrophages, followed by fusion of lysosomes to form phago-lysosome (low pH). The precise mechanisms are unclear but bacterium can survive and grow at acidic pH.

Killing phagocytes before ingestion

Other strategies of evasion of phagocytes are to attack and kill phagocytes. Some Bacteria secrete extracellular enzymes that kill phagocytes.

Killing phagocytes after ingestion

Some pathogenic microbes may grow in the phagosome and release substances which can pass through phagosomal membrane and cause discharge of lysosomal granules into the host cell thus producing damage to the host cell.

H8 CONTROL OF BACTERIAL INFECTIONS

Key Notes

The most effective approach in the control of Bacterial infections has been the improved public health measures to remove pathogenic organisms from food and water supplies, and to reduce their populations in the environment. Similarly, better housing and nutrition have raised the health of individuals, thus making them more resistant to infection.

Vaccination

Vaccines are materials originating (or artificially produced) from microbial pathogens. Whole microbes (live, attenuated, or killed), fractions of microbes or inactivated microbial products such as toxoids or their DNA (nucleic acids) encoding virulent genes may be used to induce a specific immune response in a host. This provides protection from infection with that microbe in the long term. Vaccines are not available for all microbial infections.

Antibiotics

Antibiotics are molecules produced by microbes which inhibit the growth or kill other microbes. The antibiotics normally used to treat bacterial infections are those that are selectively toxic to bacterial cells and do not harm the host. These antibiotics target sites such as peptidoglycan, ribosomes, and nucleic acid synthesis which are significantly different between prokaryotic and eukaryotic cells. The effectiveness of antibiotic therapy is endangered by the development of Bacterial antibiotic resistance and particularly by the spread of antibiotic-resistant genes between microorganisms.

Despite our efforts to control and/or eradicate pathogenic microbes, they have continued to cause million of deaths per year and remain a major contributing factor to human misery with both economical and social implications. The importance of sanitation, good housing, nutrition, and clean water supplies is paramount in reducing the number of microbial infections. For example, even before the introduction of vaccines and antibiotics, the reduction in number of infections due to tuberculosis was noticed with improved public health measures. Other than improved public health measures, vaccines and antibiotics are the major weapons in our fight against microbial diseases.

Vaccines

Vaccination is one of the most powerful means to save lives and to increase the level of health in mankind. Vaccination has proven a useful tool for protecting individuals and the population from bacterial diseases. Vaccine is a material originating from a microorganism (e.g. Bacteria, attenuated Bacteria, killed Bacteria, denatured toxin, capsule, etc.) that is introduced into individuals in a controlled way leading to the stimulation of the immune response without the symptoms of full-blown disease. Material may be produced artificially or directly obtained from the pathogen. Ultimately, this results in the production of memory cells within the host. If host encounters the pathogen again, the immune system can generate a rapid antibody response thus preventing infection. Vaccines are produced with the following objectives:

1. It should promote an effective resistance to disease.
2. It should provide sustained protection – lasts several years.
3. It induces humoral response (antibodies).
4. It induces protective T-cell responses (cellular immunity).
5. It must be safe – minimal side effects.
6. It should be stable and should remain so during transportation.
7. It is reasonably cheap and easily administered.

However, not all vaccines fit these criteria and trials have to be conducted to determine whether the beneficial effects outweigh the potential side effects. There are the following types of vaccines:

Whole cell vaccines
- Killed pathogen (ruptured or formalin-preserved), these vaccines should stimulate maximum protective immunity but may have severe side effects.
- Live attenuated or low virulence vaccines, these can be produced by knocking-out the virulent genes from the pathogen but may have side effects.

Protein subunit vaccines (single/multiple components)
These are purified proteins from the pathogen. These may induce maximum protection but may be toxic. Also, protein purification is expensive. In contrast, recombinant protein antigens are cheap and consistent. Other problems may arise from the fact that pathogen populations might be polymorphic. Vaccine may be ineffective against some strains. Simple recombinant might not stimulate all required immune components, e.g. immunogenic carbohydrate wrongly processed in production.

DNA vaccines (nucleic acid vaccines)
For these, DNA (gene) encoding a multiple vaccine protein is cloned into a plasmid (see Section F15) containing host expression promoter. Individuals are vaccinated with this plasmid. The plasmid infects the host cell and expresses foreign gene or genes and produces the protein inside the cell. Antigen will be expressed on the cell's major histocompatibilty complex (MHC) class I molecules, due to the antigen being endogenous. This stimulates the activation of cytotoxic T-cells, which are important for clearance of pathogen infected cells.

Traditional vaccines enter the MHC class II pathway, due to the fact that the antigens encountered are exogenous. This primarily stimulates antibody responses that are not effective at clearing viruses which are protected by the cells they reside in. However with DNA vaccines, once the cytotoxic T-cells have been activated, they may lyse infected cells and release the antigen allowing the antibody response to also become stimulated. These are cheaper and easier to store than purifying proteins (very stable) and may also be immunogenic but side effects remain unclear. The transient production of protein for about a month is enough to evoke a robust immune response. If any of the above vaccines are used, this process is called active immunization. In contrast, passive immunization is injection of purified antibody to produce rapid but temporary protection. Passive immunization

Table 1.   Examples of vaccines available for protection against bacterial diseases

Disease	Vaccine components
Diphtheria	Inactivated toxin
Tetanus	Inactivated toxin
Tuberculosis	Attenuated *Mycobacterium bovis* (BCG)
Whooping cough	Subcelluar fractions and pertussis toxoid
Haemophilus influenzae meningitis	Capsular polysaccharide linked to a protein carrier
Meningococcal meningitis	Capsular polysaccharide
Typhoid	Killed cells of *Salmonella typhi*
	Live oral attenuated strain Ty21A

is used to prevent a disease after known exposure, improve the symptoms of an ongoing disease, protect immunosuppressed patients, block the action of bacterial toxins and prevent disease. Some of the currently available vaccines against Bacterial diseases are indicated in *Table 1*.

Antibiotics

Molecules that inhibit the growth of, or kill, Bacteria are called antibacterial agents. These compounds are normally isolated from microbes and are called antibiotics. There are many millions of antibiotics, produced mainly by soil microorganisms that are active against Bacteria but only a few can be used to control or treat human Bacterial diseases. These are the few that, although they are toxic to Bacteria, have no significant toxic effects on the human host. The reason for this so called selective toxicity is that the site at which these antibiotics act are either unique to Bacteria (such as peptidoglycans) or very different between prokaryotes and eukaryotes (such as ribosomes and nucleic acid synthesis). *Table 2* shows the sites in the bacterial cell at which a number of clinically important antibiotics act. Some of the factors that affect antibiotic therapy are listed below:

- The antibiotic must reach the site of infection in the host.
- The antibiotic has to reach its target site. This is easier for antibiotics such as penicillin which act on peptidoglycan than for those like tetracycline which must penetrate through the plasma membrane to reach their target sites, the ribosomes.
- Gram negative Bacteria are often intrinsically resistant to the action of antibiotics due to the presence of the outer membrane, which acts as an additional barrier for the antibiotic to cross and protects the peptidoglycan.

Table 2.   Target sites for antibiotics in bacterial cells

Target site	Mode of action	Example
Cell wall (peptidoglycan biosynthesis)	Inhibition of cross-linking	Penicillin, cephalosporins (β-lactam antibiotics)
	Inhibition of polymerization	Glycopeptide antibiotics (e.g. vancomycin)
Protein synthesis	Inhibition of translocation of ribosome	Aminoglycoside antibiotics (e.g. gentamicin)
	Inhibition of binding of aminoacyl tRNAs	Tetracycline
Nucleic acid synthesis	Inhibition of tetrahydrofolic acid synthesis	Sulfonamides and trimethoprim
	Inhibition of DNA gyrase	Quinolone antibiotics (e.g.ciprofloxacin)

Table 3. Common mechanisms of antibiotic resistance in bacteria

Mechanism of resistance	Mode of action	Example
Antibiotic inactivation	β-Lactamase	Penicillin resistance
	Chloramphenicol acetyl transfease	Chloramphenicol resistance
	Aminoglycoside modifying enzymes	Aminoglycoside resistance
Reduction in permeability	Reduced uptake	Natural resistance of many Gram negative bacteria due to presence of outer membrane
	Antibiotic efflux	Tetracycline resistance
Alteration of target site	Change in target site so that it is no longer sensitive to drug	Sulfonamide resistance
	New target site produced that is not sensitive to the antibiotic	Methicillin resistance
	Overproduction of target site	Trimethoprim resistance

- Broad-spectrum antibiotics are effective against a wide range of different Gram positive Bacteria whereas other antibiotics may have only a narrow range.
- All the pathogenic Bacteria must be eradicated from the host by either inhibiting the growth of the microbe (bacteriostatic antibiotics), which can then be removed by the immune system or by killing them directly (bactericidal antibiotics).

Antibiotics have proved to be of great benefit to human kind and have ensured that people no longer need to die from diseases such as wound infections. However, recent studies have clearly shown that prokaryotes can become resistant to the action of antibiotics. The mechanisms by which they do this include the production of enzymes that break down the antibiotic, reduce the permeability to the antibiotic, and alterations to the target site as shown in *Table 3*. Antibiotic resistance may arise by mutation but more often the genes for antibiotic resistance are transferred between bacteria by conjugation, transduction and transformation. Antibiotic resistance carried by plasmids has caused particular concern as these plasmids may carry genes that confer resistance to many different antibiotics at the same time. Multiple-resistant Bacteria are therefore becoming a problem, particularly in the hospital environment and the fear is that it will not be long before there is a bacterial strain that is untreatable by all known antibiotics. Thus there is a clear need to identify novel targets in Bacterial pathogens as well as seek alternative approaches to develop preventative and therapeutic approaches.

H9 BACTERIA AS BIOLOGICAL WEAPONS

Key Notes

History	The use of biological agents as warfare tools is not novel and has been used extensively to add to human misery. At least in the recent history, British forces provided smallpox-infected blankets to American Indians, Germans used biological agents in World War I, and Japan used them in World War II, killing hundreds of thousands of people for narrow political gains. Despite efforts to quell this devastating research, the threat of germs as bioterror agents has never been greater.
Agents for bioweapons	At present, the most devastating agents of biowarfare are bacteria and their toxins including *Bacillus anthracis* (anthrax), *Yersinia pestis* (plague), *Francisella tularensis* (tularemia), *Clostridium botulinum* toxin (botulism) and viruses including Variola major virus (smallpox), viruses causing hemorrhagic fevers such as Ebola, Marburg, Lassa, Hanta, Congo, South American, Rift valley, Tick-borne and Yellow fever. Other agents (second category) are bacteria including *Coxiella burnetti* (Q fever), *Brucella* species (brucellosis), *Burkholderia mallei* (glanders), Epsilon toxin of *Clostridium perfringens*, *Staphylococcus* enterotoxin B. Foodborne or waterborne agents also are included, such as pathogenic *Salmonella* spp., *Shigella* spp., *Escherichia coli* spp., *Vibrio cholerae*; and viruses Alphaviruses (Venezuelan encephalomyelitis and eastern and western equine encephalomyelitis).
Biodefense	These include preventative measures such as international coopration and diplomacy, enforcing non-proliferation of bioweapons treaties and the availability of vaccines against biological agents. The therapeutic measures include rapid diagnostic tests, the availability of antibiotics and antivirals against bioweapons and infrastructure for environmental cleanup.

The use of germs to kill civilians is bioterrorism and their use to kill enemy forces is biowarfare. It has become clear that bioweapons are a potential threat to human kind. The most worrisome aspect is that unlike nuclear or chemical weapons, military arsenal, high-tech machinery, which are normally state-sponsored, bioweapons can be used with minimal resources by states or even by individuals.

History

The use of germs as warfare agents is nothing novel. This method of killing enemy forces or civilians has been used throughout history. For example, in 184 BC, Hannibal ordered serpent-filled pots to be thrown onto the decks of enemy ships. In the 1340s, the Tattar army used plague-infected people to spread infection to aid in their conquest. In the 1760s, the British army provided the American Indians with smallpox-infected blankets. More recently, in World War I, Germany used pathogens as agents of germ warfare in Europe. Following World War I, the biggest players in bioweapons were the U.S.A., Japan, U.S.S.R., Germany, U.K., and Canada but their reported use is unclear. It is well-established that in World War II, Japan used bioweapons (plague, typhoid, cholera, anthrax, and other agents) against the Chinese and the Soviets. During World War II, England conducted experiments on the use of bioweapons (*Bacillus anthracis*) on Gruinard Island (near the coast of Scotland). This resulted in heavy contamination of the Island, which only recently (1986) has been decontaminated with seawater and formaldehyde. Iraq's pursuits of bioweapons lasted from 1975 to 2003, while South Africa had a limited program for bioweapons from 1981 to 1993. At present, it is unclear how many nations (probably around a dozen) are actively seeking the most sophisticated bioweapons, but what is clear is that bioweapons present a clear and present danger.

Agents for bioweapons

Despite the clear devastating effects of bioweapons on humans, animals, plants and the environment, some continued to build biological weapons or pursued research to construct the most lethal microbes/toxins. As the danger of their use is real, there is a clear need to identify microbial agents/toxins that can be used as biological weapons, their mode of action and be prepared in the unfortunate event of a bioterror attack. Potentially any microbe or microbe-derived toxin capable of causing harm or human misery is a bioweapon agent. Among others, smallpox virus, anthrax and botulism toxins are considered the most significant bioweapons. To this end, the Center for Disease Control and Prevention (CDC) in Atlanta, GA, U.S.A. published a list of critical biological agents in 2000 and divided them into three categories (*Fig. 1*). The most likely entry of these pathogens/toxins is inhalation into the lungs, ingestion of contaminated food or water or absorption of toxins through the skin, presenting an easy mode of transmission to a large number of people. Of interest, the category A agents are most effectively transmitted via aerosols and are the most devastating bioweapons. In addition, our ability to genetically modify organisms may present serious consequences. For example, it is shown that genetic modification of mousepox virus can have fatal consequences for mice, which are normally resistant to this virus. Among category A, anthrax has gained the most attention in recent years. This is due to anthrax (*Bacillus anthracis*) spores enclosed in letters and mailed to U.S. media and government officials in 2001 which resulted in several fatalities. The outcome confirmed the dangers of bioweapons but more importantly the simplicity in their use and dissemination. The potential of anthrax as a bioweapon is due to its ability to readily form spores which can be stored indefinitely and can survive in the dry air for several hours. They are Gram positive Bacteria and can cause serious gut (gastrointestinal anthrax) or skin infections (cutaneous anthrax) but if inhaled, they can produce inhalational anthrax with fatal consequences (mortality rate of more than 90%). The incubation period varies from a few hours to a few days. The clinical symptoms include fever, malaise, fatigue, respiratory failure, septicaemia, and finally death within 24 hours (even with treatment). Because anthrax is non-contagious in humans, their effects are limited to the applied area. In contrast, the smallpox virus is highly contagious and can result in global pandemics with a mortality risk of 30%. *Bacillus anthracis* is normally present in soil worldwide, and there are approximately 200–2000 annual cases of anthrax (mostly cutaneous). It is the inhalation anthrax that is most associated with bioterror and has the potential of a bioweapon.

Category A
- Easily disseminated or transmitted person-to-person
- Cause high mortality
- Require special action for public health preparedness

Bacteria
- *Bacillus anthracis* (anthrax)
- *Yersinia pestis* (plague)
- *Francisella tularensis* (tularemia)
- *Clostridium botulinum* toxin (botulism)

Viruses
- Variola major virus (smallpox)
- Viruses causing hemorrhagic fevers such as Ebola, Marburg, Lassa, Hanta, Congo, South American, Rift valley, Tick-borne and Yellow fever

Category B
- Moderately easy to disseminate
- Cause moderate morbidity and low mortality
- Require special enhancements of diagnostic capacity and disease surveillance

Bacteria
- *Coxiella burnetti* (Q fever)
- *Brucella* species (brucellosis)
- *Burkholderia mallei* (glanders)
- Epsilon toxin of *Clostridium perfringens*
- *Staphylococcus* enterotoxin B
- Foodborne or waterborne agents also are included, such as pathogenic *Salmonella* spp., *Shigella* spp., *Escherichia coli* spp., *Vibrio cholerae*

Viruses
- Alphaviruses (Venezuelan encephalomyelitis and eastern and western equine encephalomyelitis)

Category C
- Pathogens that could be engineered for mass dissemination in the future because of: availability, ease of production and dissemination
- Potential for high morbidity and mortality and major health impact

Bacteria
- Multidrug-resistant *Mycobacterium tuberculosis*

Viruses
- Nipah virus
- Hantaviruses
- Tickborne hemorrhagic fever viruses
- Tickborne encephalitis viruses, Yellow fever virus

Fig. 1.   List of biological agents that can be used as bioweapons (source: www.cdc.gov).

Biodefense

Politicans aided by scientists developed bioweapons. Similarly, in defending civilization, politicians with the help of scientists must actively and urgently devise plans for eliminating this potential threat. The approaches to defending against bioweapons can be divided into two groups.

Preventative measures

These include international cooperation, monitoring all states for biological warfare programs, enforcement of non-proliferation of bioweapons treaties for all nations (including Superpowers) and the availability of vaccines against potential bioweapon agents. These approaches urgently require international diplomacy and cooperation, proactive role of independent monitoring agencies and establishments of international enforcement agencies. As with other warfare, the justification of biological warfare programs is bizarre. For example, the only purpose a bullet will serve is to kill, whether a friend or foe. The justification of making a bullet is in the eyes of the beholder, whether defense or offense. What we all must understand is that bioweapons are a common threat to human kind and need not be added on the extensive list of weapons at our disposal to kill our fellow human beings for narrow political gains. Other more applicable preventative approaches include vaccines. Vaccines are useful tools against bioweapons and entire populations must be vaccinated, but they have no value as therapeutic measures. In addition, vaccines may have potential side effects and may require many doses. For example, the present anthrax vaccine requires six shots over a period of two years. Also, the genetic manipulation of the infectious agents or use of different strains of agents may render the vaccine obsolete.

Therapeutic measures

The success of therapeutic approaches relies on the availability of rapid diagnostic tests. Accurate early detection followed by appropriate treatment, including the use of antibiotics, antiviral, and antibodies for passive protection may have protective effects. Antibiotics and antivirals have been the most effective mode of treating infectious agents. But again, with the advances in genetic approaches, the development of antimicrobial-resistant strains is relatively simple and an unfortunate opportunity to construct more deadly pathogens. Passive protection using antibodies is an effective therapeutic approach, especially in neutralizing toxins such as botulism toxin, ricin (made from waste of processed castor beans), epsilon toxin of *Clostridium perfringens*, Staphylococcal enterotoxin B. However, the use of different strains is a potential hurdle in our preparedness against bioweapons. Any bioterror attack will have serious implications on the environment. The need to develop measures for environmental cleanup is urgent. At present, we have limited if any environmentally friendly disinfectants or measures to assess the tolerable levels of contamination or infrastructure to perform such massive tasks. Future work must involve international diplomacy and cooperation, the enforcement of non-proliferation of bioweapons treaties for all nations, identification of novel antimicrobials, research in the understanding of the pathogenesis and pathophysiology of diseases caused by bioweapons, environmental assessment and cleanup. These measures may aid in our ability to combat microbial monsters created for our narrow political gains.

I1 TAXONOMY

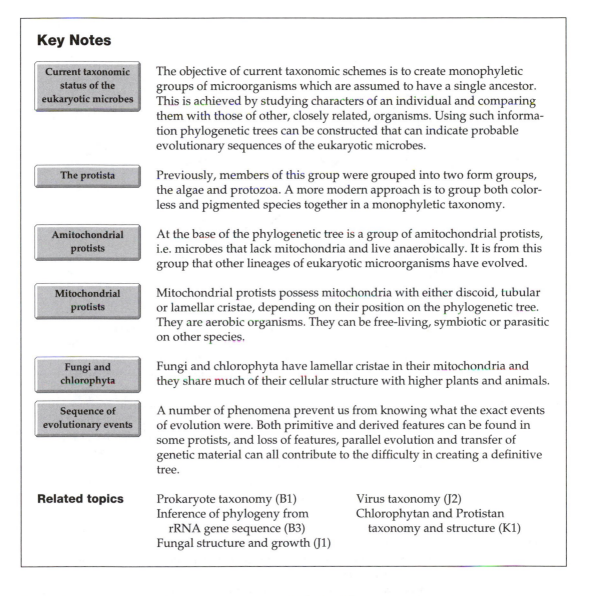

Key Notes

Current taxonomic status of the eukaryotic microbes

The objective of current taxonomic schemes is to create monophyletic groups of microorganisms which are assumed to have a single ancestor. This is achieved by studying characters of an individual and comparing them with those of other, closely related, organisms. Using such information phylogenetic trees can be constructed that can indicate probable evolutionary sequences of the eukaryotic microbes.

The protista

Previously, members of this group were grouped into two form groups, the algae and protozoa. A more modern approach is to group both colorless and pigmented species together in a monophyletic taxonomy.

Amitochondrial protists

At the base of the phylogenetic tree is a group of amitochondrial protists, i.e. microbes that lack mitochondria and live anaerobically. It is from this group that other lineages of eukaryotic microorganisms have evolved.

Mitochondrial protists

Mitochondrial protists possess mitochondria with either discoid, tubular or lamellar cristae, depending on their position on the phylogenetic tree. They are aerobic organisms. They can be free-living, symbiotic or parasitic on other species.

Fungi and chlorophyta

Fungi and chlorophyta have lamellar cristae in their mitochondria and they share much of their cellular structure with higher plants and animals.

Sequence of evolutionary events

A number of phenomena prevent us from knowing what the exact events of evolution were. Both primitive and derived features can be found in some protists, and loss of features, parallel evolution and transfer of genetic material can all contribute to the difficulty in creating a definitive tree.

Related topics

Prokaryote taxonomy (B1)
Inference of phylogeny from rRNA gene sequence (B3)
Fungal structure and growth (J1)

Virus taxonomy (J2)
Chlorophytan and Protistan taxonomy and structure (K1)

Current taxonomic status of the eukaryotic microbes

Establishing relationships within the different members of the fungi, chlorophyta and protistan microbes relies on studying **characters**, which are features or attributes of an individual organism that can be used to compare it with another organism. These features can be morphological, anatomical, ultrastructural, biochemical or based on sequences of nucleic acids. The objective of such a study is to create **monophyletic** groups which are assumed to have a single ancestor, usually extinct; a similar approach is used in the creation of classification systems for prokaryotes see Section B1. A **cladistic** approach would not assume features from an ancestor, but would merely define a monophyletic group on the basis

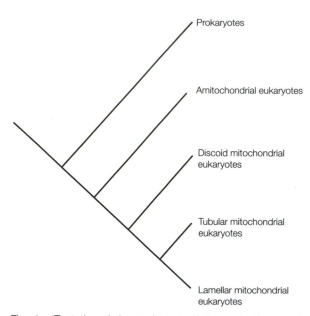

Prokaryotes

Amitochondrial eukaryotes

Discoid mitochondrial
eukaryotes

Tubular mitochondrial
eukaryotes

Lamellar mitochondrial
eukaryotes

Fig. 1.   Tentative phylogenetic tree of the probable evolutionary sequence of protistan microbes.

of shared characters. New information based on the presence and type of mitochondria, and the DNA sequencing of ribosomal RNA, place the fungal, chlorophytan and protistan members of the eukaryotic microbes into a complex **phylogenetic tree** (*Fig. 1*). For more details of these trees, see Section B3.

The protista

The protista are a paraphyletic group of organisms that are not animals, true fungi or green plants. There are approximately 200 000 named species. Traditionally they were classified into non-monophyletic, adaptive groups called the flagellates, algae and protozoa. There are about 80 different patterns of organization, clustered into 60 lineages.

In this book we have followed a classification based on current information published on the Tree of Life website.

Amitochondrial protists

At the base of the tree are amitochondrial protists, which separated from other lineages before mitochondria were acquired (see Section K1). These are anaerobic organisms that may inhabit sediments or animal intestines or they can be animal parasites. These organisms most strongly resemble the colorless, **anaerobic** cell from which, by symbiotic acquisition of mitochondria and in some cases chloroplasts, the rest of the lineages developed.

Mitochondrial protists

The next group in the protistan lineage are the **euglenozoa**, those organisms that have mitochondria with **discoid cristae**. These organisms are **aerobes**. The euglenids, ameboflagellates and acrasid slime molds share these characteristics (see Section K1). Many members of this group have two flagellae. Some are free-living while others are parasites within animals and man.

Organisms having mitochondria with **tubular** cristae, the **alveolata** appear next on the phylogenetic tree, and within this group we see the myxomycete and dictyosteloid slime molds, dinoflagellates, ciliates, apicomplexans, diatoms, oomycetes, radiolaria and heliozoans.

The fungi and chlorophyta

At the top of the tree are organisms that have mitochondria with **lamellar** cristae, the chlorophyta, chytrids, higher fungi, plants and animals (see Sections J1 and K1).

Sequence of evolutionary events

Exact orders of evolution are difficult to determine, because of morphological inconsistencies within groups. For instance, dinoflagellates have a number of less specialized features, but they also have more derived features including being biflagellate and having scales, a characteristic which brings them close to the alveolata (see Section K1).

Even sequencing of nucleic acid to provide phylogenetic **gene trees** may not answer all the questions. For instance, sequencing the small subunit nuclear-encoded rRNA reveals that one can follow the evolution of an individual gene, but not necessarily the whole organism, as different regions of rRNA genes evolve at different rates. Phylogenetic patterns may also be confused because of transfer of genetic information between lineages as a consequence of **endosymbiosis** and other mechanisms. **Parallel** evolution within these organisms is also likely to occur, and superficial similarities within distinct lineages may arise because of loss of features, which is likely to be a significant factor within these apparently simple organisms.

12 EUKARYOTIC CELL STRUCTURE

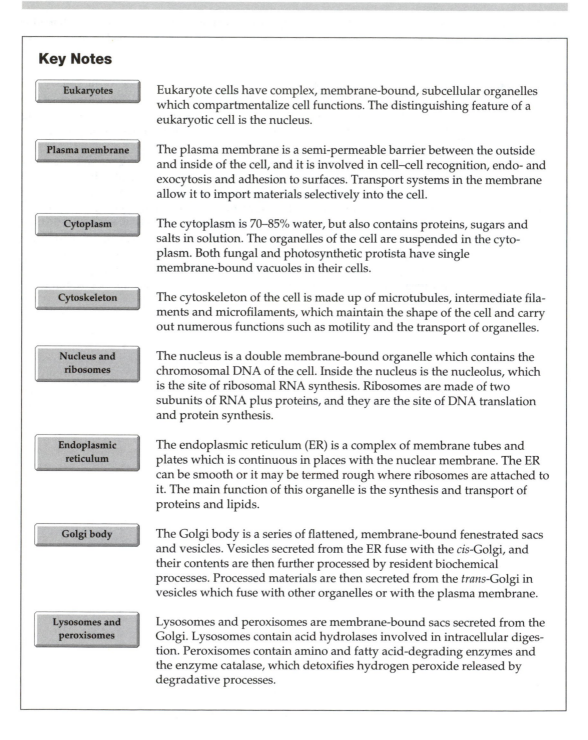

Key Notes

Eukaryotes
Eukaryote cells have complex, membrane-bound, subcellular organelles which compartmentalize cell functions. The distinguishing feature of a eukaryotic cell is the nucleus.

Plasma membrane
The plasma membrane is a semi-permeable barrier between the outside and inside of the cell, and it is involved in cell–cell recognition, endo- and exocytosis and adhesion to surfaces. Transport systems in the membrane allow it to import materials selectively into the cell.

Cytoplasm
The cytoplasm is 70–85% water, but also contains proteins, sugars and salts in solution. The organelles of the cell are suspended in the cytoplasm. Both fungal and photosynthetic protista have single membrane-bound vacuoles in their cells.

Cytoskeleton
The cytoskeleton of the cell is made up of microtubules, intermediate filaments and microfilaments, which maintain the shape of the cell and carry out numerous functions such as motility and the transport of organelles.

Nucleus and ribosomes
The nucleus is a double membrane-bound organelle which contains the chromosomal DNA of the cell. Inside the nucleus is the nucleolus, which is the site of ribosomal RNA synthesis. Ribosomes are made of two subunits of RNA plus proteins, and they are the site of DNA translation and protein synthesis.

Endoplasmic reticulum
The endoplasmic reticulum (ER) is a complex of membrane tubes and plates which is continuous in places with the nuclear membrane. The ER can be smooth or it may be termed rough where ribosomes are attached to it. The main function of this organelle is the synthesis and transport of proteins and lipids.

Golgi body
The Golgi body is a series of flattened, membrane-bound fenestrated sacs and vesicles. Vesicles secreted from the ER fuse with the *cis*-Golgi, and their contents are then further processed by resident biochemical processes. Processed materials are then secreted from the *trans*-Golgi in vesicles which fuse with other organelles or with the plasma membrane.

Lysosomes and peroxisomes
Lysosomes and peroxisomes are membrane-bound sacs secreted from the Golgi. Lysosomes contain acid hydrolases involved in intracellular digestion. Peroxisomes contain amino and fatty acid-degrading enzymes and the enzyme catalase, which detoxifies hydrogen peroxide released by degradative processes.

Mitochondria	Mitochondria are the site of respiration and oxidative phosphorylation in aerobic organisms. They are bound by a double membrane, the inner one being in-folded to form plates or tubes called cristae. ATP production is located in particles attached to the cristae.
Hydrogenosomes	Hydrogenosomes are organelles found in some amitochondrial, anaerobic groups. Their function is energy production. They contain enzymes of electron transport which use terminal electron acceptors that generate hydrogen.
Glycosomes	Glycosomes contain the enzymes of glycolysis and are found only in the apicomplexa.
Chloroplasts	Chloroplasts are double membrane-bound organelles which contain the photosynthetic pigment chlorophyll. Within the chloroplast are stacks of flattened sacs called thylakoids where the photosynthetic systems are located.
Cell walls	Cell walls are found in the photosynthetic protista (cellulose-based) and the fungi (chitin-based). They delimit the outside of the cell from the environment and are important in maintaining cell rigidity and controlling excess of water influx due to osmosis.
Flagella	Flagella are microtubule-containing extensions of the cell membrane. They provide the cell with motility by their flexuous bending, which is controlled by the microtubule motor protein dynein.
Cilia	Cilia have the same internal structure as flagellae but they are smaller and more numerous. Co-ordinated movement, which can be seen as wave-like beating, is required for motility.
Contractile vacuoles	Contractile vacuoles are found in free-living freshwater protozoa. They expel water absorbed into the cell by osmosis through pores in the cell surface.

Related topics

The microbial world (A1)
Prokaryote cell structure (C7)
Bacterial cell envelope and cell wall synthesis (C8)
Heterotrophic pathways (E1)
Electron transport, oxidative phosphorylation and β-oxidation of fatty acids (E3)

Autotrophic reactions (E4)
DNA replication (F5)
Transcription (F6)
Translation (F8)
Taxonomy (I1)
Cell division and ploidy (I3)
Chlorophytan and protistan taxonomy and structure (K1)

Eukaryotes

Eukaryotic cells are compartmentalized by **membranes**. The cell contains several different types of membrane-bound organelle in which different biochemical and physiological processes can occur in a regulated way (*Fig. 1*). Membranes also transport information, metabolic intermediates and end-products from the site of biosynthesis to the site of use.

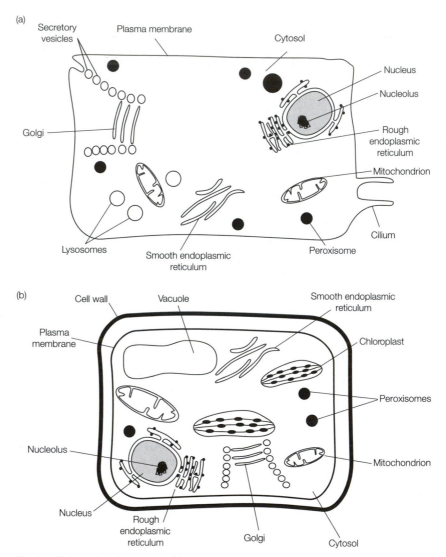

Fig. 1.   *Eukaryotic cell structure: (a) structure of a typical animal cell, (b) structure of a typical plant cell.*

Plasma membrane

The plasma membrane of eukaryotes is a **semi-permeable** barrier that forms the boundary between the outside and the inside of the cell. It is similar to that of the prokaryotes (see Section C7 and C8), except that it contains **sterols**, flat molecules that give the membrane a greater rigidity and which stabilize the eukaryotic cell. There are transport systems in the membrane that selectively import materials into the cell, and it is also involved in **endo-** and **exo-cytosis**, where food particles are engulfed and waste products are expelled from the cell in membrane-bound vesicles. The plasma membrane is involved in key interactive processes between cells, like **cell–cell recognition** systems, as well as adhesion of the cell to solid surfaces.

Cytoplasm

The cytoplasm is a dilute solution (70–85% water) of proteins, sugars and salts in which all other organelles are suspended. It has **sol–gel** properties, which means it can be liquid or semi-solid depending on its molecular organization. Vacuoles act as storage sites for nutrients and waste products in a very weak solution. The high water content of the vacuole maintains a high cell turgor pressure.

Cytoskeleton

The eukaryotic cell is further stabilized by a cytoskeleton made up of **tubulin** containing **microtubules** (25 nm diameter), and **microfilaments** (4–7 nm diameter) and **intermediate fibers** (8–10 nm diameter). The cytoskeleton is a dynamic structure, providing support for the cell but also the machinery for ameboid movement, cytoplasmic streaming and nuclear and cell division.

Nucleus and ribosomes

The nucleus contains the DNA of the microbe. In eukaryotic microbes the nucleus usually contains more than one **chromosome**, in which the DNA is protected by **histone** proteins. In diploid organisms these chromosomes are paired. However, in some protista there are two distinct types of nuclei within one cell, and they may be polyploid. The larger of the two nuclei are termed **macronuclei** and this nucleus is associated with cellular function. The smaller **micronucleus** functions in controlling reproduction.

The nucleus is surrounded by the **nuclear membrane**, a double membrane which is perforated by many pores where the two membranes fuse. It is via these pores that the nucleus remains in constant control of the rest of the cell machinery via **mRNA** and **ribosomes**. In places the nuclear membrane is also continuous with the endoplasmic reticulum. Within the nucleus is the **nucleolus**, an RNA-rich area where rRNA is synthesized (see Sections F6 and F8 for DNA transcription and translation). Eukaryotic ribosomes are essentially very similar to those of prokaryotes (see Section F8) but they are slightly larger, their two subunits are of 60S and 40S, making a dimer of 80S. Their function is the same as that of prokaryotes (see Section F8).

Endoplasmic reticulum

The outer membrane of the nucleus is in places continuous with a complex, three-dimensional array of membrane tubes and sheets, the endoplasmic reticulum (ER). Tubular ER can be studded with ribosomes, and described as **rough ER** (**RER**), where **ribosomal translation** and **protein modification** takes place (see Section F8). These proteins can either be secreted into the lumen of the ER or inserted into the membrane. Plates of **smooth ER** are associated with **lipid synthesis** and **protein** and **lipid transport** across cells.

Golgi

The Golgi is composed of stacks of a flattened series of membrane-bound sacs or **cisternae**, surrounded by a complex of tubes and vesicles. There is a definite **polarity** across the stack, the *cis* or forming face receiving vesicles from the ER, the contents of these vesicles then being processed by the Golgi, to be budded from the sides or the *trans* (maturing) face of the organelle. The Golgi apparatus processes and packages materials for secretion into other subcellular organelles or from the cell membrane. Golgi in fungi are less well developed than in algae, and tend to have fewer or single cisternae. They are sometimes termed **dictyosomes.**

Lysosomes and peroxisomes

The Golgi body generates these single-membrane-bound organelles which contain enzymes (**acid hydrolases** in the lysosome, **aminases, amidases** and lipases plus catalase in the peroxisomes) needed in the digestion of many different macromolecules. The internal pH of the lysosome is **acidic** (pH 3.5–5) to enable the enzymes to work at the optimum pH, and this pH is maintained by proton pumps present on the membrane. The breakdown of amino and fatty acids by the peroxisomes generates **hydrogen peroxide**, a potentially cytotoxic by-product. The enzyme **catalase**, also present in the peroxisome, degrades the peroxide into water and oxygen, protecting the cell.

Mitochondria

Mitochondria are double-membrane-bound organelles where the processes of **respiration** and **oxidative phosphorylation** occur (see Sections E2 and E3). They are approximately 2–3 mm long and 1 mm in diameter. Their numbers in a cell vary. They contain a small, circular DNA molecule which encodes some of the mitochondrial proteins, and 70S ribosomes (see Sections F7 and F8). The inner membranes of the mitochondria contain an **ATP/ADP transporter** that moves the ATP, which is synthesized in the organelle, outwards into the cytoplasm. ATP production is located on particles attached to the cristae, the inner infolded mitochondrial membrane (*Fig. 2*). Their structures differ slightly between the three protistan groups. Not all protista have mitochondria (see Section K2) and in these cells metabolism is essentially anaerobic (see Section K4). In the aerobic, very primitive eukaryotes' mitochondrial **cristae** are discoid. Mitochondria of the fungi are large and highly lobed with flat, plate-like cristae, while those of the chlorophyta have much more inflated cristae.

Hydrogenosomes

Hydrogenosomes are unique organelles found in anaerobic protista that lack mitochondria. They are membrane-bound organelles containing **electron-transport pathways** in which **hydrogenase** enzymes transfer electrons to terminal electron acceptors which generate molecular hydrogen (see Section K4) and ATP.

Glycosomes

Glycosomes are unique to the protistan group Apicomplexa. This organelle is surrounded by a single unit membrane and contains the **enzymes of glycolysis** (see Topic B1).

Chloroplasts

Chloroplasts are **chlorophyll**-containing organelles that can use light energy to fix carbon dioxide into carbohydrates (**photosynthesis**) (see Section E4 and K4). They are bound by double membranes and contain flattened membrane sacs called **thylakoids** where the light reaction of photosynthesis takes place (*Fig. 3*). In photosynthetic protista and the chlorophyta these organelles are large, almost filling the cell. The **pyrenoid** is a proteinaceous region within the chloroplast where polysaccharide biosynthesis takes place.

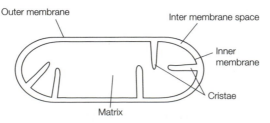

Fig. 2.   Structure of a mitochondrion.

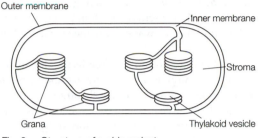

Fig. 3.   Structure of a chloroplast.

Cell walls

The protoplasts of fungal and photosynthetic protistan cells are in most cases surrounded by rigid cell walls. In the fungi the cell wall is composed of a microcrystalline polymer of **chitin** (repeating units of β1–4 linked NAG) and amorphous β-glucans, while in the chlorophyta and some other photosynthetic protista, cell walls are composed of **cellulose** (repeating units of β1–4 linked glucose) and **hemicelluloses** (see Sections C8 and K1).

Flagella

Flagellae are membrane-bound extensions of the cell, which contain microtubules (see Section K2). The microtubules are arranged as a bundle of nine doublets around the periphery of the flagellum, with a pair of single microtubules running within them. This structure is called the **axoneme**. Flagella provide cells with motility, because they flex and bend when supplied with ATP. Each outer pair of microtubules has arms projecting towards a neighboring doublet (*Fig. 4*) and a spoke extending to the inner pair of microtubules. Microtubules are formed from **tubulin,** a self-assembling protein. Tubulin is composed of two subunits α and β arranged in a helical fashion. The projecting arms between outer subunits are made up of the protein, **dyenin**. This protein is involved in converting the energy released from ATP hydrolysis into mechanical energy for flagellar movement. Movement is produced by the interaction of the dyenin arms with one of the microtubules of adjacent doublets. A basal body (**kinetosome**) anchors the flagellum within the cytoplasm.

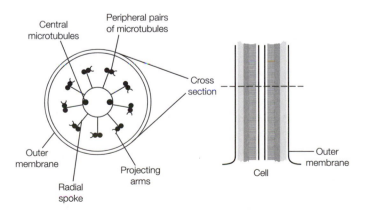

Fig. 4.   General structure of a flagellum or cilium.

Cilia

Cilia have the same internal structure as flagellae but they are typically shorter in length. They are usually present on a cell in great numbers and propel the cell by co-ordinated beating seen as waves over the surface of the organism.

Contractile vacuoles

Contractile vacuoles are found in free-living freshwater non-photosynthetic protistans. Their function is to regulate osmotic pressure within the cell by expelling water from a central vacuole through a pore in the outer surface. The simplest contractile vacuoles consist of a vacuole that can form anywhere in the cell, to a fixed structure that is surrounded by bands of microtubules and surrounded by collecting canals that collect fluid from the cytoplasm (*Fig. 5*).

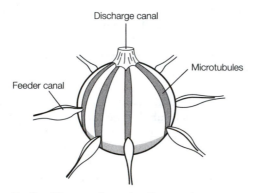

Fig. 5.   Diagram of a contractile vacuole.

I3 CELL DIVISION AND PLOIDY

Key Notes

Cell cycle	The cell cycle describes all the events that occur in a cell from the end of one cell division to the end of the next. There are four phases in the eukaryotic cell cycle, which include cell growth (G1), DNA synthesis (S), a second gap or growth phase (G2) and finally nuclear division (N).
Mitosis and asexual cell division	Mitosis is nuclear division which results in progeny nuclei that are identical to the parent. It is usually followed by cytokinesis, cell division, which produces cells that have the same phenotype as the parent. In some eukaryotic microbes multiple nuclear divisions may occur without cytokinesis, giving rise to large, multinucleate cells. There are four stages of mitosis, prophase, metaphase, anaphase and telophase. During prophase chromosomes duplicate to form chromatids, joined at the centromeres. Centromeres are attached to spindle microtubules. During metaphase, chromatids are arranged across center of the cell, and during anaphase the microtubules pull the sister apart to the poles of the dividing cell. During telophase the microtubules disappear and the nuclear envelopes fuse.
Meiosis and sexual cell division	Meiosis is a nuclear division where there is a halving of chromosome numbers from a diploid to a haploid state. In organisms with an extended diploid phase of their life cycle, meiosis produces haploid gametes, and it is immediately followed by gamete fusion and formation of a new diploid organism. In organisms with an extended haploid stage, diploid formation is immediately followed by meiosis and produces the new haploid organism. There are eight stages to meiosis; prophase 1, metaphase 1, anaphase 1 and telophase 1, and prophase 2, metaphase 2, anaphase 2 and telophase 2. During prophase 1, homologous chromosomes associate and duplicate. At this point there may be recombination. During metaphase 1, homologous chromosomes assemble on the spindle across the cell, and they are separated during anaphase 1 and telophase 1. Prophase and metaphase 2 are transient and the chromatids assemble across the cell to be separated during anaphase 2. During telophase 2 cell division usually occurs.
Chromosomes	Eukaryotic microorganisms package the large amount of DNA they contain into chromosomes. Chromosomes contain a single, linear double strand of DNA tightly bound with histone proteins. There are usually between four and eight chromosomes per cell.
Histones	Histones are basic proteins that bind to DNA to condense and fold it. They are vital to the structure of DNA and have been highly conserved during evolution.

Related topics	Genomes (F2)	Recombination (F13)
	DNA replication (F5)	Reproduction in fungi (J3)
	Cell Division (C9)	Life cycles in the Chlorophyta and
		Protista (K3)

Cell cycle

The cell cycle consists of an **interphase** during which growth of the cell (G_1) occurs, the cell increases in volume to maximum size and there is synthesis of cytoplasmic constituents and RNA (*Fig. 1*). Synthesis of DNA occurs next, during the **S** phase, as chromosomes are duplicated in preparation for nuclear division. During the final part of the cell cycle a second gap or growth phase occurs (G_2), when specific cell division-related proteins are synthesized. Nuclear division (**N**) then follows to complete the cell cycle.

Mitosis and asexual cell division

Asexual cell division in unicellular eukaryotic organisms is synonymous with growth. Division is usually by **binary fission** after a single nuclear division. The parent cell divides, usually longitudinally, into two even-sized identical progeny cells. Division can be by **multiple fission** (see Section K5), where many nuclear divisions occur, producing either a large multinucleate **coenocyte** or many uninucleate progeny cells. All cells produced from mitosis are genetically identical to their parent.

In both cases **somatic** cell division is preceded by the mitotic division of the cell nucleus. In mitosis the replicated DNA from the S phase of the life cycle is separated equally into two progeny cells. The events of mitosis can be separated into four stages for convenience, but each flows into the other as a continuous process.

The first phase in mitosis is the **prophase**. In this phase microtubules form from the **microtubule organizing centers** (**MTOCs**). In fungi these structures are known as **spindle pole bodies** (**SPBs**), located close to the nuclear envelope. In the motile species of protista the MTOC is called a **centriole** and it becomes surrounded by microtubules in a process termed **aster formation** (*Fig. 2a*). The MTOCs begin to move towards opposite poles of the nucleus, and spindle microtubules appear between them. Single chromosomes, which have been

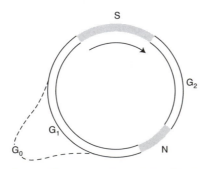

Fig. 1.   The eukaryotic cell cycle. The S phase is typically 6–8 h long, G2 is a phase in which the cell prepares for mitosis and lasts for 2–6 h, and nuclear division (N) takes only about 1 h. The length of G1 is very variable and depends on the cell type. Cells can enter G0, a quiescent phase, instead of continuing with the cell cycle. From Hames, B.D. et al., Instant Notes in Biochemistry. © BIOS Scientific Publishers Limited, 1997.

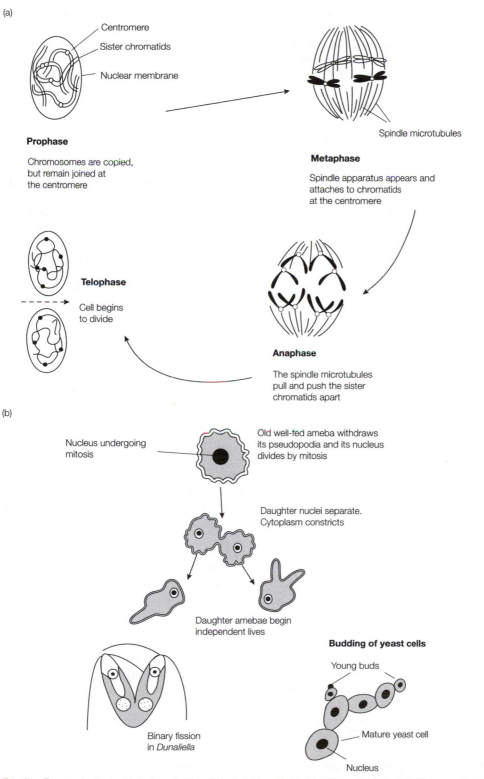

Fig. 2.   Events of mitosis: (a) nuclear division, (b) cell division. Reproduced from Beckett, B.S, Biology, with kind permission.

duplicated to form **chromatids**, are joined together at their **centromeres**. These centromeres are also attached to spindle microtubules. By late prophase–early **metaphase**, MTOCs are opposite each other and the spindle is complete, with chromosomes aligned across the center in a **metaphase plate**. In many species of fungi, chromosomes remain extremely indistinct during mitosis, and they do not appear to assemble across a metaphase plate.

The nuclear envelope may disappear between the prophase and the beginning of the metaphase in some eukaryotic microbes, but in some protista the nuclear envelope remains intact throughout the process, the microtubules of the spindle penetrating through it.

In the next phase of mitosis, the **anaphase**, pairs of chromatids that were held together at the centromere begin to separate simultaneously, and spindle microtubules begin to pull them towards the two poles of the cell. By the end of the anaphase the chromatids have been pulled close to the MTOCs. In many species of the fungi there is an asynchronous chromosome separation during the anaphase.

In the final phase of mitosis, the **telophase**, the aster microtubules disappear, and the nuclear envelope reforms if it has disintegrated. In the two progeny nuclei the MTOC duplicates, and **cell division** commences with the division of the cytoplasm by an invaginating plasma membrane or the formation of a **cell plate** by Golgi-derived vesicles across the midline between the two nuclei (*Fig. 2b*). In some protista separation can be by **budding**, producing a progeny cell that is much smaller than the parent.

Meiosis and sexual cell division

Most protista are **haploid** for most of their life cycle. They have only one set of chromosomes. The **diploid** stage (two sets of chromosomes) is often very transient and found only in resting structures such as spores. The life cycle ends with meiosis, which returns the new cell to its haploid state. In organisms with a dominant diploid vegetative phase, meiosis occurs just before cell division, producing haploid gametes which then fuse to reform the diploid.

At the end of the interphase and before meiosis begins, the duplication of chromosomes to chromatids occurs just as it does in mitosis. However, as the cell is diploid, **homologous** pairs of chromosomes associate during **prophase 1** (*Fig. 3*), and at this point it is possible for genetic recombination to occur (see Section F13).

During metaphase 1 chromosomes assemble across the metaphase plate, and in anaphase 1 homologous chromosomes are separated. Telophase 1 is very transient, and the chromosomes rapidly move into the prophase and metaphase 2, where pairs of chromatids assemble across the metaphase plate. The chromatids separate from each other at anaphase 2, and in telophase 2 cytoplasm begins to separate around the four progeny nuclei, each containing a haploid complement of chromosomes.

In some circumstances **multiple sets** of chromosomes can exist in a cell and this is termed **polyploidy**. Nuclear division in polyploid organisms is complex and often results in the loss of single chromosomes, leading to odd numbers of chromosomes in some progeny cells. This is termed **aneuploidy**. Polyploid and aneuploid cells are usually unable to participate in meiosis because of their odd chromosome numbers.

Chromosomes

Compared with the prokaryotes, eukaryotic microbes contain much more DNA, and therefore have had to evolve structures which pack, store and present DNA

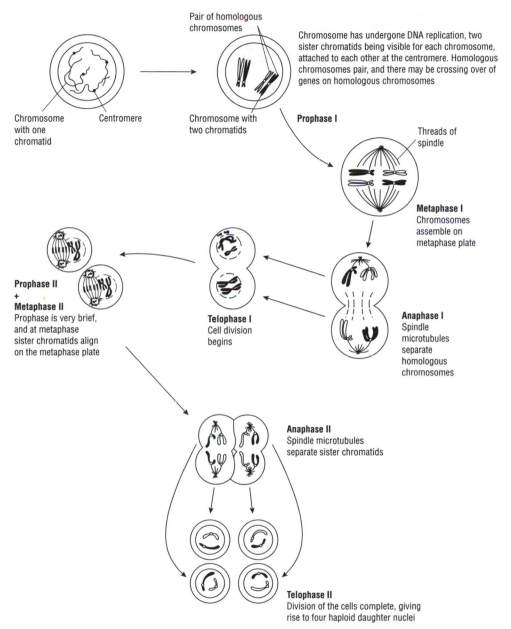

Pair of homologous chromosomes

Chromosome has undergone DNA replication, two sister chromatids being visible for each chromosome, attached to each other at the centromere. Homologous chromosomes pair, and there may be crossing over of genes on homologous chromosomes

Chromosome with one chromatid   Centromere

Chromosome with two chromatids

Prophase I

Threads of spindle

Metaphase I
Chromosomes assemble on metaphase plate

**Prophase II
+
Metaphase II**
Prophase is very brief, and at metaphase sister chromatids align on the metaphase plate

Telophase I
Cell division begins

Anaphase I
Spindle microtubules separate homologous chromosomes

Anaphase II
Spindle microtubules separate sister chromatids

Telophase II
Division of the cells complete, giving rise to four haploid daughter nuclei

Fig. 3. The events of meiosis. Redrawn from Gross, T., Faull, J.L., Kettridge, S. and Springham, D. Introductory Microbiology, 1995, Stanley Thornes Ltd., Cheltenham, UK.

during different parts of the cell cycle (see Section C9). There are usually between four and eight chromosomes per cell, but there can be more or less. Chromosomes contain a single linear double strand of DNA, tightly condensed with **histone** proteins. The combined DNA and histone is often termed **chromatin**. The histone proteins bind to the DNA and create three levels of folding. During the prophase, chromatin is very condensed but, in the interphase, chromatin is described as dispersed.

Histones   Histones are very basic proteins because they contain many basic amino acids (lysine and arginine), with positively charged side chains. These side chains associate with the negatively charged phosphate groups on the DNA molecule. The histones are vital to the structure of DNA and have been **highly conserved** during evolution (see Section B3).

J1 FUNGAL STRUCTURE AND GROWTH

Key Notes

Fungal structure	Fungi are heterotrophic, eukaryotic organisms with a filamentous, tubular structure, a single branch of which is called a hypha. A network of hyphae is called a mycelium. Hyphae are bound by firm, chitin-containing walls and contain most eukaryotic organelles. Not all fungi are multicellular, some are single-celled and are termed yeasts.
Fungal taxonomy	There are four phyla within the fungi, divided from each other on the basis of differences in their mechanisms of sexual reproduction. The four phyla are the Zygomycota, Chytridiomycota, Ascomycota and Basidiomycota. A fifth group exists which contains fungi where sexual reproduction is not known but where asexual reproduction is seen. These fungi are placed in the phylum Deuteromycota.
Fungal wall structure and growth	Fungal walls are formed from semi-crystalline chitin microfibrils embedded in an amorphous matrix of β-glucan. In the Ascomycota and Basidiomycota, hyphae grow by tip growth followed by septation. In the Chytridiomycota and Zygomycota fungal hyphae grow by tip growth but remain aseptate.
Colonial growth	Colonial growth is characterized by the radial extension of mycelium over and through a substrate, creating a circular or spherical fungal colony.
Kinetics of growth	Fungal growth can be measured by measuring mycelial mass changes with time under excess of nutrient conditions. From this information the specific growth rate can be calculated. After a lag phase, a brief period of exponential growth follows as hyphal tips are initiated. As the new hypha extends, it grows at a linear rate until nutrient depletion causes a retardation phase, followed by a stationary phase.
Hyphal growth unit	Hyphal growth may also be measured by microscopy and by counting the total numbers of hyphal tips, and dividing that number by the total length of mycelium in the colony, the average length of hypha required to support a growing tip can be calculated. This is termed the hyphal growth unit.
Peripheral growth zone	The peripheral growth zone is the region of mycelium behind the tip, which permits radial extension at a rate equal to the specific growth rate.
Related topics	Composition of typical prokaryotic cell (C7) Taxonomy (I1)
	Cell division (C9) Eukaryotic cell structure (I2)
	Measurement of microbial growth (D1) Reproduction in fungi (J3)
	Batch culture in the laboratory (D2) Chlorophytan and Protistan taxonomy and structure (K1)

Fungal structure   Fungi are filamentous, non-photosynthetic, eukaryotic microorganisms that have a **heterotrophic** nutrition (see Section E2). Their basic cellular unit is described as a **hypha** (*Fig. 1*). This is a tubular compartment which is surrounded by a rigid, **chitin**-containing wall. The hypha extends by tip growth, and multiplies by branching, creating a fine network called a **mycelium**. Hyphae contain nuclei, mitochondria, ribosomes, Golgi and membrane-bound vesicles within a plasma–membrane bound cytoplasm (see Section I2). The subcellular structures are supported and organized by microtubules and endoplasmic reticulum. The cytoplasmic contents of the hypha tend to be concentrated towards the growing tip. Older parts of the hypha are heavily vacuolated and may be separated from the younger areas by cross walls called **septae**. Not all fungi are multicellular, some are **unicellular** and are termed **yeasts**. These grow by binary fission or budding.

Fungal taxonomy   In the past the fungi were a **polyphyletic** group which contained microorganisms that had very different ancestors. Current thinking now prefers a **monophyletic** classification where all groups within a phylum are descendants of one ancestor (see Section I2). Fungi are currently divided into four major phyla on the basis of their morphology and sexual reproduction.

In the **Zygomycota** and the **Chytridiomycota** the vegetative mycelium is non-septate, and complete septa are only found in reproductive structures. Asexual reproduction is by the formation of **sporangia**, and sexual reproduction by the formation of non-motile **zygospores** or motile **zoospores** respectively.

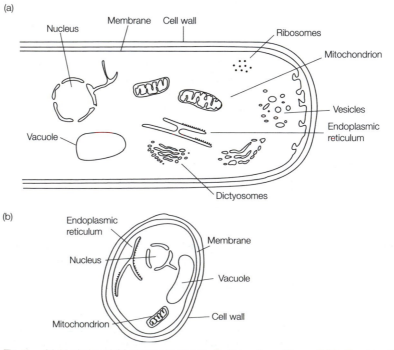

Fig. 1. (a) Hyphal and (b) yeast structures. Redrawn from Grove, S.M., Bracker, C.E. and Morré, B.J., American Journal of Botany, vol. 57, pp. 245–266, with kind permission from the Botanical Society of America.

Table 1. Features of the main groups of fungi

Group	Perforate septae present or absent	Asexual sporulation	Sexual sporulation
Zygomycota	Absent	Non-motile sporangiospores	Zygospore
Chytridiomycota	Absent	Motile zoospores	Oospore
Ascomycota	Present	Conidiospores	Ascospores
Basidiomycota	Present	Rare	Basidiospores
Deuteromycota	Present	Conidiospores	None

The other two phyla have a more complex mycelium with elaborate, perforate septa. They are divided into the **Ascomycota** and the **Basidiomycota**. Members of the Ascomycota produce asexual **conidiospores** and sexual **ascospores** in sac-shaped cells called **asci**. Fungi from the **Basidiomycota** rarely produce asexual spores, and produce their sexual spores from club-shaped **basidia** in complex fruit bodies.

A fifth group exists in the higher fungi which contains all forms that are not associated with a sexual reproductive stage, and these are termed the **Deuteromycota**.

Within each of the major phyla are several classes of fungi, indicated by names ending with *-etes*, for example Basidiomycetes, and within classes can be sub-classes or orders, indicated by names ending in *-ales*, within which there are genera and then species (see Section B). The important differences between fungi used to distinguish taxonomic groups are summarized in *Table 1*.

Fungal wall structure and growth

Fungal walls are rigid structures formed from layers of semi-crystalline chitin **microfibrils** that are embedded in an amorphous matix of β-**glucan**. Some protein may also be present. Growth occurs at the hyphal tip by the fusion of membrane-bound vesicles containing wall-softening enzymes, cell-wall monomers and cell-wall polymerizing enzymes derived from the Golgi with the hyphal tip membrane (*Fig. 2*). The fungal wall is softened, extended by turgor pressure and then rigidified.

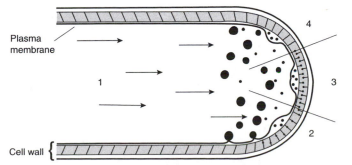

Fig. 2. Hyphal tip growth. Step 1, vesicles migrate to the apical regions of the hyphae. Step 2, wall lysing enzymes break fibrils in the existing wall, and turgor pressure causes the wall to expand. Step 3, amorphous wall polymers and precursors pass through the fibrillar layer. Step 4, wall synthesizing enzymes rebuild the wall fibrils. Redrawn from Isaac, S, Fungal–plant interactions, 1991, with kind permission of Kluwer Academic publishers.

Septae are cross walls that form within the mycelium. Growth in the Zygomycota and Chytridiomycota is not accompanied by septum formation, and the mycelium is cenocytic. Septae only occur in these groups to delimit reproductive structures from the parent mycelium, and they are complete (see Section I3). In the Ascomycota and Basidiomycota growth of the mycelium is accompanied by the formation of incomplete septae. Septae in the ascomycetes are perforate, and covered by endoplasmic reticulum membranes to limit movement of large organelles such as nuclei from compartment to compartment. This structure is called the **dolipore** septum. In dikaryotic basidiomycetes, septum formation is co-ordinated with divisions of the two mating-type nuclei, maintaining the dikaryotic state by the formation of **clamp connections**. These septae resemble crozier formation in the formation of asci (see Section I3).

Colonial growth

Hyphal tip growth allows fungi to extend into new regions from a point source or **inoculum**. Older parts of the hyphae are often emptied of contents as the cytoplasm is taken forwards with the growing tip. This creates the radiating colonial pattern seen on agar plates (*Fig. 3*), in ringworm infections of skin and fairy rings in grass lawns.

Kinetics of growth

When fungi are filamentous their growth rate cannot be established by cell counting using a hemocytometer or by turbidometric measurements (which can be used to measure bacterial and yeast growth) (see Sections D1 and D2). However, by measuring mass (*M*) changes with time (*t*) under excess nutrient conditions the **specific growth rate** (m) for the culture can be calculated using the formula:

$$\frac{dM}{dt} = mM$$

Fungal growth in a given medium follows the growth phases of lag, acceleration, exponential, linear, retardation, stationary and decline (see Section D1, *Fig. 3*).

Exponential growth occurs only for a brief period as hyphae branches are initiated, and then the new hypha extends at a linear rate into uncolonized regions of

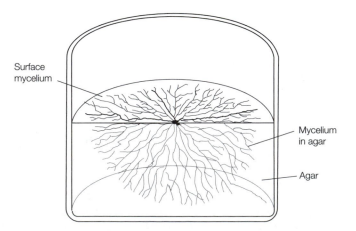

Fig. 3. Colonial growth patterns of filamentous fungi. Reproduced from Ingold, C.T. and Hudson, H.J., The Biology of Fungi, 1993, with kind permission from Kluwer Academic publishers.

substrate. Only hyphal tips contribute to extension growth. However, older hyphae can grow aerially or differentiate to produce sporing structures.

Hyphal growth unit

It is also possible to observe hyphal growth by microscopy, measuring tip growth and branching rates of mycelium. From these data the hyphal growth unit and the **peripheral growth zone** can be calculated. The hyphal growth unit (G), which is the average length of hypha that is required to support tip growth, is defined as the **ratio** between the total length of mycelium and the total number of tips:

$$\frac{\text{Total number of tips}}{\text{Total length of mycelium}} = \text{Hyphal growth unit}$$

In most fungi the hyphal growth unit includes the tip compartment plus two or three sub-apical compartments. The ratio increases exponentially from germination, but stabilizes to give a constant figure for a particular strain under any given set of environmental conditions.

Peripheral growth zone

The peripheral growth zone is the region of mycelium behind the tip, needed to support maximum growth of the hyphal tip. This zone permits radial extension at a rate equal to that of the **specific growth rate** of unicells in liquid culture. The mean rate of hyphal extension (E) is a function of the hyphal growth unit and the specific growth rate

$$E = G\mu$$

The **radial extension rate** (K_r) is a function of the peripheral growth zone (v) and the specific growth rate (μ):

$$K_r = \omega\mu$$

J2 FUNGAL NUTRITION

Key Notes

Carbon nutrition	Fungi require organic carbon compounds to satisfy their carbon and energy requirements. They obtain this carbon by saprotrophy, symbiosis or parasitism. The carbon must be available in a soluble form in order to cross the rigid cell wall, or must be broken down by enzymes secreted by the fungal cells.
Carbon metabolism	Fungi normally utilize glycolysis and aerobic metabolism of carbohydrates. Some can use fermentative pathways under reduced oxygen levels. A few fungi are truly anaerobic.
Nitrogen nutrition	Fungi cannot fix gaseous nitrogen but can utilize nitrate, ammonia and some amino acids as nitrogen sources.
Macro-, micro-nutrients and growth factors	Most macro- and micro-nutrients that fungi require are present in excess in their environments. Phosphorus and iron may be in short supply, and fungi have specific mechanisms to obtain these nutrients. Some fungi may require external supplies of some vitamins, sterols and growth factors.
Water, pH and temperature	Fungi require water for nutrient uptake and are therefore restricted to damp environments. They occupy acidic environments between pH 4 and 6, and by their activity further acidify it. Most fungi are mesophilic, growing between 5° and 40°C, but some can tolerate high or low temperatures.
Secondary metabolism	Secondary metabolites, derived from many different metabolic pathways, are produced by fungi when vegetative growth becomes restricted by nutrient depletion or stress. Such compounds may offer a competitive advantage to the producer.
Related topics	Measurement of microbial growth (D1) Eukaryotic cell structure (I2) Heterotrophic pathways (E2) Chlorophytan and Protistan Autotrophic reactions (E4) taxonomy and structure (K1) Biosynthetic pathways (E5) Chlorophytan and Protistan Control of bacterial infections (H8) nutrition and metabolism (K2)

Carbon nutrition

Fungi are heterotrophic for carbon (see Topic B1). They need organic compounds to satisfy energy and carbon requirements. There are three main modes of nutrition: **saprotrophy**, where fungi utilize dead plant, animal or microbial remains; **parasitism**, where fungi utilize living tissues of plants and animals to the detriment of the host; and **symbiosis**, where fungi live with living tissues to the benefit of the host.

Carbohydrates must enter hyphae in a soluble form because the rigid cell wall prevents endocytosis. Soluble sugars cross the fungal wall by diffusion, followed by active uptake across the fungal membrane (see Topic E2). This type of nutrition is seen in the symbiotic and some parasitic fungi. For the saprophytic fungi most carbon in the environment is not in a soluble form but is present as a complex polymer like cellulose, chitin or lignin. These polymers have to be broken down enzymically before they can be utilized. Fungi release **degradative enzymes** into their environments. Different classes of enzyme can be produced, including the **cellulases**, **chitinases**, **proteases** and multi-component **lignin-degrading enzymes**, depending on the type of substrate the fungus is growing on. Regulation of these enzymes is by **substrate induction** and **end-product inhibition**.

Carbon metabolism

Once within the hypha, carbon and energy metabolism is by the processes of **glycolysis** and the **carboxylic acid cycle** (see Section E2). Fungi are usually aerobic, but some species, for example the yeasts, are capable of living in low oxygen-tension environments and utilizing **fermentative** pathways of metabolism (see Section E5). Recently, truly **anaerobic** fungi have been discovered within animal rumen and in anaerobic sewage-sludge digesters.

Nitrogen nutrition

Fungi are heterotrophic for nitrogen (see Section E5). They cannot fix gaseous nitrogen, but they can utilize nitrate, ammonia and some amino acids by direct uptake across the hyphal membrane. Complex nitrogen sources, such as peptides and proteins, can be utilized after extracellular proteases have degraded them into amino acids.

Macro-, micro-nutrients and growth factors

Phosphorus, potassium, magnesium, calcium and sulfur are all macronutrients required by fungi. All but phosphorus are usually available to excess in the fungal environment. Phosphorus can sometimes be in short supply, particularly in soils, and fungi have the ability to produce extracellular **phosphatase** enzymes which allow them to access otherwise unavailable phosphate stores.

Micronutrients include copper, manganese, sodium, zinc and molybdenum, all of which are usually available to excess in the environment (see Section D1). Iron is relatively insoluble and therefore not easily assimilated, but fungi can produce **siderophores** or organic acids, which can **chelate** or alter iron solubility and improve its availability.

Some fungi may require pre-formed vitamins, for example, thiamin and biotin (see Section D1). Other requirements can be for sterols, riboflavin, nicotinic acid and folic acid.

Water, pH and temperature

Fungi require water for nutrient uptake and they are therefore restricted to fairly moist environments such as host tissue if they are parasites or symbionts, or soils and damp substrates if they are saprophytes. Desiccation causes death unless the fungus is specialized, as they are in the lichens (see Section J4). Some fungi are wholly aquatic.

Fungi tend to occupy acidic environments, and by their metabolic activity (respiration and organic acid secretion) tend to further acidify it. They grow optimally at pH 4–6.

Most fungi are **mesophilic**, growing between 5° and 40°C. Some are **psychrophilic** and are able to grow at under 5°C, others are **thermotolerant** or **thermophilic** and can grow at over 50°C.

Secondary metabolites

Nutrient depletion, competition or other types of metabolic stress which limit fungal growth promote the formation and secretion of secondary metabolites. These compounds can be produced by many different metabolic pathways and include compounds termed **antibiotics** (active against bacteria, protista and other fungi) (see Section H8), **plant hormones** (gibberellic acid and indoleacetic acid (IAA)) and **cytotoxic** and **cytostimulatory** compounds.

J3 REPRODUCTION IN FUNGI

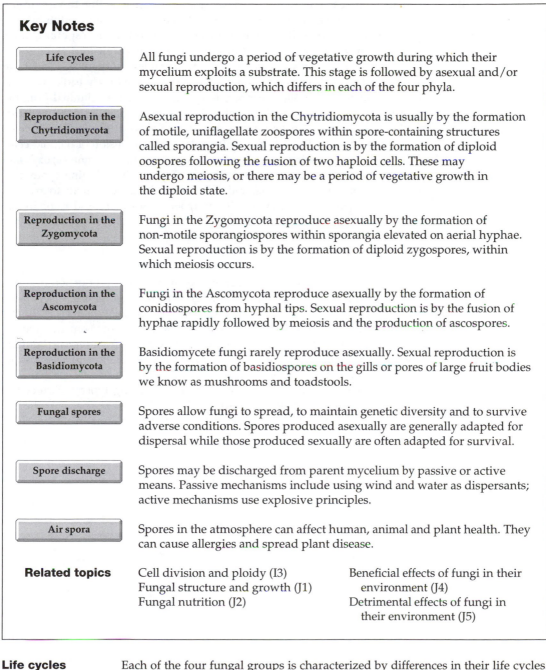

Key Notes

Life cycles

All fungi undergo a period of vegetative growth during which their mycelium exploits a substrate. This stage is followed by asexual and/or sexual reproduction, which differs in each of the four phyla.

Reproduction in the Chytridiomycota

Asexual reproduction in the Chytridiomycota is usually by the formation of motile, uniflagellate zoospores within spore-containing structures called sporangia. Sexual reproduction is by the formation of diploid oospores following the fusion of two haploid cells. These may undergo meiosis, or there may be a period of vegetative growth in the diploid state.

Reproduction in the Zygomycota

Fungi in the Zygomycota reproduce asexually by the formation of non-motile sporangiospores within sporangia elevated on aerial hyphae. Sexual reproduction is by the formation of diploid zygospores, within which meiosis occurs.

Reproduction in the Ascomycota

Fungi in the Ascomycota reproduce asexually by the formation of conidiospores from hyphal tips. Sexual reproduction is by the fusion of hyphae rapidly followed by meiosis and the production of ascospores.

Reproduction in the Basidiomycota

Basidiomycete fungi rarely reproduce asexually. Sexual reproduction is by the formation of basidiospores on the gills or pores of large fruit bodies we know as mushrooms and toadstools.

Fungal spores

Spores allow fungi to spread, to maintain genetic diversity and to survive adverse conditions. Spores produced asexually are generally adapted for dispersal while those produced sexually are often adapted for survival.

Spore discharge

Spores may be discharged from parent mycelium by passive or active means. Passive mechanisms include using wind and water as dispersants; active mechanisms use explosive principles.

Air spora

Spores in the atmosphere can affect human, animal and plant health. They can cause allergies and spread plant disease.

Related topics

Cell division and ploidy (I3)
Fungal structure and growth (J1)
Fungal nutrition (J2)

Beneficial effects of fungi in their environment (J4)
Detrimental effects of fungi in their environment (J5)

Life cycles

Each of the four fungal groups is characterized by differences in their life cycles. All fungi are characterized by having a period of vegetative growth where their biomass increases. The length of time and the amount of biomass needed before spores, but a few have lost all sporing structures and are referred to as *mycelia sterilia*. Different types of spore are produced in different parts of the life cycle.

Reproduction in Chytridiomycota

Fungi in the Chytridiomycota are quite distinct from other fungi as they have extremely simple thalli and motile zoospores. Some species within this group can be so simple that they consist of a single vegetative cell within (**endobiotic**) or upon (**epibiotic**) a host cell, the whole of which is converted into a **sporangium**, a structure containing spores. These types are termed **holocarpic** forms.

Other members of this group have a more complex morphology, and have **rhizoids** and a simple mycelium. Asexual reproduction in the chytridiomycota is by the production of motile **zoospores** in sporangia that are delimited from the vegetative mycelium by complete septae. The zoospores have a single, posterior flagellum. Sexual reproduction occurs in some members of the chytridiomycota by the production of **diploid** spores after either somatic fusion of haploid cells, either two different mating-type mycelia, fusion of two motile **gametes**, or fusion of one motile gamete with a **non-motile egg** (*Fig. 1*). The resulting spore may undergo meiosis to produce a haploid mycelium or it may germinate to produce a diploid vegetative mycelium, which can undergo asexual reproduction by the production of diploid zoospores. The diploid mycelium can also produce resting sporangia in which meiosis occurs, generating haploid zoospores that germinate to produce haploid vegetative mycelium.

Reproduction in Zygomycota

In the Zygomycota, asexual reproduction begins with the production of aerial hyphae. The tip of an aerial hypha, now called a **sporangiophore**, is separated from the vegetative hyphae by a complete septum called a **columella** (*Fig. 2*). The cytoplasmic contents of the tip differentiate into a sporangium containing many asexual spores. The spores contain haploid nuclei derived from repeated mitotic divisions of a nucleus from the vegetative mycelium (see Section I3). Dispersal of the spores is by wind or water.

In sexual reproduction, two nuclei of different mating types fuse together within a specialized cell called a **zygospore** (*Fig. 2*). In some species the different mating-type nuclei may be within one mycelium (**homothalism**). In other species,

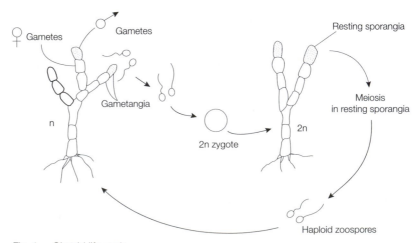

Fig. 1.   Chytrid life cycle.

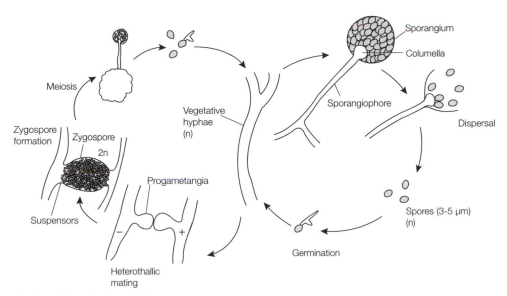

Fig. 2.   Life cycle of a typical zygomycete.

two mycelia with different mating-type nuclei must fuse (**heterothalism**). In both cases, fusion occurs between modified hyphal tips called **progametangia**, which once fused are termed the zygospore. Within the developing zygospore meiosis occurs; usually three of the nuclear products degenerate, leaving only one nuclear type present in the germinating mycelium (see Section I3).

Reproduction in Ascomycota

The vegetative stage of the Ascomycota life cycle is accompanied or followed by asexual sporulation by the production of single spores called conidia from the tips of aerial hyphae called conidiophores (*Fig. 3*). The spores can be delimited by a complete transverse wall formation followed by spore differentiation (*Fig. 3a*) termed **thallic** spore formation, or more usually by the extrusion of the wall from the hyphal tip, termed **blastic** spore formation (*Fig. 3b*). These spores can be single-celled and contain one haploid nucleus, or they can be multicellular and contain several haploid nuclei produced by mitosis (see Section I3).

Spores can be produced from single, unprotected conidiophores or they can be produced from **aggregations** that are large enough to be seen with the naked eye (*Fig. 3c*). The conidiophores can aggregate into stalked structures where the spores produced are exposed at the top (**synnema** or **coremia**). Alternatively, varying amounts of sterile fungal tissue can protect the conidia, as in the flask-shaped **pycnidia**. Some species produce conidia in plant tissue, and the conidial aggregations erupt through the plant epidermis as a cup-shaped **acervulus** or a cushion-shaped **sporodochium**.

Sexual reproduction in this group occurs after **somatic** fusion of different mating-type mycelia. A transient diploid phase is rapidly followed by the formation of **ascospores** within sac-shaped **asci** differentiated from modified hyphal tips. In the initial stages of ascal development hooked hyphal tips form, called **croziers** or **shepherds' crooks** because of their shape. They have distinctive septae at their base which ensure that two different mating-type nuclei are maintained

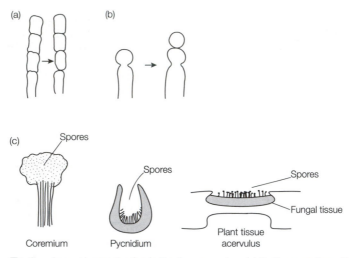

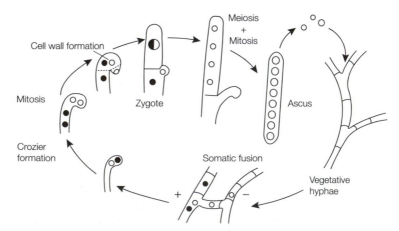

Fig. 3.   Asexual reproduction in the Acsomycetes. (a) thallic spore formation; (b) blastic spore formation; (c) aggregations of conidiophores. Redrawn from Ingold, C.T. and Hudson, H.J., The Biology of Fungi, 1993, with kind permission from Kluwer Academic Publishers.

in the terminal cell. Formation of the septae is coordinated with nuclear division (*Fig. 4*). In yeasts all these events occur within one cell, after fusion of two mating-type cells, the whole cell being converted into an ascus.

In more complex Ascomyceta many asci form together, creating a fertile tissue called a **hymenium**. In some groups the hymenium can be supported or even enclosed by large amounts of vegetative mycelium. The whole structure is called a **fruit body** or **sporocarp** and is used as a major taxonomic feature (*Fig. 5*). They can become large enough to be seen with the naked eye. Flask-shaped sexual reproductive bodies are called **perithecia**, cup-shaped bodies are called **apothecia** and closed bodies are called **cleistothecia**. These structures have

Fig. 4.   Sexual reproduction in the Ascomycetes.

(a)

(b)

(c)

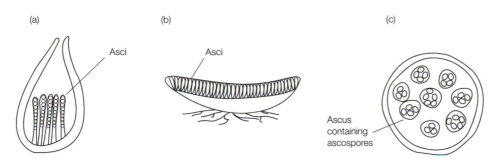

Asci

Asci

Ascus
containing
ascospores

Fig. 5. Structures of sexual sporocarps in the Ascomycetes. (a) perithecium; (b) apothecium; (c) cleistothecium. Redrawn from Ingold, C.T. and Hudson, H.J., The Biology of Fungi, *with kind permission from Kluwer Academic Publishers.*

evolved to protect the asci and assist in spore dispersal, but the hymenium itself is unaffected by the presence of water.

Reproduction in Basidiomycota

This group of fungi are characterized by the most complex and large structures found in the fungi. They are also distinctive in that they very rarely produce asexual spores. Much of the life cycle is spent as vegetative mycelium, exploiting complex substrates. A preliminary requisite for the onset of sexual reproduction is the acquisition of two mating types of nuclei by the fusion of compatible hyphae. Single representatives of the two mating-type nuclei are held within every hyphal compartment for extended periods of time. This is termed a **dikaryotic** state, and its maintenance requires elaborate septum formation during growth and nuclear division (see Section J1).

Onset of sexual-spore formation is triggered by environmental conditions and begins with the formation of a **fruit body primordium**. Dikaryotic mycelium expands and differentiates to form the large fruit bodies we recognize as mushrooms and toadstools. Diploid formation and meiosis occur within a modified hyphal tip called a **basidium** *(Fig. 6)*.

Four spores are budded from the basidium. Basidia form together to create a hymenium which is highly sensitive to the presence of free water. The hymenium

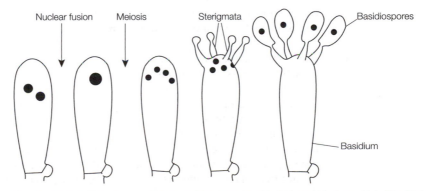

Nuclear fusion   Meiosis   Sterigmata   Basidiospores

Basidium

Fig. 6.   Basidium formation. Redrawn from from Ingold, C.T. and Hudson, H.J., The Biology of Fungi, *with kind permission from Kluwer Academic Publishers.*

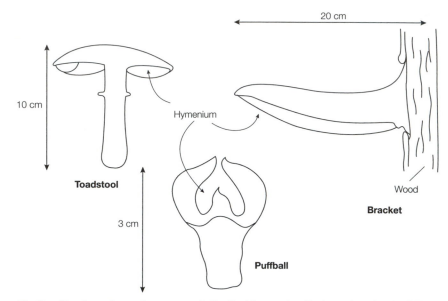

Fig. 7.   Structure of sexual sporocarps in the Basidiomycetes. Redrawn from Ingold, C.T. and Hudson, H.J., The Biology of Fungi, with kind permission from Kluwer Academic Publishers.

is distributed over sterile, dikaryotic-supporting tissues which protect it from rain. The hymenium can be exposed on **gills** or **pores** beneath the fruit body, seen in the **toadstools** and **bracket fungi**, or enclosed within chambers as in the **puff-balls** and **truffles** (*Fig. 7*).

Fungal spores

There are two conflicting requirements fungi have for their spores. Spores must allow fungi to spread, but they must also allow them to survive adverse conditions. These requirements are met by different types of spores. Small, light spores are carried furthest from parent mycelium in air and these are the dispersal spores. They are usually the products of asexual sporulation, the sporangiospores and the conidiospores, and so spread genetically identical individuals as widely as possible. Genetic diversity is maintained by sexual reproduction, and the spore products are often large **resting spores** that withstand adverse conditions but remain close to their site of formation. Spores therefore vary greatly is size, shape and ornamentation, and this variation reflects specialization of purpose.

Spore discharge

Spores that have a dispersal function can be released from their parent mycelium by **active** or **passive** mechanisms (*Fig. 8*). As many spores are wind-dispersed, they are produced in dry friable masses which are passively discharged by wind. Other spores are passively discharged by water droplets splashing spores away from parent mycelium.

Fungal spores can be actively discharged by **explosive** mechanisms. These mechanisms use a combination of an increasing turgor pressure within the spore-bearing hypha, combined with an in-built weak zone of the hyphal wall. This insures that when the hypha bursts the spore discharge is directed for maximum distance. Asci are usually dispersed in this way, and a few sporangia too. Basidiospores are also actively discharged.

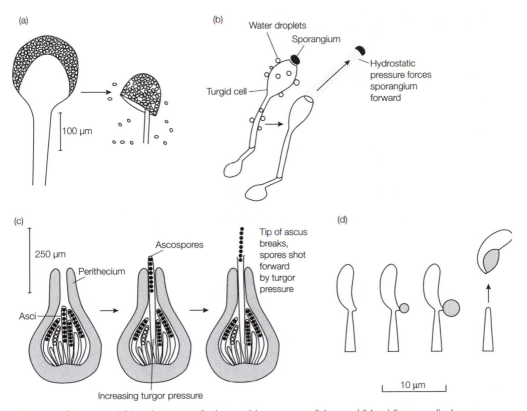

Fig 8.   (a) Passive and (b) active spore discharge; (c) ascospore disharge; (d) basidiospore discharge.

Air spora   Air-borne fungal spores can be carried great distances. Their presence in the air can have an impact on human health as they can cause **allergic rhinitus** (hay fever) and **asthma**. Many plant diseases that cause significant economic losses are air-borne (see Section J5). Spore clouds can be tracked across continents, and epidemic disease forecasts can be made, depending on weather conditions and air-spora counts.

J4 BENEFICIAL EFFECTS

Key Notes

Bread and brewing

The products of yeast fermentation (CO_2 and alcohol) are exploited in bread making and alcohol brewing. Both processes enhance the value of the substrate but contribute little to its nutritional value.

Symbioses

Fungi can enter into specialized and intimate, mutually beneficial associations with higher plants, other microbes and animals. The associations can be external to the host cell, as in the ectomycorrhizae and lichens, or inside the cell as in the endomycorrhizae and endophytic fungi.

Decomposers

Fungi are the main agents of decay of plant wastes in the environment, decomposing substrates to CO_2, H_2O and fungal biomass, and releasing other nutrients back to the biosphere.

Biological control

Fungi can be used to control insect pests, weed plants and plant diseases by exploiting their natural antagonistic, competitive and pathological attributes.

Bioremediation

The degradative abilities of fungi can be exploited to decompose man-made pollutants such as hydrocarbons, pesticides and explosives. They may decompose substrates into CO_2 and H_2O by respiratory pathways, or they may reduce toxicity by co-metabolic activity. The toxicity of some compounds can also be increased in this manner.

Industrially important natural products of fungi

Fungi naturally produce antibiotics, immunosuppressants, acids, enzymes and several other classes of useful natural products. They also can be used to produce large quantities of protein, including the popular meat substitute Quorn.

Related topics

Heterotrophic pathways (E2)
Fungal structure and growth (J1)

Fungal nutrition (J2)
Reproduction in fungi (J3)
Detrimental effects (J5)

Bread and brewing

The metabolic products of yeast metabolism are exploited by humans for **bread making** and **alcohol brewing**. Yeast metabolism of flour starch by respiration generates CO_2 which is trapped within the gluten-rich dough and forces bread to rise (see Section E2). In the brewing of alcohol, yeasts are forced into fermentative metabolism in the sugar-rich, low-oxygen environments of beer wort or crushed grape juice. The fermentative metabolism is inefficient, and only partially metabolizes the available substrates, yielding CO_2 and **ethanol**. Both processes considerably enhance the value of the original substrate while contributing little to its nutritional status!

Symbioses

Fungi can enter into close associations with other microbes and with higher plants and animals. These beneficial associations are termed **symbioses**, and in most cases the symbiotic fungus gains carbohydrates from its associate, while the associating organism gains nutrients and possibly protection from predation and herbivory or plant pathogens.

Fungal symbioses are common on plant roots, and the symbiotic roots are termed **mycorrhizae** (*Fig. 1*). Their presence enhances plant-root nutrient uptake and plant performance. The fungal association can be predominantly external to the root tissue. These associations are called **ecto**mycorrhizae and can be seen on beech and pine trees, for example. Other associations are predominantly within the plant root, and are termed **endo**mycorrhizae. These associations are seen on the roots of herbaceous species like grasses, but are also found in the roots of tropical species of tree and shrub.

Other associations between fungi and plants can occur within leaves or stems, and these are called **endophytic** fungi. They live almost all their life cycle within the host, grow very slowly and do not cause any signs of infection. They appear to protect their host from herbivory and fungal infection by the production of metabolites. However, these products can have dramatic effects on herbivorous animals, causing symptoms of fungal toxicosis similar to St Anthony's Fire (see Section J5).

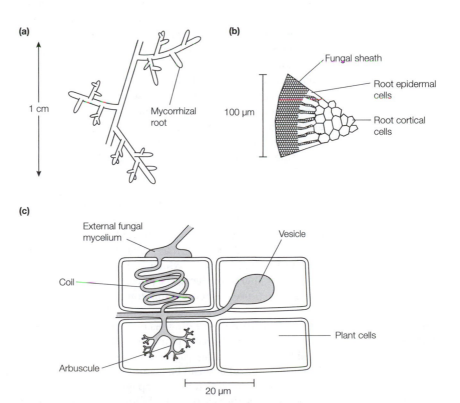

Fig. 1. Structure of ecto- and endo-mycorrhizae. (a) Macromorphology of ectomycorrhiza; (b) micromorphology of ectomycorrhizae; (c) micromorphology of endomycorrhizae. Redrawn from Isaac, S., Fungal-Plant Interactions, 1991, with kind permission from Kluwer Academic Publishers.

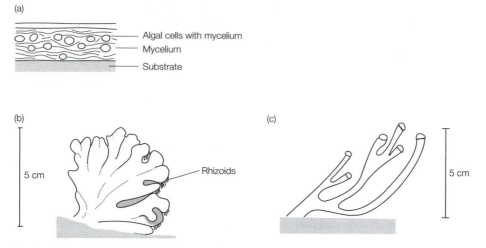

Fig. 2.   Lichen structure. (a) Crustose, tightly attached to substrate; (b) foliose, loosely attached to substrate; (c) fruticose, attached only at base.

Some fungi can form very intimate associations with algal species. These associations are termed **lichens**, and they have a form quite distinct from that of either component species, with **crustose, foliose** or **fruticose** thalli (*Fig. 2*). They are slow-growing, and are adapted to occupy extreme or marginal environments, like bare rock faces, walls and house roofs. As they are under quite extreme stress they are highly sensitive to pollution, for instance, from acid rain or heavy metals, and their presence or absence in an environment has become a useful indicator of urban and industrial pollution.

Fungi can also associate with insect species in symbioses of varying intimacy. Some species of ants culture specific fungi on cut plant remains within their nests, and then browse on the fungal mycelium that develops. **Termites** have symbiotic fungi within their guts, which, in association with a consortium of other microbes, including protista and bacteria, help the termite digest its woody gut contents (see Section K4).

Decomposers

The degradative processes that fungi perform with their extracellular enzymes are essential to the terrestrial biosphere (see Section J2). They are the main agents of decay of cellulosic wastes produced by plants, which in the tropical rain forest can amount to 12 000 kg hectare^{-1} year^{-1}. They decompose this material into CO_2, H_2O and fungal biomass, which in its turn is decomposed by other microbes, returning mineral nutrients like phosphorus, nitrogen and potassium to the biosphere. This process is termed mineralization.

Biological control

The natural attributes of fungi as disease-causing organisms can be exploited by humans to control weed plant populations and insect pests. They are even capable of parasitizing plant disease-causing fungi. This process is termed **biological control** and can be an alternative to the application of chemical pesticides. Applications of fungal propagules to targeted problem populations can cause epidemic disease or overwhelm and monopolize a niche.

Bioremediation

The degradative processes that fungal enzymes catalyze on their natural substrates can be used on other, man-made substrates to provide **biological cleanup**.

Hydrocarbons like oils can be degraded by fungi and other microbes to CO_2 and H_2O by aerobic respiration. These activities are termed bioremediation, and cont-aminated areas of land can be actively bioremediated by the addition of fungal propagules.

Pesticides, explosives and other recalcitrant molecules can be changed by **co-metabolic** activities of fungi, where enzymes normally used for one metabolic process within the fungus coincidentally catalyze another reaction. The products of co-metabolic reactions are not utilized any further by the fungus, but can some-times be used by other microbes in the ecosystem. Reactions like these can lead to reduced toxicity of some contaminants, but in other cases can lead to **activation** of the pollutant, leading to an increase in the toxicity of the compound.

Industrially important natural products of fungi

Fungi are able to produce many different types of metabolite that are of commer-cial importance. These include **antibiotics** (e.g. Penicilliins and Cephalosporins) and **immunosuppressants** (cyclosporins), important in medicine, **enzymes** that are used in the food industry (e.g. α-amylase, renin), other enzymes (e.g. cellu-lases, catalase), **acids** (e.g. lactic, citric) and several other products.

Protein extracted from fungi is also an important commercial product. In the 1960s protein from yeasts (single-cell protein) was developed. Currently, fungal protein extracted from *Fusarium graminearum*, known as **Quorn**, is a useful meat alternative.

J5 DETRIMENTAL EFFECTS

Key Notes

Biodeterioration	The degradative activities of fungi in unwanted situations cause significant economic losses. Materials containing large quantities of cellulose, leather and hydrocarbons can be used as substrates by fungi, providing there is an adequate water supply.
Plant disease	Fungi are capable of causing significant losses to crops both before and after harvest. However, this can be countered by the use of fungicides, storage conditions that do not favor fungal growth, and the development of resistant plant varieties.
Animal and human disease	Fungi can cause both superficial and deep, life-threatening infections of both man and animals. The latter are particularly dangerous in immunocompromised individuals.
Fungal toxicosis	Ingestion of fungi or their secondary metabolic products, accidentally or deliberately, can cause intoxication and occasionally death in both humans and animals.
Related topics	Fungal structure and growth (J1) Reproduction in fungi (J3) Fungal nutrition (J2)

Biodeterioration

The same extracellular enzymes that are important in the degradation of leaf litter and the recycling of nutrients in the biosphere can cause massive economic losses when they occur in circumstances where they are not wanted. Fungi can attack and utilize as substrates paper, cloth, leather and hydrocarbons, but also can cause degradative change in other materials, for instance, glass and metal, because of their ability to produce acid as they grow. The supply of water is a key control point in these processes, and keeping substrates dry is an effective way of avoiding these changes.

Plant disease

Fungi are capable of attacking all plant species, causing serious damage and in some circumstances even death. In crop production over half of potential crop yield is lost to plant pathogens, most of it to **fungal disease**. In storage, up to one-third of the harvested product can be lost to post-harvest disease, again mostly as a result of the activities of fungi. Use of **fungicides** can reduce both pre- and post-harvest disease, and **plant-breeding** programs can introduce disease-resistant strains of crop plant. Post-harvest losses can be reduced by storage of products at low temperatures and low moisture levels.

Animal and human disease

Human and animal epidermis can be attacked by fungi, causing superficial damage and discomfort like **ringworm** infections, **athletes foot** and **thrush**. Other deeper, systemic fungal infections of the lung and central nervous and lymphatic systems cause much more serious diseases, for example, **aspergillosis**, **coccidiomycosis**, **blastomycosis**, **histoplasmosis** and **pneumocystis** pneumonia are all caused by fungi. Although most humans experience superficial fungal infections and survive, these deeper diseases are especially dangerous for the immuno-compromised patient after transplantation, and the HIV-positive population.

Fungal toxicosis

Accidental or deliberate consumption of wild fungi or fungally contaminated food can lead to poisoning or **toxicosis** of the consumer because some fungi naturally contain toxic metabolites called **mycotoxins**. Deliberate toxicosis can arise from consumption of mushrooms and toadstools that are known to contain naturally hallucinatory drugs like **psilocibins**, which lead to euphoric states followed by extreme gastrointestinal distress. Accidental consumption of mis-identified fungal fruit bodies can lead to fatal mushroom poisoning from fungal toxins, causing total liver failure between 8 and 10 h. Consumption of food accidentally contaminated by fungal metabolites also leads to human and animal death. For instance, rye flour contaminated by the ergots of the fungus *Claviceps purpurea* leads to the symptoms of **St Anthony's fire**, where peripheral nerve damage is caused by the presence of **ergometrine** in the fungal tissue. This can be followed by gangrene of the limbs and death. Detection of fungal mycotoxins such as **ochratoxin** in apple juice and **aflatoxin** in peanuts have also caused problems for food producers and consumers and forced improvements in product processing.

K1 TAXONOMY AND STRUCTURE

Key notes

Chlorophytan and protistan taxonomy	Until fairly recently, the photosynthetic and non-photosynthetic protistan genera were divided into two polyphyletic form groups, the algae and protozoa based on the presence or absence of chloroplasts. Currently, a monophyletic taxonomy of the protista is being developed, based on molecular data.
Chlorophytan and protistan structure	Many members of the Chlorophyta are unicellular, photosynthetic organisms. Some have a filamentous or membranous morphology. They have a cellulose cell wall. Many species are flagellate, and they contain chloroplasts, which vary in structure and pigment content. Members of the protista are heterotrophic or photosynthetic, unicellular eukaryotes. They vary greatly in shape and size, and contain most eukaryotic cell organelles. They also have some unique organelles. There are three major groups within the protista, the Euglenozoa (containing the euglenids and kinetoplastids), the Alveolata (containing the ciliates, dinoflagellates and the apicomplexans) and the Stramenopila (containing the diatoms, Chrysophytes, Oomycetes, opalines and the amoebae and slime molds).
Chlorophytan and protistan growth	Growth in most unicellular Chlorophyta and protista is synonymous with longitudinal binary fission. Cenocytic, tubular or filamentous Chlorophyta grow by tip growth like the fungi. Other filamentous or membranous algae grow by intersusception of new cells into the filament. The kinetics of growth are similar to those of bacteria, but in addition to estimations of growth by mass measurement, cell counts and chlorophyll content can be assessed. Rapid cell division can lead to very high cell populations, only limited by nitrogen, phosphate or silicon availability.

Related topics	The microbial world (A1)	Chlorophytan and protistan
	Composition of a typical	nutrition and metabolism (K2)
	prokaryotic cell (C7)	Beneficial effects of the
	Cell division (C4)	Chlorophyta and protista (K4)
	Eukaryotic cell structure (I2)	
	Cell division and ploidy (I3)	
	Fungal structure and growth (J1)	

Chlorophytan and protistan taxonomy

In the past the eukaryotic photosynthetic and non-photosynthetic microorganisms were divided into **form groups**, the algae and the protozoa, based on the presence or absence of chloroplasts. The algae were sub-divided into groups based on pigmentation, the number and type of flagellae and other structural characteristics. Protozoa were divided on similar structural characteristics into four **polyphyletic** form groups, the ciliates, flagellates, sporozoans and amoebas. Advances in molecular biology now allow us to begin to create a monophyletic taxonomy of the Chlorophyta and protista and such a scheme includes many former members of the algae, fungi and protista. A suggested taxonomic scheme is shown in *Fig. 1* and characteristics of the major monophyletic groups are shown in *Table 1*. Only those organisms included in the microbial world will be discussed here, the red and brown algae are multicellular organisms and are omitted from consideration.

Chlorophytan and protistan structure

The Chlorophyta are a monophyletic group. They range in complexity from unicellular motile or non-motile organisms to **sheets**, **filaments** and **cenocytes** (*Fig. 2*). They are found in fresh and salt water, in soil and on and in plants and animals. Most Chlorophytan cell walls are formed from cellulose and they may

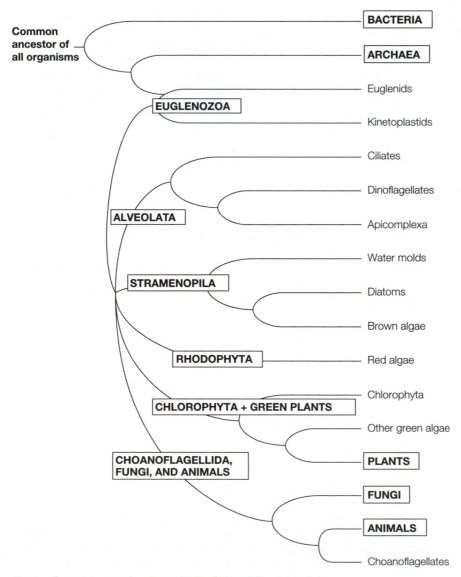

Fig. 1.   Current taxonomic scheme for the Chlorophyta and protista.

Table 1. Characteristics of the major monophyletic groups of the Chlorophyta and protista

Group	Common name	Characteristics	Examples
Euglenoza		**Flagellate unicells**	
Euglenoids		Mostly photosynthetic	*Euglena*
Kinetoplastids		Single large mitochondrion	*Trypanosoma*
Alveolata		**Unicellular; sacs (alveolata) below cell surface**	
Pyrrophyta	Dinoflagellates	Golden brown algae	*Peridinium*
Apicomplexa		Apical complex aids host cell penetration	*Plasmodium*
Ciliophora	Ciliates	Cilia; macro- and micro-nucleus	*Paramecium*
Stramenopila		**Motile stages with two unequal flagellae, one with hairs**	
Bacillariophyta	Diatoms	Unicellular, photosynthetic, 2 silicon containing frustules	*Navicula*
Chrysophytes	Golden Brown algae	Multicellular photosynthetic marine species	*Ceratium*
Oomycota	Water molds	Heterotrophic coenocytes	*Saprolegnia*
Chlorophyta	Green algae	Chlorophyll a and b, cellulose cell walls	*Chlamydomonas*

be fibrillar, similar to those of the fungi, and sometimes impregnated with silica or calcium carbonate.

Chlorophytan cells contain nuclei, mitochondria, ribosomes, Golgi and chloroplasts (Section I2). The internal cell structure is supported by a network of microtubules and endoplasmic reticulum. Chloroplasts in the Chlorophyta are very variable structures and can be large and single, multiple, ribbon-like or stellate chloroplasts with chlorophylls *a* and *b* and carotenoids and they store **starch**. They have a vegetative phase that is haploid, and sexual reproduction occurs when cells are stimulated to produce gametes instead of normal vegetative cells at binary fission.

They often possess flagella that have a 9 + 2 microtubule arrangement within them (see Section I2) and there may be one or two per cell. They may be inserted **apically, laterally** or **posteriorly** and trail or girdle the cell. The flagellum can be a single whiplash or it can have hairs and scales. The presence of **eyespots** near the flagellar insertion point allows the cell to swim towards the light. Movement may be by lateral strokes or by a spiral movement that can push or pull the cell through the water.

Members of the protista are unicellular, heterotrophic or photosynthetic, organisms. They may be motile or non-motile. They contain many of the organelles found in other eukaryotes, including nuclei, mitochondria (absent in some groups), chloroplasts (absent in some groups), ribosomes, endoplasmic reticulum, Golgi vesicles, microtubules and microfibrils. Chloroplasts can be very variable in this group, and their shape and pigment content are useful distinguishing taxonomic features. The numbers of ER membranes that surround the chloroplast, and the numbers of thylakoid stacks within them, are also important indicators of phylogeny. Unique organelles found in the protista include **contractile vacuoles** that control influx of water to the cells due to osmosis. Some of the protista have **hydrogenosomes** where fermentative reactions take place (see Topic E2). Some protista have a cell covering of scales or plates formed of pseudochitin, and this may be impregnated with calcium or silica scales.

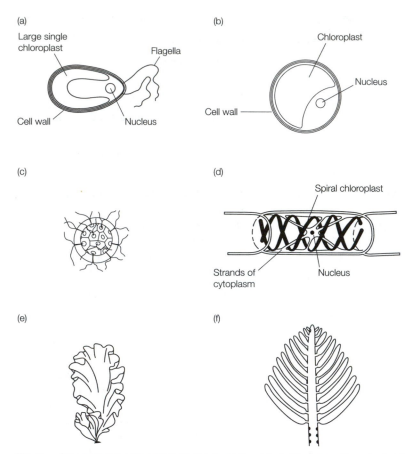

Fig. 2.   Chlorophytan cells and thalli. (a) A motile unicell; (b) non-motile unicell; (c) colonial forms; (d) filamentous forms; (e) membranous (only two cells thick); and (f) tubular. Reproduced from Clegg, C.J., Lower Plants, 1984, with kind permission from John Murray (Publishers) Ltd.

These scales are secreted from the Golgi bodies within alveoli. Some species have the ability to from thick-walled cysts during a resting phase in the life cycle. These cysts can be very resistant to adverse conditions.

The group is characterized by a considerable variation in morphology (*Fig. 3*). Free-floating species are radially or laterally symmetrical and streamlined into bullet or kidney shapes. At its simplest, the outer surface of protistan cells may be naked plasma membrane. This is seen in the gametes of some species, in amoebae and in some of the intracellular parasites. The simple shape of the amoeba, a simple naked cell with locomotion based on pseudopodia and a phagocytic nutrition, appears to have arisen on a number of occasions in all of the major taxonomic groups. Such a naked cell has advantages in nutrient uptake that outweigh its vulnerability. The naked membrane can be extended with lobes of cytoplasm that are called pseudopodia. These structures are important in locomotion and feeding. This type of movement requires that the cell is in contact with a solid surface. The cell is partitioned into two regions, the viscous ectoplasm and the more liquid endoplasm. Within the ectoplasm are myosin and actin filaments whilst in the endoplasm only un-polymerized actin is present. To produce ameboid movement the microfilaments in the ectoplasm

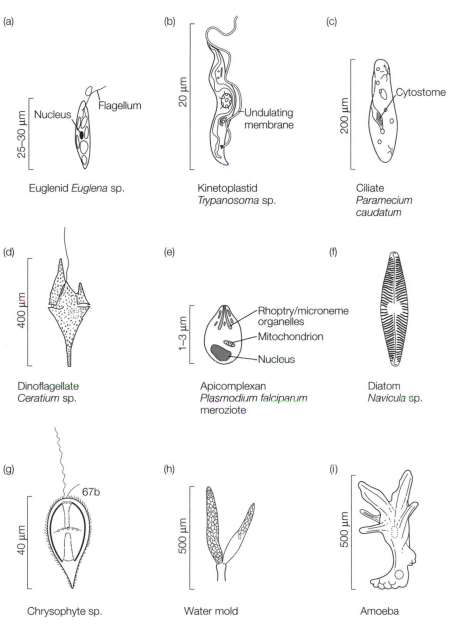

(a)

25–30 μm

Nucleus

Flagellum

Euglenid *Euglena* sp.

(b)

20 μm

Undulating
membrane

Kinetoplastid
Trypanosoma sp.

(c)

200 μm

Cytostome

Ciliate
*Paramecium
caudatum*

(d)

400 μm

Dinoflagellate
Ceratium sp.

(e)

1–3 μm

Rhoptry/microneme
organelles

Mitochondrion

Nucleus

Apicomplexan
Plasmodium falciparum
meroziote

(f)

Diatom
Navicula sp.

(g)

40 μm

67b

Chrysophyte sp.

(h)

500 μm

Water mold

(i)

500 μm

Amoeba

Fig. 3.   Structure of the unicellular protista.

produce a pressure on the endoplasm (*Fig. 4*) moving the endoplasm to one part of the cell, and away from another. Motility can also be via flagellae or cilia that propel the cells through water by rhythmic beating. Flagellae typically exhibit a sinusoidal motion in propelling water parallel to their axis (*Fig. 5a*). The undulating action of the flagellum either propels water away from the surface of the cell body or draws water towards and over the cell body. Cilia exhibit an oar-like motion, propelling water parallel to the cell surface (*Fig. 5b*).

Within the protista the major groups currently considered to be monophyletic are the **Euglenozoa**, **Alveolata** and **Stramenopila**. However, there are a large

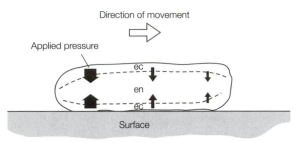

Fig. 4. A proposed mechanism for amoeboid movement. Symbols: ec = ectoplasm; en = endoplasm.

number of taxa that have yet to be assigned, and there are new groupings emerging as more molecular data are published.

Euglenozoa
The Euglenozoa contains the **flagellate** forms, the **Euglenids** and **Kinetoplastids**.

Euglenids
Euglenids are unicellular, motile organisms, with two flagellae that originate from a pocket at the anterior end of the cell (*Fig. 3a*). They may have a heterotrophic or photosynthetic nutrition. When photosynthetic they contain chloroplasts with chlorophylls *a* and *b* and they store **paramylon** starch. These features suggest that the chloroplasts in this group originated from a Chlorophytan ancestor. They inhabit nutrient-rich (**eutrophicated**) fresh and brackish water. Many species are capable of phagotrophy, and some are wholly saprophytic and devoid of chloroplasts. A few species are parasitic. They have a proteinaceous flexible pellicle that lies underneath the cell membrane very like that of the closely related alveolatan species. The flexibility of this pellicle allows euglenids a characteristic flexibility termed **euglenoid movement** when cells are on solid substrates. The pellicle allows the organism to move through muds and sands. The nuclei appear to contain many chromosomes, and polyploidy appears common.

Kinetoplastids
The sister group to the euglenozoa is the kinetoplastids, parasitic flagellates with a single large mitochondrion and a naked cell membrane (*Fig. 3b*). They share many of the characteristics with the euglenoids. The group are named after a

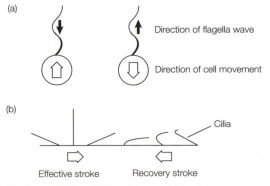

Fig. 5. Motility in (a) flagellates and (b) ciliates.

large structure within the mitochondria called the **kinetoplast** which contains DNA and associated proteins. The DNA is in the form of **maxicircles** and **minicircles**. Maxicircles encode oxidative metabolism enzymes, minicircles encode unusual RNA-editing enzymes (see Section K6).

Alveolata

The second major group is the alveolata. They have a unique cell surface where the cell plasma membrane is underlain by a layer of vesicles called alveoli, which contain cellulose plates or scales generated from the Golgi. These vesicles fuse with the plasma membrane and provide the cell with an effective cell wall. The plates can be impregnated with silica or calcium carbonate. Within this group are the ciliates, dinoflagellates and the apicomplexa. Some species in the Alveolata have a plasma membrane supported by a proteinaceous pellicle, very similar to that seen in the euglenids.

The ciliates

The ciliates are a largely free-living group, that are characterized by their dense covering of cilia and complex nuclear arrangements (*Fig. 3c*). Many species are photosynthetic, having acquired chloroplasts from photosynthetic symbionts (see Section K6).

Dinoflagellates

Dinoflagellates are predominantly unicellular, marine, free-living, motile organisms (*Fig. 3d*) Although unicellular, the dinoflagellates are structurally an extremely diverse group of unicellular organisms. Most have two flagellae inserted into the cell at right angles to each other, around the midline of the cell. One is wrapped around the waist of the cell in a groove, the other extends from the posterior of the cell. Their chlorolasts contain chlorophylls *a*, *c*1 and *c*2 and the carotenoid **fucoxanthin**. They store **chrysolaminarin**, a β1–3 linked glucose. Some are capable of phagotrophy; others live within marine invertebrates and are termed **zooxanthallae** (see Section K4).

Apicomplexans

Apicomplexans are a wholly parasitic group and have a body-form much like an amoeba, but they have an apical complex, a mass of organelles contained within the apical end of the cell (*Fig. 3e*). This structure helps the cell invade the host tissue (see Section K6).

The Stramenopila

The Stramenopila contain the **diatoms**, **chrysophytes**, **oomycetes**, and **opalines** plus other large multicellular groups. Also within the stramenopila are several microbial genera with the amoebic body form including the **slime molds.** Many of the Stramenopila have arisen after the symbiotic association between a non-photosynthetic protistan and **red** or **brown** photosynthetic species some time ago in evolutionary history. The pigment content, the numbers of ER layers around the chloroplast, and the numbers of thylakoid lamellae confirm this hypothesis. Those species that contain red pigments from the cryptophytes can occupy the very deepest layers of the photic zone (see Topic I2). Most species in this group are unicellular, but some are colonial or filamentous. Life cycles are similar to those of the chlorophytes, but the dominant vegetative stage is diploid.

The diatoms

The diatoms (golden brown species of protista) contain chorophylls *a* and *c* and various accessory pigments. Diatoms differ from other members of the golden brown protista because their vegetative stage often lacks flagellae, but they have a **gliding motility** on solid surfaces. They have a silica-containing cell wall composed of a pair of 'nested' shells called **frustules** with a **girdle band** around them. The large half of the shell is termed the **epitheca**, the smaller the **hypotheca** (*Fig. 3f*).

Chrysophytes

The chrysophytes contain chlorophylls *a*, *c*1 and *c*2 and have two flagellae inserted into the cell at near right angles to each other. Some species are covered in radially or bilaterally symmetrical scales (*Fig. 3g*). They are also character- ized by the formation of spores. Spore formation is termed **intrinsic** and it is independent of external conditions. Around 10% of the population will encyst as a zygotic spore in any one generation, allowing populations in optimal growth conditions to maintain genetic diversity and reduce intraspecific competition.

The water molds, Oomycetes and Hyphochytrids

Members of the Oomycota are common water molds, saprophytes or parasites of animals and plants. They share many morphological characters of the chytrids (see Section J1), but differences include cellulose-containing cell walls, tubular mitochondrial cristae and a life cycle where the dominant somatic phase is diploid rather than haploid or dikaryotic (*Fig. 3h*). They also possess biflagel- late zoospores instead of the single flagellate zoospores found in the chytrids (see Section K3). From DNA-sequence analysis their closest relatives appear to be the dinoflagellates. These organisms are capable of causing devastating disease in both plants and fish.

Members of the Hyphochytrids also have cellulose-containing cell walls, and these organisms have only recently been separated from the chytrids. A haploid vegetative stage is dominant in the life cycle. Members of this phyla of organisms are destructive **intracellular parasites** of plants with an absorptive nutrition. They cause enlargement and multiplication of host cells, creating large and unsightly clubbed roots and often death of the plant. The best known disease is **club root** of crucifers. Like the Oomycota, though they share morphological and nutritional similarities with the chytrid fungi, they have some cellulose in their cell walls, and DNA-sequence data also indicate a closer relationship with the dinoflagellate algae than that with the fungi.

Amoebae and slime molds

Amoebae are naked protistan cells that have an absorptive nutrition (*Fig. 3i*) They have a wide distribution, living free in soil and water, and as parasites of animals and man. They should not be considered as a monophyletic group as there are amoebic groups aligned closely to the Euglenozoa and the Alveolata. Reproduction in the amoebae is by binary fission of the cell after mitotic division of the nucleus. A single cell splits to form two identical progeny.

Often grouped together, the phyla termed **Dictyosteliomycota** and the **Acrasiomycota** are phyla that contain the cellular slime molds. They exist for most of their life cycle as haploid amoebae that feed within soil by engulfing bacteria. They are uninucleate for most of their life cycle but form a plasmodium at sporulation.

Myxomycota are acellular slime molds. They exist as haploid amoebae for their vegetative stage, but fuse in pairs to form a diploid cell that undergoes repeated mitotic nuclear divisions without cell division, forming a plasmodium. The life cycle is then very similar to the cellular slime molds.

Opalines
A group of specialized ciliate-like organisms, found living saprophytically in the bowels of amphibians. They are covered with closely spaced flagellae (see Section K6).

Chlorophytan and protistan growth

Growth in the unicellular Chlorophyta and protista is synonymous with binary fission. In most unicells, haploid or diploid nuclei undergo mitosis, and the cell then divides longitudinally to form two daughter cells. In some species there are two haploid divisions within the parent cell, followed by the formation of four motile daughter cells (see Section I3). Some cenocytic filamentous algae grow from the tip of the filament in a way very similar to that of hyphal growth (see Section J1). Others grow by division of vegetative cells within filaments or sheets.

Accurate estimates of Chlorophytan and protistan growth rates can be made by cell counting or by estimating chlorophyll content of a culture. Kinetics of growth are similar to those seen in the bacteria (see Topic D1), but for photosynthetic species, depletion of nutrients other than carbon leads to culture limitations and the stationary and death phases. Nitrogen, phosphates or silicon are frequently limiting.

K2 NUTRITION AND METABOLISM

Key notes

Carbon and energy metabolism	The Chlorophyta and pigmented protista are photosynthetic organisms and obtain their carbon and energy requirements by the fixation of CO_2, using photosynthesis. In a terrestrial habitat, light levels are usually adequate to support photosynthesis, but in the aquatic habitat light energy is rapidly absorbed in the top 0.5 m of the water column. Aquatic species have evolved three chlorophylls, *a*, *b* and *c*, and a large number of accessory pigments to allow them to extend the depth to which they can grow. Photosynthetic species use aerobic respiration via glycolysis and the citric acid cycle. Non-photosynthetic protista can obtain their carbon and energy requirements from the environment by diffusion, pinocytosis and phagocytosis. In aerobic species, glycolysis, the citric acid cycle and mitochondrial respiration provide the cell with energy and metabolites. Anaerobic protista may utilize fermentative pathways within the hydrogenosomes.
Oxygen and carbon dioxide	In the terrestrial environment, photosynthetic CO_2 and O_2 requirements are almost always satisfied by atmospheric gases. In the aquatic environment, the solubility of O_2 decreases with temperature and increasing dissolved CO_2 levels, and availability of O_2 therefore becomes limiting in warm waters.
Nitrogen nutrition	No Chlorophytan or protistan species can fix nitrogen and they all must therefore obtain it in a fixed, inorganic or organic form. Most can utilize nitrate or ammonia; some require organic compounds. Nitrogen levels may be limiting to growth in marine environments.
Macronutrients, micronutrients and growth factors	Most nutrients are available to excess in the aquatic environment, but phosphates and silicon are only poorly soluble in water and are often limiting to growth in fresh water. Some species are predominantly autotrophic, but many require an external supply of amino acids, vitamins, nucleic acids and other growth factors; such species are described as auxotrophic.
Water, pH and temperature optima	The members of the Chlorophyta and protista do not survive severe desiccation and are therefore found mostly in damp terrestrial habitats or in water. Phagocytic species have an absolute requirement for liquid water. Most species can tolerate a wide range of pH and temperature. Some are specialized and can inhabit extremely acidic, hot springs while others can complete their entire life cycle below 0°C.

Related topics

Heterotrophic pathways (E2)
Autotrophic reactions (E4)
Measurement of microbial
   growth (D1)
Eukaryotic cell structure (I2)
Beneficial effects of fungi in their
   environment (J4)
Biosynthetic pathways (E5)

Chlorophytan and protistan
   taxonomy and structure (K1)
Beneficial effects of the Chlorophyta
   and protista (K4)

Carbon and energy metabolism

All species of the Chlorophyta and many species of the protista are photosynthetic and therefore gain carbon and energy from the fixation of atmospheric or dissolved carbon dioxide using photosynthesis.

The photosynthetic reactions take place in the chloroplasts (see Section K1), the light reactions occurring in the chloroplast thylakoids, and the light-independent reactions occurring in the stroma (see Section E4). The chlorophyll pigments are membrane-bound within the thylakoids, and their properties include the ability to be excited by light. Different chlorophylls accept light energy of different wavelengths, depending on their structure. Accessory pigments, the **carotenoids, phycobilins** and **xanthophylls**, also absorb light energy of different wavelengths and pass their excitation to chlorophyll, maximizing the breadth of wavelength over which light energy can be absorbed. The role of accessory pigments is particularly important as they allow photosynthetic organisms to occupy different parts of the **photic zone**, the shallow layer of water where sufficient light penetrates to support photosynthesis. The actual depth to which light will penetrate varies with turbidity and dissolved organic matter content (see Topic I4). Furthermore, light of different wavelengths penetrates water to

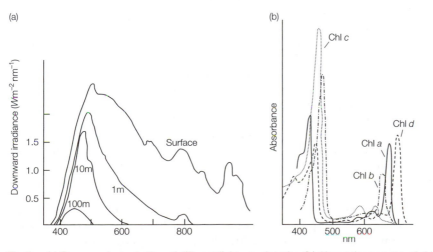

Fig. 1.   (a) Downward penetration of different light wavelengths; (b) Absorption spectra of chlorophylls.

different degrees (*Fig. 1a*). Red light is absorbed rapidly by water, blue light least. The members of the Chlorophyta occupy only the shallowest of waters.

Chlorophylls *a*, *b* and *c* are the photosynthetic pigments, with chlorophyll *a* being the primary photosynthetic pigment. The absorption **maxima** of chlorophyll *a* are at 430 and 663 nm (*Fig. 1b*). Other chlorophylls are accessory pigments and have slightly different absorption maxima: chlorophyll *b* at 435 and 645 nm, chlorophyll *c*1 at 440, 583 and 634 nm, and *c*2 at 452, 586 and 635 nm. Phycobilins absorb light energy at 565 nm. There are also other accessory pigments found in the different algal groups, the carotenes and xanthophylls. Depending on the different accessory pigments present in individual species, light of different wavelengths can be absorbed and utilized in photosynthesis.

Nutrient uptake in non-photosynthetic protista may be by **diffusion**, **pinocytosis**, or **phagocytosis**. Within nutrient-rich tissues, such as those that intracellular parasites or blood-dwelling parasites live in, nutrient uptake is by diffusion and pinocytosis, the formation and digestion of small **vesicles** from the plasma membrane. These vesicles capture soluble nutrients from the environment.

In most larger, free-living species, particulate organic matter and bacteria are engulfed by phagocytosis. In this process large vacuoles can form anywhere over the cell surface in the amoebae, or from specialized sites such as the cytostome, in ciliates and flagellates. Digestion of the vesicle content occurs by the fusion of lysozyme-containing enzyme vesicles with the phagocytic vesicles. Once digestion is complete enzymes are recycled by the cell and sequestered into small vesicles, whilst cell debris is expelled from the cell by **reverse phagocytosis**, where the food vacuole re-fuses with the plasma membrane.

A large number of symbiotic species can be found in the protista, in association with many different photosynthetic symbionts. Species of Alveolata and Euglenozoa with these symbionts have an autotrophic nutrition, relying on photosynthetic products for carbon and energy sources (see Topic B4). In some cases the cytostome seen in non-photosynthetic protistans is not present. Further evidence of the close taxonomic relationships between the photosynthetic and non-photosynthetic groups is provided by the fact that it is possible to 'cure' some species of their symbionts, and once cured, these protista return to their holozoic nutrition and form food vesicles from a reformed cytostome.

Almost all members of the Chlorophyta are aerobic and use mitochondrial respiration with oxygen as the terminal electron acceptor. In most of the protista, glycolysis and the citric acid cycle provide energy and intermediates for cellular metabolism (see Section E2). Aerobic protista use mitochondrial respiration, with oxygen as the terminal electron acceptor, generating ATP.

Anaerobic species have a fermentative metabolism (see Section E2), which results in the incomplete oxidation of substrates. A few species that live in anaerobic lake sediments can use alternative electron acceptors such as nitrate.

There are several modifications to the usual glycolytic pathway (see Topic B1) found in some members of the protista. Some can utilize **inorganic pyrophosphate** rather than ATP, replacing enzymes like pyruvate kinase with pyrophosphate kinase. The advantage of this is that pyrophosphate kinase activity is reversible and can be used to synthesize glucose from other substrates.

In parasitic kinetoplastids glycolysis does not occur in the cytoplasm but in its own organelle, the **glycosome** (see Section K6). Within this organelle a modified form of glycolysis can give rise to glycerol which is the substrate of respiration in this group (see *Fig. 2*). In anaerobic protista fermentative reactions occur within the **hydrogenosome**. Substrates like pyruvate and malate from

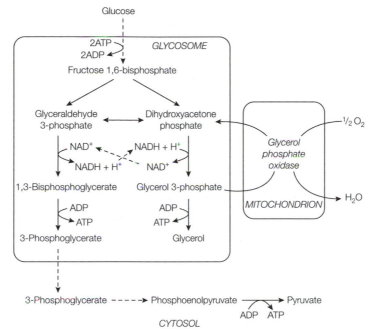

Fig. 2.   Aerobic glycolysis in bloodstream forms of trypanosomes, showing compartmental-ization of much of the pathway in the glycosome. Dotted lines indicate several reaction steps.

glycolysis are taken up by this organelle and incompletely oxidized to end products like acetate, in the process forming ATP. Hydrogen is another characteristic end-product of hydrogenosomal metabolism (see *Fig. 3*).

In many protista mitochondrial metabolism resembles that of bacteria respiration, having a branched pathway, part of which is sensitive to cyanide (mammalian-like) and part of which has an alternative terminal oxidase which allows the mitochondrion to operate at low oxygen levels. These modifications are most developed in the specialized parasites, where both cellular and mitochondrial structure change as the parasite passes though different stages of the life cycle (see Section K6).

Oxygen and carbon dioxide

In an aquatic environment temperature influences levels of dissolved oxygen. Water is saturated with oxygen at 14 mg $O_2 l^{-1}$ at 0°C, but only 9 mg l^{-1} at 20°C. Oxygen utilization by living organisms increases with a rise in temperature; thus, availability of oxygen is likely to limit growth in warmer waters.

Carbon dioxide, the levels of which in water vary inversely with dissolved oxygen, provides carbon to autotrophic species for photosynthesis. Dissolved carbon (as **carbonic acid**, HCO_3^-) is usually present at between 2.2 and 2.5 $nmol^{-1}$, while CO_2 is present at only 10 mmol l^{-1}. Most Chlorophyta and photosynthetic protista utilize carbonic acid. Anaerobic photosynthesis occurs in a few species, using hydrogen sulfide or carbon dioxide as terminal acceptors (see Section E3).

The absolute requirement for light in photosynthetic species of Chlorophyta and protista means that they must have adaptations, which allow them to remain in the photic zone of their environment. On land this is not problematic, but in the aquatic ecosystem there is a natural tendency for cells to sink, and thus

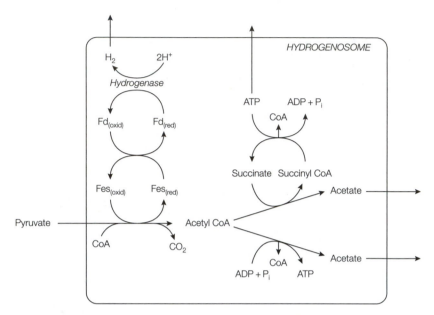

Fig. 3.   Hydrogenosomal metabolism in the trichomonads. Fes = Iron-sulfur proteins; Fd = ferrodoxin; oxid = oxidized state; red = reduced state.

there are cell modifications that counter this. Many Chlorophytan and protistan cells, zoospores and gametes are flagellate and can swim towards the light (see Section K1). Other species have structural modifications like spines that increase the resistance of the cell to sinking, and catch water uplift. Some have **gas vacuoles** that increase bouyancy.

Nitrogen nutrition   Members of the Chlorophyta and the protista are heterotrophic for nitrogen and must obtain it in a fixed form such as nitrate, ammonia or amino acids. Amino acid requirement is common amongst the algae (see Growth factors). Poor availability of nitrogen is a common limiting factor to growth in the marine environment. Complex nitrogen sources such as peptides and proteins can be utilized after either extracellular or intracellular enzyme secretion has digested the polymers. Amino acids can also be stored as a nitrogen store.

Macronutrients, micronutrients and growth factors   Carbon is rarely limiting for growth of photosynthetic species, but nitrogen and phosphorus often are. Fresh-water productivity is often limited by phosphate availability. Silicon is often limiting to diatom growth in nutrient poor, **oligotrophic** lakes. Most other nutrients are present to excess in the aquatic environment.

Although some photosynthetic species are completely **autotrophic**, many require an external supply of **vitamins**, frequently thiamin, biotin, B_{12} and riboflavin, purines, pyrimidines and other classes of growth factors. This requirement is termed **auxotrophy**, and it reflects the abundance of dissolved organic matter to be found in their environment, which allows the selection of populations that do not synthesize all their own metabolic requirements (see Section D1).

Water, pH and temperature optima   Almost all Chlorophyta and protista are limited to damp environments. A few photosynthetic species that are in lichen symbiosis (see Section J4) are protected from desiccation and can survive extremely dry conditions. Most members of the Chlorophyta and protista are tolerant of a wide range of pH; some are specialized and can inhabit highly acidic environments like those of hot sulfur springs. In highly illuminated, oligotrophic lakes, large changes in pH can be caused by variations in dissolved CO_2 concentrations and these changes can have detrimental effects on the populations of photosynthetic species.

Some dormant stages of protista are capable of withstanding 100°C for several hours. Other species can grow and divide at –2°C in seawater, and specialized 'snow algae' have growth optima between 1 and 5°C. Most species have a temperature optimum between 5 and 50°C.

Osmolarity

Seawater and fresh water have very different **osmolarities**. Fresh water species of Chlorophytes and protists have an internal osmotic pressure of 50–150 mOs ml^{-1} while fresh water osmolarity is <10 mOsml^{-1}. Those species that do not have a rigid cell wall to prevent excessive water uptake have contractile vacuoles that collect and expel excess water from the cell. Marine organisms have cytoplasm that is roughly isotonic with seawater and therefore usually do not require contractile vacuoles.

K3 LIFE CYCLES

Key notes

Life cycles in the Chlorophyta	Many members of the Chlotrophyta have a haploid vegetative phase, and gametes (motile or non-motile) are formed by the differentiation of a vegetative cell. Zygote formation is followed by meiosis, producing haploid progeny cells. Other species have a diploid vegetative phase and produce haploid gametes by meiosis.
Life cycles in the Protista	Life cycles can be dominated by haploid or diploid phases. Life cycles in the Euglenozoa are characterized by asexual reproduction in flagellate species by longitudinal binary fission. Sexual reproduction is seen in the parasitic flagellates. In the Alveolata, asexual reproduction is by homothetogenic cell division. Sexual reproduction occurs and the dominant phase of the life cycle is diploid. Asexual reproduction in the apicomplexa is characterized by multiple fission. Sexual reproduction occurs in alternative hosts.
Life cycles in the Stramenopila	The diploid vegetative cells of diatoms reproduce asexually by mitosis. For sexual reproduction in round centric diatoms, one or several macrogametes are formed in an oogonium which is then fertilized by motile microgametes produced from the other mating-type diatom to generate a zygote within an auxospore. Pennate diatoms do not display motile microgametes but after meiosis fuse to form a zygote.
Reproduction in the Oomycota and Hyphochytridiomycota	Except for a few terrestrial species, asexual reproduction in Oomycetes is via flagellate motile diploid zoospores released from terminal sporangia. Sexual reproduction involves haploid gametes produced by terminal or sub-terminal antheridia and oogonia. Hypochytrids reproduce via a multinucleate protoplast that is formed in the host cell, then releasing cysts into the soil that germinate into biflagellate zoospores able to invade new host plant roots.
Life cycles in the amoebae and slime molds	Amoebae undergo mitosis and reproduce asexually by binary fission. Sexual reproduction is by fusion of haploid amoebae of different mating types to produce a diploid, followed by meiosis to generate haploid amoebae once more. Amoebae of the cellular slime molds (Dictyosteliomycota and Acrasiomycota) aggregate upon starvation to yield a multicellular pseudoplasmodium that differentiates into a fruiting body and generates haploid spores. Amoebae of the acellular slime molds (Myxomocota) fuse to form a diploid cell (or this is produced via haploid flagellated cells) which grows into a large multinucleate plasmodium. Sporangia arise and meiosis within them generates haploid spores that germinate to release amoebae.
Related topics	Cell division and ploidy (I3) Fungal structure and growth (J1)

Life cycles in the Chlorophyta

The dominant vegetative phase of members of the Chlorophyta can be haploid or diploid. Vegetative growth is associated with mitotic cell division. In chlorophytes with a haploid life history, meiosis occurs at zygote germination and the cells remain haploid for the whole of their vegetative life. Diploid formation occurs only in the zygote. Chlorophytes with a diploid life history are only haploid at gamete formation and after zygote formation continue their life cycles as diploids.

Motile compatible gametes first entangle flagellae, cells then **conjugate** and nuclei fuse to form a zygote. The zygote may remain motile or it may form a thick-walled resting cyst. Meiosis occurs within the zygote and haploid, flagellate cells are released (*Fig. 1*). Similar events are seen in the colonial forms of the Chlorophyta; all cells of a colony may develop into free-swimming gametes after breakdown of the colony structure.

Filamentous Chlorophyta may reproduce sexually by conjugation. In this process, two vegetative cells form a **conjugation tube** between them and fuse. The cellular content from one cell then moves into the other, where nuclear fusion and zygote formation occur. The zygote encysts and meiosis occurs before the emergence of a haploid new filament (*Fig. 2*). Other filamentous species produce motile gametes of two mating types from different vegetative cells. Often one gamete is considerably larger (**macrogamete**) than the other (**microgamete**). In some species, only one motile microgamete is produced and it fuses with a non-motile gamete cell called an **oogonium** to form the zygote.

Life cycles in the protista

Vegetative growth is associated with mitotic nuclear division followed by cell division. Cell division can be by budding, binary or multiple division. Each of the different groups of protista are characterized by differences in their life cycle. The dominant vegetative phase of protista can be haploid or diploid.

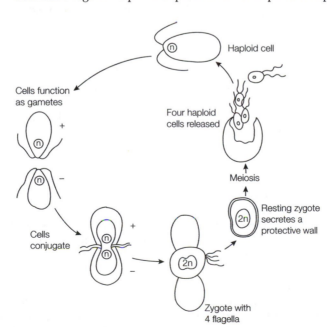

Fig. 1. Sexual reproduction in the unicellular Chlorophyta. Reproduced from Gross, T., Faull, J.L., Kettridge, S. and Springham, D., Introductory Microbiology, *1995, with kind permission from Kluwer Academic Publishers.*

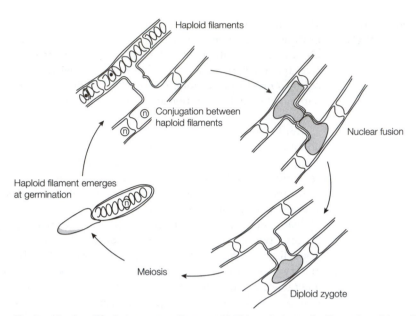

Fig. 2. Conjugation between two filamentous Chlorophytan cells. Reproduced from Gross, T., Faull, J.L., Kettridge, S. and Springham, D., Introductory Microbiology, *1995, with kind permission from Kluwer Academic Publishers.*

In protista with a haploid life cycle, meiosis occurs at the germination of the diploid zygote. The cells remain haploid for the rest of the vegetative phase, forming haploid gametes that fuse to form the transient diploid. Protista with a diploid life history are haploid only at gamete formation. After formation of the diploid zygote they continue their life cycle as diploids until gamete formation necessitates meiosis.

Euglenozoa
Reproduction in the flagellates is characterized by mitosis of the nucleus followed by cell division, which occurs along the longitudinal axis of the cell. This produces mirror image progeny (see Section I3). Sexual reproduction is seen in some species of this group. The group includes many of the parasitic species including trypanosomes (see Section K6).

Alveolata
Dinoflagellates are haploid and have unusual chromosomes, which are condensed throughout their life cycle. The chromosomes contain very little histone. Sexual reproduction commences when motile cells differentiate to become **macro-** or **microgametes** *(Fig. 3)*. The gametes fuse to from a zygote. Meiosis occurs followed by the degeneration of three of the four nuclear products. A haploid, motile cell then emerges from the zygote.

Apicomplexa
Apicomplexan protists are parasites with a complex life cycle (see Section K6). There are both diploid and haploid phases and often two host species are infected in a life cycle. The group is characterized by a type of cell division that is called multiple division or **shizogony**. During this process, multiple division of haploid nuclei occurs, producing many progeny. These progeny are then released into

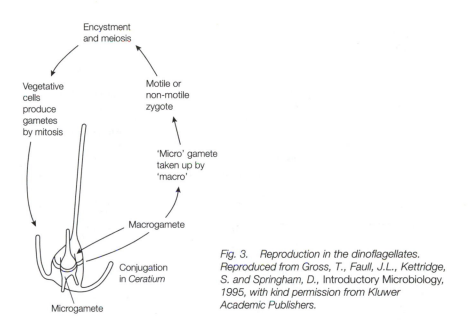

Fig. 3.   Reproduction in the dinoflagellates. Reproduced from Gross, T., Faull, J.L., Kettridge, S. and Springham, D., Introductory Microbiology, 1995, with kind permission from Kluwer Academic Publishers.

body fluids like blood where they rapidly enter new host cells and establish themselves as intracellular parasites. The malarial parasite *Plasmodium* spp. is an example of an apicomplexan parasite (see Section K6).

Ciliates

Ciliate asexual reproduction is also by binary fission after mitosis of the nucleus. Cell division is described as **homothetogenic**, across the narrow part of the cell. Ciliates have two types of nuclei, a **micronucleus** that is diploid, contains little RNA and a lot of histone, and a **macronucleus** which is polyploid and controls day-to-day cellular activities. There may be hundreds of macronuclei in one ciliate, and up to 80 micronuclei. During asexual mitotic cell division, both of these nuclei divide to provide progeny cells with at least one copy of both nuclei. During sexual reproduction the macronucleus degenerates and the micronucleus undergoes meiosis to form gametes. Gametes fuse to form a new diploid nucleus. This new nucleus divides, one copy of the nucleus remaining as the new micronucleus, and the others differentiating to provide a new polyploid macronucleus (*Fig. 4*).

This unusual state of nuclear dualism appears to confer an advantage to the ciliates by having a separate genetic store in the micronucleus, and enhances RNA synthesis by the macronucleus. This allows them to be very adaptable to changing environmental conditions.

Life cycles in the Stramenopila

Within the Stramenopila there are a number of life cycle patterns.

The diatoms

Vegetative cells are diploid, and repeated mitotic divisions lead to a reduction in cell volume as daughter cells synthesize new frustules that fit within the inherited parent frustule. Once a 30% reduction in volume has been reached diatoms either produce a resting spore (or **auxospore**) to regain cell size or they reproduce sexually (*Fig. 5*). Sexual reproduction begins by the formation of

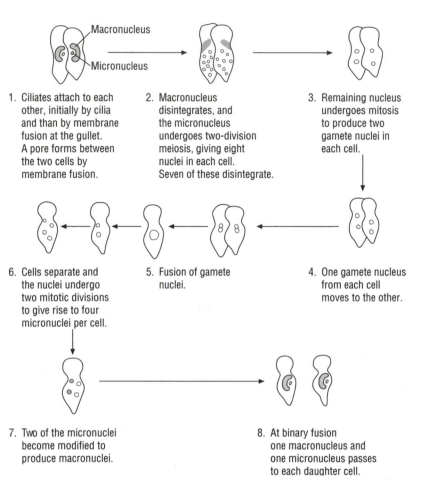

1. Ciliates attach to each other, initially by cilia and than by membrane fusion at the gullet. A pore forms between the two cells by membrane fusion.

2. Macronucleus disintegrates, and the micronucleus undergoes two-division meiosis, giving eight nuclei in each cell. Seven of these disintegrate.

3. Remaining nucleus undergoes mitosis to produce two gamete nuclei in each cell.

6. Cells separate and the nuclei undergo two mitotic divisions to give rise to four micronuclei per cell.

5. Fusion of gamete nuclei.

4. One gamete nucleus from each cell moves to the other.

7. Two of the micronuclei become modified to produce macronuclei.

8. At binary fusion one macronucleus and one micronucleus passes to each daughter cell.

Fig. 4.   Sexual reproduction in the ciliates. Redrawn from Gross et al., Introductory Microbiology, 1993.

gametes after meiosis. In round, centric diatoms a single or multiple macro-gamete forms within an **oogonium**. Motile microgametes are formed within the other mating-type diatom. These microgametes are released and fertilization of the oogonium leads to the formation of a zygote within an auxospore. The auxospore enlarges and secretes a new pair of full-size frustules. The diploid, vegetative life cycle then continues. The long, thin diatoms, termed pennate, do not form motile gametes, but after a meiotic division fuse somatically to form a zygote.

Reproduction in the Oomycota and Hyphochytri-diomycota

Oomycetes are filamentous, coenocytic organisms. Asexual reproduction is characterized by the production of flagellate, motile, diploid zoospores from terminal sporangia. Sexual reproduction occurs after meiosis and gamete pro-duction in terminal or sub-terminal antheridia and oogonia. A few species are terrestrial and produce non-motile gametes from sporangia and Oogonia. The life cycle of the plant pathogen *Phytophthora infestans* is typical of this terrestrial group (*Fig. 6*).

Hyphochytrids reproduce by the formation of multinucleate, unwalled proto-plasts within an enlarged host cell. Cyst formation within the host cell occurs, and

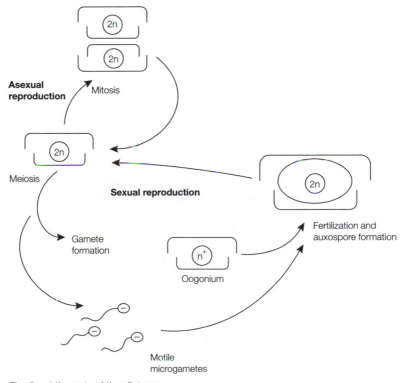

Fig. 5.   Life cycle of the diatoms.

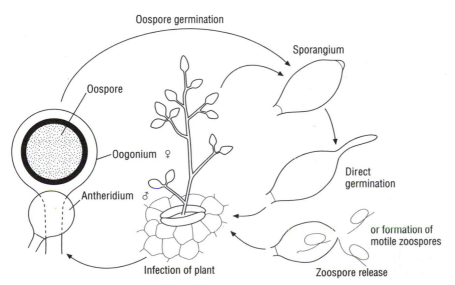

Fig. 6.   Life cycle of a water mold, Phytophthora infestans. Redrawn from Deacon, Modern Mycology, with permission of Blackwell Science.

these are released into the soil on breakdown of the plant root where they can persist for years. The presence of a suitable host plant root breaks dormancy, and the cyst germinates by producing anteriorly biflagellate zoospores that swim to the host root and actively penetrate it.

Life cycles in the amoebae and slime molds

Amoebae

Amoebae have a very wide distribution, living free in soil and water, and as parasites of man and animals. They should not be considered as a mono-phyletic group. Reproduction in amoebae is by binary fission of the cell after mitotic division of the nucleus. A single cell splits to form two identical progeny. Sexual reproduction is seen in some species, the haploid amoebae differentiating into gametes of different mating types which can then fuse to form the diploid zygote. Meiosis then occurs and the haploid phase of the life cycle continues.

Cellular slime molds

Starvation of amoebae of the Dictyosteliomycota and Acrasiomycota cellular slime molds leads to their aggregation into multicellular **pseudoplasmodium**, which eventually differentiates into a fruiting body consisting of a foot, stalk and sporangium. Haploid spores are formed which are disseminated and give rise to haploid amoebae (*Fig. 7*).

Acellular slime molds

In the Myxomycota, the acellular slime molds, under certain conditions, repro-duction will commence by the fusion of amoebae, or amoebae can produce flagellate cells which fuse, to form a diploid cell (*Fig. 8*). This diploid cell grows, feeds and undergoes repeated mitotic nuclear divisions without cell division, forming a large plasmodium. Sporangia are produced within which meiosis occurs to generate haploid spores. Each spore germinates to release an amoeba to complete the life cycle.

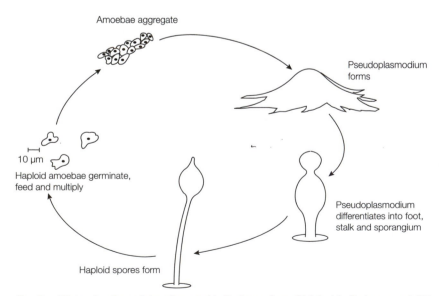

Fig. 7.   Life cycle of a cellular slime mold. Redrawn from Sleigh, M., Protozoa and Other Protists, 1991, with kind permission from Cambridge University Press.

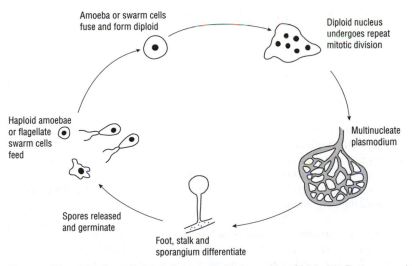

Fig. 8. Life cycle of an acellular slime mold. Redrawn from Sleigh, M., Protozoa and Other Protists, 1991, with kind permission from Cambridge University Press.

K4 BENEFICIAL EFFECTS

Key notes

Primary productivity

Primary productivity in the oceans due to Chlorophytan and photosynthetic protistan photosynthesis is estimated to be ~50 × 109 tonnes per annum. Carbon can be released as dissolved organic matter (80–90%) or as particulates (10–20%). The photosynthetic cells, associated saprophytic Bacteria and dissolved organic matter together form the basis of the aquatic food chain.

Symbiosis

Symbiotic associations can occur between species of Chlorophyta, protista, fungi and animals. In these associations, carbohydrates from photosynthesis are exchanged for nutrients from the fungus or animal partner. Protista can have internal symbionts that can be prokaryotes or other protists. Photosynthetic symbionts provide photosynthate or other factors for the heterotrophic protistan partner. Bacterial endosymbionts can provide vital cell metabolites. Ectosymbiotic prokaryotes may provide protista with motility or alternative electron acceptors. Some protista are important within gut and rumen microflora of large animals, as part of a complex cellulolytic community.

Diatomaceous earth

Diatomaceous earth is formed from the silica-containing shells of diatoms. It has several commercial uses that take advantage of its chemically inert but physically abrasive qualities.

Bioluminescence

Bioluminescence is found in several of the protistan phyla and is associated with luciferin–luciferase reactions that occur within scintillons in the protistan cell. The precise function of this reaction is unknown, although it may provide defense against predators.

Related topics

Beneficial effects of fungi in their environment (J4)
Chlorophytan and protistan taxonomy and structure (K1)

Chlorophytan and protistan nutrition and metabolism (K2)
Life cycles in the Chlorophyta and protista (K3)

Primary productivity

In the ocean the total plant biomass (as **phytoplankton**) is estimated to be 4×10^9 tonnes (dry weight). Annual net **primary production** is around 50×10^9 tonnes. Productivity ranges from 25 g carbon m^{-3} in oligotrophic waters to 350 g carbon m^{-3} in eutrophic waters. This is released as **dissolved organic matter (DOM)**, which accounts for 80–90% of organic matter present in the sea. The DOM is used as a nutrient source by heterotrophic Bacteria; populations between 10^4 and 10^6 ml^{-1} are commonly found in nutrient-poor (**oligotrophic**) waters, and much higher populations are found in nutrient-rich (**eutrophic**) waters (see Topic I5). The cells (both alive and as dead particulates), DOM and prokaryotic populations form the base of the aquatic food chain.

Symbiosis

Protista can form beneficial associations with prokaryotes, fungi, Chlorophyta, other protista and insects and higher animals. These symbioses can be specific and physically and physiologically intimate or they can be relatively non-specific and merely loose ecological associations.

Endosymbionts are symbionts that live inside other organisms. Photosynthetic species can occur within Alveolata and Euglenozoa (and others) and their presence allows the protistan dual organism to adopt a phototrophic habit. The photosynthetic partner is confined to a membrane-bound vacuole, but it is capable of cell division. There is a two-way exchange of materials where the products of nitrogen metabolism of the heterotrophic protistan are utilized by the photosynthetic partner and the products of photosynthesis are utilized by the heterotrophic partner.

The **zooxanthellae** are symbiotic dinoflagellates that are found as coccoid cells within animal cells. They are enclosed in intracellular double membrane-bound vacuoles, which remain undigested. They are found in protista, hydroids, sea anemones, corals and clams where they provide glycerol, glucose and organic acids for the animal, and the symbiotic alga gains CO_2, inorganic nitrogen, phosphates and some vitamins from the animal. **Reef-building corals** are only able to build reefs if they have their symbiotic photosynthetic partner, and many of the animal hosts are at least partly dependent on the algal partner for carbohydrates. **Radiolaria**, responsible for massive primary productivity in the oceans, are wholly dependent on their photosynthetic symbiont for carbohydrate.

Many aerobic protista contain bacteria as endosymbionts. *Amoeba proteus* has a symbiotic Gram-negative Bacterium which is essential to the survival of the amoeba, and *Paramecium aurelia* has specific symbiotic algae that are responsible for the secretion of killer factors that are important in competition within environments and mating.

There are a large number of Bacterial ectosymbionts. Spirochetes and many other species of Bacteria are often attached to the euglenozoal pellicle, arranged in very specific patterns. For example *Myxotricha paradoxa* is an inhabitant of termite guts. Its motility within this environment depends on the co-ordinated movement of adherent spirochetes.

Ecto- and endosymbiotic Bacteria are also found associated with anaerobic alveolata from sulfur-rich environments. *Kentrophorous lanceolata* has a dense mat of sulfur Bacteria on its dorsal surface. These symbionts provide alternative electron acceptors to the protistan.

Protista can be symbiotic with insects and higher animals, and are typically found within fermentative guts as mentioned above. They are part of a complex ecosystem where cellulose is broken down by fungi and prokaryotes, and their metabolic products are fermented by the protista and prokaryotes to provide fatty acids for their hosts. Some of the protistan population also predate the prokaryotic population to provide control of numbers.

Diatomaceous earth

At death, diatom cells fall through the water column to the sea-bed. The inert nature of the frustule silica, silicon dioxide (SiO_2), means that it does not decompose but accumulates, eventually forming a layer of diatomaceous earth. This material has many commercial uses to humans, including filtration, insulation and fire-proofing and as an active ingredient in abrasive polishes and reflective paints. Recently, it has been used as an insecticide, where the abrasive qualities of diatomaceous earth are used to disrupt insect cuticle waxes, causing desiccation and death.

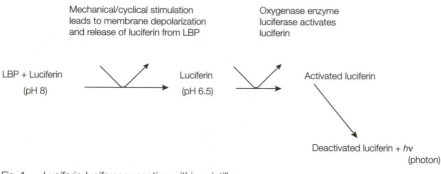

Fig. 1.   Luciferin-luciferase reaction within scintillon.

Bioluminescence   Many of the marine dinoflagellates are capable of **bioluminescence**, where chemical energy is used to generate light. The light is in the blue–green range, 474 nm, and can be emitted as high-intensity short flashes (0.1 s) either spontaneously, after stimulation, or continuously as a soft glow.

Bioluminescence is created by the reaction between a tetrapyrrole (**luciferin**), and an oxygenase enzyme (**luciferase**). The entire reaction is held within membrane-bound vesicles called **scintillons**, where the luciferin is sequestered by **luciferin-binding protein** (**LBP**) and held at pH 8. Release of light is stimulated by either a cyclical or mechanical stimulation of the scintillon, which leads to pH changes within it. Luciferin is released from LBP, and luciferase is able to activate it. Activated luciferin exists for a very brief time before it returns to its inactivated form, releasing a photon of light energy (*Fig. 1*).

The distribution of scintillons within the photosynthetic cell varies over 24 h. During the night they are distributed throughout the cytoplasm, but in daylight they are tightly packed around the nucleus. The function of bioluminescence for the organism is uncertain. It appears to be useful in defense against predators. It has become important in medicine and science because the coupled system of luciferin/luciferase can be used to mark cells. Once tagged with the bioluminescent marker, marked cells can be mechanically sorted from other cells, visualized by microscopy or targeted for therapy.

K5 DETRIMENTAL EFFECTS

Key Notes

Parasitic relationships	A wide range of organisms can be parasitized by members of the protista. They may enter via the mouth and colonize the gut where they can cause asymptomatic, mild disease. More specialized parasites cause greater effects on the gut, and cause disease symptoms. Some highly specialized parasites can enter via the gut but move into other tissues to complete their life cycles. Blood parasites are transferred via vectors from previously infected hosts. They have a highly specialized life cycle. A colorless Chlorophytan, Prototheca, can also parasitize damaged skin tissue.
Toxic blooms	Toxic blooms occur when nutrient-rich water supports population explosions of some photosynthetic species of protista. Some of these species can produce toxins that affect marine animals and humans that come into contact with them.
Eutrophication	The presence of excess nutrients in an aquatic ecosystem will support high levels of bacterial and eukaryotic microbial growth. The oxygen demand for this growth is great and will exceed supply, creating anoxic conditions. This is called eutrophication.
Related topics	Chlorophytan and protistan taxonomy and structure (K1) Chlorophytan and protistan nutrition and metabolism (K2) Lifecycles in the Chlorophyta and protista (K3) Beneficial effects of the Chlorophyta and protista (K4)

Parasitic relationships

Parasitism of other organisms by protista is a very common nutritional strategy. Hosts can range from other members of the protista, including aquatic and higher plants, to many species of multicellular animals (see Section K6).

The lowest level of specialization is seen in those that inhabit the intestines of animals as commensals. They gain entry via food or water and exit via the feces. Many species live on gut prokaryotes. They are transient visitors, able to encyst before they are shed to ensure they survive adverse conditions outside the host. Many of these species have a fermentative metabolism to enable them to survive the low oxygen levels of the gut. A more specialized group attaches itself to the host gut wall to remain within the gut for longer periods. They again gain their nutrition from the passing gut contents and prokaryotes. These may cause some irritation to the gut epithelium.

The third group derive their nutrition from the gut epithelium and cause considerable damage in doing so. They have a number of pathogenicity factors including toxins and proteolytic enzymes called proteolysins. A more specialized form of parasitism is found in species that infect orally but then migrate into other tissues for part of their life cycle.

Parasites that have a life cycle that includes a blood-borne phase are highly specialized. They require vectors to reach their chosen niche, usually the blood-sucking insects. Part of the life cycle is completed in the vector. Trypanosomes and the malarial parasites (see Topic K6) are two of the most significant human protistan parasites. During the blood-borne phase of disease, parasite metabolism can be anaerobic and mitochondrial metabolism is switched off (see Topic K6).

Prototheca is a common soil-dwelling alga that has lost its chlorophyll and lives saprophytically. It is an opportunist pathogen that can enter wounds on the feet, where it can cause a subcutaneous infection. The initially small lesion can spread through the skin, producing a crusty, warty lesion. It may spread to lymph nodes and the disease can become debilitating. In rare cases it can grow rapidly in the bloodstream, causing rapid death in ailing or immunosuppressed animals and humans.

Toxic blooms

When environmental conditions for growth are optimal, photosynthetic protista can grow exponentially, leading to high local populations (see Topic I1). This can occur in several species of marine dinoflagellates that contain poisons which are toxic to fish, invertebrates or mammals, depending on the species of alga and the class of compound they produce. The toxins are accumulated in the digestive glands of shellfish and when consumed by man cause **paralytic shellfish poisoning**. Many of these compounds are neurotoxic. **Saxitoxin**, accumulated in shellfish and accidentally consumed, causes numbness of the mouth, lips and face which reverses after a few hours.

Toxins can also accumulate in higher animals like fish. For example **ciguatoxin** from *Gambierdiscus toxicus* accumulates in muscle tissue of grouper and snapper and when eaten causes gastric problems, central nervous system damage and respiratory failure.

Toxins can also be formed by other groups of algae. Some species of diatoms can produce **domoic acid**, which can accumulate in mussels and, when consumed by humans, causes **amnesic shellfish poisoning**, a short-term loss of memory but which can occasionally cause death. Some of the golden-brown protista that give rise to these types of poisoning are highly pigmented members of the Chrysophycae, and high cell densities can be seen as so-called **red tides** in the sea. Other toxins from species of the Prymnesiophycae can affect gill function of fish and molluscs.

Eutrophication

The presence of high populations of photosynthetic protista, their DOM products and the high numbers of Bacteria that can be supported by large amounts of organic matter lead to a condition known as **eutrophication**. Water is in a nutrient-rich state, biomass is high and oxygen demand exceeds supply. Anoxic conditions rapidly develop, leading to the death of aerobic organisms.

The decomposition of dead material by prokaryotes leads to further demands for oxygen, and the entire environment becomes anaerobic, allowing for the growth of anaerobic Bacteria and Archaea with the production of methane, hydrogen sulfide, hydrogen and many other products of anaerobic metabolism (see Section K4). Eutrophication is commonly seen around untreated sewage outfalls and dairy farm waste run-off, and in fresh-water streams polluted by nitrate-rich agricultural field run-off.

K6 PARASITIC PROTISTA

Key Notes

Protista as human pathogens	Protista are the largest group of single-cell organisms and include deadly parasites. These include the causative agents of some of the most notorious and deadly diseases, e.g. malaria. Here, the major protista pathogens causing human infections are divided into six major groups: flagellates, amoebae, sporozoa, ciliates, myxozoa, and microsporidia.
Flagellates (phylum mastigophora)	Protists in this phylum are characterized by the presence of a flagellum, at least during some phases of their life cycles. These organisms may be parasitic or free-living in anoxic environments and lack mitochondria and Golgi apparati. Examples include *Giardia, Trypanosome, Leishmania, Trichomonas*.
Amoebae (phylum sarcodina)	The majority of protista in this phylum exhibit movement using characteristic pseudopodia (moving of the protoplasm) into a direction. They are typically uninucleate and possess mitochondria. Reproduction is by asexual fission and they may be parasitic or free living. Examples include *Entamoeba, Acanthamoeba, Naegleria, Balamuthia*.
Sporozoans (phylum apicomplexa)	Organisms in this group are all parasitic and characterized by the presence of an apical complex, located at one end of the organism. They exhibit both sexual and asexual reproduction. Examples include *Plasmodium, Toxoplasma, Babesia, Isospora, Cryptosporidium*.
Ciliates (phylum ciliophora)	Members of this phylum are ciliated during at least one stage of their life cycle, and typically move by the beating of cilia. They exhibit both asexual and sexual reproduction, i.e. asexual reproduction occurs by budding, binary, and multiple fission, as well as sexual reproduction by conjugation, autogamy or cytogamy. They may be parasitic, commensal, or free-living. Examples are *Balantidium, Isotricha, Sonderia*.
Microsporidia (phylum microspora)	These include unicellular spores that are intracellular parasites. They do not possess mitochondria, peroxisomes, or hydrogenosomes. The sporoplasm is injected into the host cells via an extruded hollow tube (polar filament), where reproduction takes place by binary fission. Examples include Microsporidia (usually in AIDS patients), *Pleistophora, Glugea, Amblyospora*.
Phylum myxozoa	Members of this phylum are parasites of cold-blooded vertebrates, mostly parasites of fish, but some are reported from amphibians and reptiles. They exhibit both sexual and asexual reproduction. Examples include *Myxobolus, Triactinomyxon, Myxidium, Chloromyxum*.
Locomotion	Protista locomotion is usually mediated by pseudopodia, flagella, cilia or gliding movements.

Pseudopodia	Pseudopodia are involved in amoeboid movement and involve the streaming of endoplasm from one end of a cell to the other. This is driven by unequal pressure applied by a viscous and filamentous ectoplasm (a viscous region of cytoplasm close to the outer cell surface). This type of movement requires that the cell is in contact with a surface.
Cilia and flagella	Cilia exhibit an oar-like motion. During the power stroke, the stiffened cilium propels water lateral to its surface and parallel to the cell surface. For structures as small as cilia, water is highly viscous and thus cilia exert a considerable propelling force on the surrounding water. Flagella exhibit a wave-like motion that propels water either away or towards the cell. Movement away from the cell causes movement of the cell in a direction opposite to the flow of water.
Gliding	In these movements, cell contacts the surface of a substratum and slides over it with the aid of microtubules.
Locomotory proteins	The major proteins involved in protista movement include microtubules that are formed of tubulin molecules and microfilaments composed of actin molecules. The polymerization and depolymerization of these molecules just beneath the plasma membranes pushes the cell surface and causes movement.
Asexual reproduction	Protista exhibit either fission (binary or multiple) or budding as mechanisms of asexual reproduction. Binary fission produces two daughter cells from a parent cell. This requires mitotic division (division produces two identical sets of chromosomes). In multiple fission, the mitotic division occurs several times producing large number of nuclei giving rise to several daughter cells. Budding is a specialized type of fission with daughter nuclei produced within the parent cell, which migrate into the cytoplasmic buds that are released by fission.
Sexual reproduction	The majority of protista with sexual reproduction exhibit gamete formation by meiosis (reduction division). Gametes fuse to form zygote, which undergoes meiosis and numbers of chromosomes are halved followed by asexual reproduction to produce large populations of the organisms. If both gametes arise from the same clone, the species is called monoecious but if they arise from different clones, species is called diecious. However, if self-fertilization occurs, the species is called hermaphrodite. If the new population of an organism develops from unfused gamete without fertilization, it is known as parthenogenesis. The factors that induce gamete formation are not clearly known but may involve environmental conditions such as salinity, pH, temperatures, nutrients, etc. Some ciliated protista are unable to produce gametes. Instead they possess dual nuclei and during their sexual reproduction, the nuclei from two organisms (instead of gametes) fuse together yielding a zygotic nucleus.
Protista life cycles	The life cycle of pathogenic protista can be simple, involving two parasite forms, or one host and two parasite forms or highly complex involving several hosts and several parasite forms. The pathogenic protista with complex life cycles include bloodborne parasites, e.g. *Plasmodium*, that are human parasites and are transmitted via insects. The life cycle in insects may be necessary for parasite development or insects may simply act as carriers

to transmit parasite to the new host. In contrast, protista with simple life cycles include pathogens which are transmitted via contaminated food/water (e.g. *Giardia*, *Acanthamoeba*) and those which are sexually transmitted (e.g. *Trichomonas*). These are usually acquired by new host through exposure to the contaminated water/food or through sexual intercourse respectively. Pathogens thrive on excess food and grow vegetatively and produce infections, however, they switch to resistant forms under harsh conditions, i.e, lack of food, extremes in pH, extremes in temperatures, salinity, etc.

Strategies of immune evasion

The immune system is highly professional in controlling and/or eradicating pathogens. To overcome this, protozoan pathogens employ various strategies to evade the immune system.

Non-induction of immune system

Protozoan pathogens may target organs and/or tissues that have a limited immune response. For example, *Acanthamoeba* produces infection of the avascular cornea or *Naegleria* produces infection of the brain tissues, both of which have limited immune responses.

Anatomical seclusion

Some protozoan pathogens invade non-immune cells and remain intracellular, thus hide from the immune cells. For example, some stages of *Plasmodium* spp. live inside liver cells.

Antigenic variation

Some protozoan parasites evade the immune response by changing the expression of their surface antigens every few generations. Thus antibodies produced against the former antigens become redundant.

Host mimicry

This is achieved by expressing proteins/glycoproteins on the parasite surface that mimic host proteins/glycoproteins or by adsorbing host components on their surfaces. Thus the parasite molecules are not recognized as foreign antigens.

Interference with host immune signalling molecules

Host immune signalling molecules such as cytokines play a crucial role in building an effective immune response. Some protozoan pathogens interfere with their production or mode of action, thus interfering with the build up of an overwhelming immune response.

Strategies against protozoan pathogens

Despite advances in antimicrobial therapy and supportive care, the burden of protozoa infections has remained significant. Thus future research must employ a variety of approaches in the prevention, control, and treatment of parasitic infections.

Chemotherapy

The use of chemicals (natural products or synthetic compounds) has remained the most common and cheapest approaches to treat parasitic infections. A complete understanding of the parasite metabolism will help identify novel targets for i) the rationale development of drugs and/or ii) their role as vaccines.

Control measures

The majority of parasitic protozoa have complex life cycles, thus a complete understanding of their life cycles, vectors, ecological distribution will aid in the development of control measures. Controls can be applied by improvements in sanitation, controlling the vector, and reducing the parasite reservoir.

Protista as human pathogens   Protista are responsible for the most notorious and deadly diseases, killing millions of animals and humans. Among the protista, the following are the important human pathogens (*Fig. 1*).

Flagellates (phylum mastigophora)

Organism	*Trichomonas vaginalis*
Biology	Trophozoites range from 10–25 mm in length, contain a single nucleus, phagocytose Bacteria and leukocytes as food source and reproduce by binary fission but do not form cysts.
Disease	Infection of the urogenital systems.
Symptoms	Usually asymptomatic but when acquired these parasites attach to epithelial cells and can produce severe inflammation or swelling of the sexual organs.
Transmission	Sexual intercourse with infected individuals.
Treatment	Oral application of metronidazole is effective.
Occurrence	Worldwide.

Organism	*Giardia lamblia*
Biology	Trophozoites are approximately 12–15 µm long, contain two nuclei, feed on nutrients obtained from the intestinal fluid, and reproduce by binary fission. They form cysts for transmission.
Disease	Diarrhea (infection of the gut).
Symptoms	Parasites stick to the intestinal epithelium but do not lyse host cells. May cause diarrhea, which is observed within few days to weeks but without blood. Other symptoms include intestinal pain, dehydration and weight loss but non-fatal.
Transmission	Oral uptake of *Giardia* cysts from human faeces in contaminated food or drinking water.

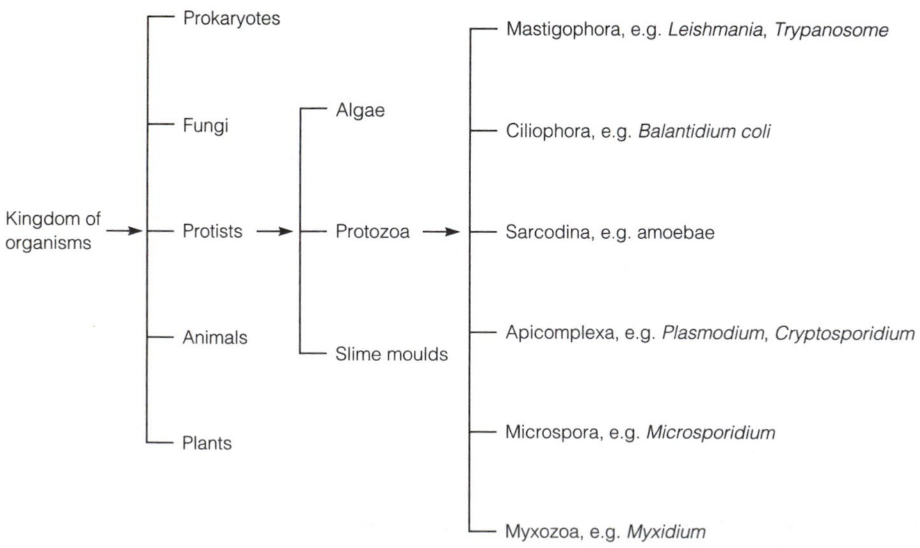

```
                    ┌── Prokaryotes            ┌── Mastigophora, e.g. Leishmania, Trypanosome
                    │                          │
                    │            ┌── Algae     │
                    ├── Fungi    │             ├── Ciliophora, e.g. Balantidium coli
                    │            │             │
Kingdom of          │            │             │
organisms    ──▶    ├── Protists ──▶ Protozoa ──▶── Sarcodina, e.g. amoebae
                    │            │             │
                    │            │             │
                    ├── Animals  │             ├── Apicomplexa, e.g. Plasmodium, Cryptosporidium
                    │            └── Slime moulds
                    │                          │
                    │                          ├── Microspora, e.g. Microsporidium
                    └── Plants                 │
                                               │
                                               └── Myxozoa, e.g. Myxidium
```

Fig. 1.   The classification scheme for major protista.

Treatment	Oral application of metronidazole is effective against the trophozoite.
Occurrence	Worldwide.
Organism	*Trypanosoma brucei gambiense/Trypanosoma brucei rhodesiense*
Biology	Trophozoites are approximately 18–29 µm long, contain a single nucleus, feed on nutrients obtained from the host and reproduce by binary fission but do not form cysts.
Disease	Sleeping sickness or African Trypanosomiasis.
Symptoms	Symptoms observed within weeks and include fever, headache, swollen lymph nodes, weight loss, and heart involvement (usually with *T. b. rhodesiense*). It is important to note that *T. b. rhodesiense* causes acute disease and the host often dies before the disease can develop fully. In contrast, *T. b. gambiense* produces chronic disease and parasite invades the CNS and produces typical symptoms, such as tremors, sleepiness, paralysis, and finally death within months. Of interest, *T. b. brucei* causes fatal infections in non-human mammals, horses, sheep, etc.
Transmission	Disease transmission occurs via insect bite, i.e. tsetse fly (vector) where parasites are present in the salivary glands.
Treatment	Use of suramin, pentamidine, Berenil and difluoromethy-lornithine in early disease may have beneficial effects.
Occurrence	Africa.
Organism	*Trypanosoma cruzi*
Biology	Trophozoites are approximately 20 µm long, contain a single nucleus, feed on nutrients obtained from the host and reproduce by binary fission but do not form cysts.
Disease	Chagas' disease or American Trypanosomiasis.
Symptoms	Following invasion, parasites infect susceptible tissues in particular the heart. Parasites penetrate the myocardial fibers, multiplying for several days and producing a cavity in the invaded tissue and escape into the bloodstream and invade other susceptible tissues including liver, spleen, muscles, intestinal mucosa, resulting in organ failure, and finally death within months to years
Transmission	Disease transmits via insects, i.e. triatomid bugs (vector). While feeding, these bugs defecate on the host, and parasites in the fecal material gain entry into the human body through the bite wound.
Treatment	No effective treatment but use of nifurtimox, ketoconazole, and benznidazole may be effective.
Occurrence	South and Central America.
Organism	*Leishmania tropica / Leishmania major*
Biology	Trophozoites are approximately 2–5 µm long, contain a single nucleus, feed on nutrients obtained from the host and reproduce by binary fission but do not form cysts.
Disease	Cutaneous leishmaniasis.

Symptoms	Following invasion, parasites multiply within reticuloen-dothelial and lymphoid cells. Symptoms include cutaneous lesions within days to months at the site of insect bite.
Transmission	Disease transmits via insects, i.e. sandfly. While feeding, these bugs excrete on the host, which contain parasites. These parasites gain entry into the human body through site of the bite.
Treatment	Injections of Pentostam and Glucantime provide effective treatment.
Occurrence	Africa, Middle East, Asia.

Organism	*Leishmania donovani*
Biology	Trophozoites are approximately 2–5 μm long, contain a single nucleus, feed on nutrients obtained from the host and reproduce by binary fission but do not form cysts.
Disease	Kala-azar or visceral leishmaniasis.
Symptoms	Following invasion, parasites are taken up by macrophages and multiply eventually killing the macrophages and severely affecting the host defenses. Symptoms appear within a few weeks to months and include fever, malaise, anemia, enlarged liver and spleen, and finally death. Of interest, *L. donovani* taken up by neutrophils are killed but this does not have a major effect on the outcome of the disease.
Transmission	Disease transmits via insects, i.e. sandfly.
Treatment	Injections of Pentostam and Glucantime provide effective treatment.
Occurrence	Americas, Africa, Asia, Mediterranean.

Amoebae

Organism	*Entamoeba histolytica*
Biology	Trophozoites are about 20–30 μm long, contain a single nucleus, feed on nutrients obtained from the host, reproduce by binary fission and form cysts for transmission.
Disease	Diarrhea or amoebic dysentery and liver abscesses.
Symptoms	Intestinal lesion develop in the gut and with increasing number of parasites the mucosal destruction becomes extensive with abdominal pain, diarrhea (bloody), cramps, vomiting, malaise, weight loss, with extensive scarring of the intestinal wall. Death can occur with gut perforation, exhaustion, and liver abscesses.
Transmission	Oral uptake of *Entamoeba* cysts from human faeces with contaminated food or drinking water.
Treatment	Treatment includes oral application of metronidazole, diiodohydroxyquin, iodoquinol.
Occurrence	Worldwide (most prominent in warm countries).

Organism	*Acanthamoeba* **spp. causing blinding keratitis**
Biology	Trophozoites are about 15–35 μm long, contain a single nucleus, feed on bacteria, reproduce by binary fission and form cysts under harsh conditions as well as for transmission.
Disease	Blinding keratitis, usually associated with contact lens use.

Symptoms	Initially irritation of the eye, inflammation, photophobia, epithelial defects, excruciating pain, and can result in blindness.
Transmission	Washing contact lenses with tap water or home-made saline is a common mode of transmission. Swimming/sleeping with contact lenses or extended wear of contact lenses are additionally predisposing factors.
Treatment	Successful treatment requires early diagnosis with aggressive application of polyhexamethylene biguanide (PHMB) or chlorhexidine digluconate (CHX) together with propamidine isethionate, also known as Brolene.
Occurrence	Worldwide.

Organism	*Acanthamoeba* **spp. causing fatal encephalitis**
Biology	Trophozoites are about 15–35 µm long, contain a single nucleus, feed on bacteria, reproduce by binary fission and form cysts under harsh conditions as well as for transmission.
Disease	Granulomatous encephalitis involving the central nervous system (limited to immunocompromised patients).
Symptoms	Infection initiates with amoebae entry into the bloodstream via lower respiratory tract or directly through skin lesions followed by their crossing of the blood-brain barrier into the CNS to produce disease. Alternative routes of entry involve olfactory neuroepithelium. The clinical syndromes include meningitis-like symptoms such as headache, fever, behavioral changes, lethargy, stiff neck, vomiting, nausea, increased intracranial pressure, seizures and finally leading to death.
Transmission	Limited to immunocompromised patients. Their exposure (especially in the presence of skin lesions) to standing water, soil, mud, swimming in contaminated water, i.e. lakes, untreated pools may attribute to this fatal infection.
Treatment	No recommended treatment, usually fatal, current therapeutic agents include a combination of ketoconazole, fluconazole, sulfadiazine, pentamidine, amphotericin B, azithromycin, or itraconazole. Alkylphosphocholine compounds, such as hexadecylphosphocholine, have shown promise.
Occurrence	Worldwide.

Organism	*Balamuthia mandrillaris*
Biology	Trophozoites are about 20–45 µm long, contain a single nucleus, feed on eukaryotic cells, reproduce by binary fission and form cysts under harsh conditions as well as for transmission.
Disease	Granulomatous encephalitis involving the central nervous system. Unlike *Acanthamoeba*, it can cause infections in relatively immunocompetent individuals.
Symptoms	Infection initiates when amoebae enter the bloodstream via the lower respiratory tract or directly through skin lesion followed by their crossing of the blood–brain barrier into the

CNS to produce disease. Alternative routes of entry is via olfactory neuroepithelium. Symptoms include headache, fever, behavioral changes, nausea, stiff neck, photophobia, seizures, and finally leading to death within few weeks to months.

Transmission Exposure (especially in the presence of skin lesions) to soil, mud, during gardening, as well as swimming in contaminated water may attribute to this fatal infection.

Treatment No recommended treatment, usually fatal, current therapeutic agents include a combination of ketoconazole, fluconazole, sulfadiazine, pentamidine, amphotericin B, azithromycin, or itraconazole.

Occurrence Worldwide.

Organism *Naegleria fowleri*

Biology Trophozoites are about 15–35 µm long, contain a single nucleus, feed on bacteria, reproduce by binary fission and form cysts under harsh conditions as well as for transmission.

Disease Primary amoebic meningoencephalitis involving the central nervous system.

Symptoms Infection initiates when amoebae enter the nasal passage (during swimming/diving into the water). Amoebae migrate along the olfactory nerves, through the cribiform plate into the cranium. Death occurs within days.

Transmission Exposure to contaminated water, i.e. swimming/diving.

Treatment No recommended treatment, usually fatal, current therapeutic agents include amphotericin B or qinghaosu.

Occurrence Worldwide.

Organism *Pneumocystis carinii* (taxonomic status unclear, protozoa/fungi)

Biology Trophozoites are about 1–5 µm long, contain a single nucleus, reproduce both by asexual and sexual reproduction and form cysts under harsh conditions as well as for transmission.

Disease Pneumonia (limited to immunocompromised patients).

Symptoms Parasites infect the alveolar epithelial cells in lungs. Symptoms include fever, cough, and breathing problems. Parasites may disseminate to spleen, lymph nodes, bone marrow, and may result in death if not treated.

Transmission Through aerosol droplets, sputum or direct contact with infected individuals.

Treatment Pentamidine, or combination of trimethoprim-sulfamethoxazole.

Occurrence Worldwide.

Organism *Blastocystis hominis*

Biology Trophozoites vary greatly in size from 5–40 µm, contain a single nucleus, reproduce by asexual reproduction and form cysts under harsh conditions as well as for transmission.

Disease	Blastocystosis (especially in patients with weak immune system).	
Symptoms	Usually asymptomatic, however in minority individuals it can cause diarrhea, loose stools, anal itching, abdominal pain, and weight loss but its not proven whether symptoms come from *Blastocystis* sp. or a result of other (Bacterial/viral) infections.	
Transmission	Faecal-oral route through contaminated food/water. Parasite divides by binary fission in the gastrointestinal tract.	
Treatment	Treatment involves oral application of metronidazole for prolonged periods of time.	
Occurrence	Worldwide.	

Spoozoa
Apicomplexa

Organism | *Toxoplasma gondii*
Biology | Trophozoites are about 7–10 μm long, contain a single nucleus, reproduce both by asexual (humans as well as other mammals) and sexual reproduction (cats) and form cysts for transmission.
Disease | Toxoplasmosis.
Symptoms | Usually asymptomatic, it is an intracellular parasite of intestinal epithelial cells (first point of contact) as well as macrophages and muscle cells. But especially in immuno-compromised patients, it can cause fever, inflammation, swelling of lymph nodes, and disseminate to other organs including lungs, liver, heart, and brain, and may cause death. Pregnant women with this toxoplasmosis may pass it to their child, a form known as congenital toxoplas-mosis.
Transmission | Through eating oocysts in cat faeces or by eating raw or uncooked meat of pigs, sheep, cattle, etc.
Treatment | Treatment involves oral application of a combination of pyrimethamine and sulfadiazine.
Occurrence | Worldwide.

Organism | *Cryptosporidium parvum*
Biology | The reproductive stage is about 7 μm, contain a single nucleus, reproduce both by asexual fission and sexual reproduction with male and female gametes and form oocysts for transmission.
Disease | Cryptosporidiosis (commonly occurs in immunocompro-mised patients).
Symptoms | Often in AIDS patients, it infects intestinal epithelial cells causing watery diarrhea lasting for several months. The frequency of bowel-movement ranges from 5–25 per day and could result in death if not treated. In normal individ-uals it may be asymptomatic or, if it occurs, lasts few days and is self-limiting.
Transmission | Oral uptake of *Cryptosporidium* oocysts from faeces with contaminated food or drinking water.

Treatment	Treatment involves oral application of nitazoxanide with limited effects.
Occurrence	Worldwide.

Organism	*Isospora belli*
Biology	Oocyst containing sporocyst (18–30 µm) are taken up orally followed by sporozoite release and invasion of intestinal epithelial cells. Sporozoite divide asexually and produce merozoites, which invade epithelial cells and produce male and female gametocytes. Fertilization results in the development of oocysts that are excreted in the faeces.
Disease	Diarrhea (mostly limited to immunocompromised patients).
Symptoms	Usually in AIDS patients, it infects intestinal epithelial cells causing heavy diarrhea lasting for up to 24 months.
Transmission	Faecal-oral route through food/water contaminated with *Isospora* oocysts.
Treatment	Treatment includes oral application of co-trimoxazole for couple of weeks.
Occurrence	Worldwide.

Organism	*Plasmodium spp.*
Biology	Parasite differentiate into different forms in the vertebrate host and depending on the stage, the size varies from 2.5–15 µm, reproduce both by asexual and sexual reproduction.
Disease	Malaria.
Symptoms	Infects liver endothelial cells and red blood cells. Symptoms include abdominal pain, headache, and typical intermittent fever-chills and anemia associated with destruction of the red blood cells (periods vary depending on species). *Plasmodium falciparum* may produce continuous fever and results in death.
Transmission	Disease transmission via insect bite, i.e. *Anopheles* mosquito (vector), as parasites are present in the salivary glands.
Treatment	Use of chloroquine, primaquine and qinghaosu in early disease is effective.
Occurrence	Warm countries (mostly tropical and subtropical countries).

Ciliates

Organism	*Balantidium coli*
Biology	The trophozoite stage is about 50–60 µm long, contains a macro and micronuclei, reproduce by asexual binary fission and by conjugation and form cysts under harsh conditions as well as for transmission.
Disease	Balantidiasis.
Symptoms	Eighty percent of infections are asymptomatic and 20% infects intestinal epithelial cells producing ulcer. Advanced cases show symptoms similar to amoebic dysentery, i.e. vomiting, diarrhea, nausea and could lead to death.
Transmission	Faecal-oral route via food/water contaminated with *Balantidium* cysts.

Treatment   Treatment involves oral application of metronidazole, diiodohydroxyquin and tetracycline.

Occurrence   Worldwide.

Microsporidia   Microsporidia is a general term used to describe obligate intracellular protozoan parasites (>1200 species). They produce resistant spores that infect both humans and animals. Spores associated with human infections are 1–4 μm, and possess a unique organelle, i.e. polar tubule coiled inside the spore. The human pathogens of the microsporidia include *Brachiola algerae, B. connori, B. vesicularum, Encephalitozoon cuniculi, E. hellem, E. intestinalis, Enterocytozoon bieneusi, Microsporidium ceylonensis, M. africanum, Nosema ocularum, Pleistophora* sp., *Trachipleistophora hominis, T. anthropophthera, Vittaforma corneae. Enterocytozoon bieneusi* is the most common human pathogen causing 40% of all human microsporidian infections. They mostly produce disease in AIDS patients. Various organs can be affected, including kidneys and intestine, and infectious spores are found in urine and faeces. Symptoms of disease include diarrhoea and dysfunction of infected organs.

Transmission   Oral uptake of spores from human urine or faeces with contaminated food, drinking water, hand–mouth contact or by unprotected homosexual intercourse.

Occurrence   Worldwide.

Treatment   No satisfactory drug available; albendazole has promising effects.

Locomotion   Movement in protista is usually mediated by pseudopodia, flagella, or cilia. Other modes of protista locomotion involve gliding movements in which no changes in body shape are observed (*Fig. 2*). The various protista have evolved to exhibit

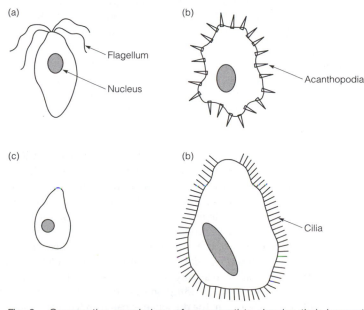

Fig. 2. Comparative morphology of some protista showing their locomotory organelles. (a) flagellates, e.g. Trichomonas vaginalis; (b) amoeba, e.g, Acanthamoeba; (c) apicomplexan, e.g. Toxoplasma gondii; (d) ciliate, e.g. Balantidium coli.

distinct movements depending on where they normally live. For example, protista with amoeboid movements using pseudopodia are normally present in the environments with abundant organic matter or in flowing water with plant life. Cilia or flagella are used to travel longer distances *per se* and maximize the possibility of encountering food particles.

Pseudopodia   Pseudopodia are not permanent structures present on the surface of organisms but are formed upon a stimulus. Pseudopodia are observed in amoeboid movement. They are characterized as a flow of cytoplasm in a particular direction. The cytoplasmic membrane temporarily attaches to the substratum and cytoplasm is drawn into the new attachment protruding as a foot-like structure, hence it is called pseudopodia. These extensions may exhibit following distinct phenotypes: i) broad, round-tipped pseudopodia are known as lobopodia, ii) extensively branched forming a net-like structure is rhizopodia, iii) sharp, pointed projections are filopodia, and iv) axopodia, i.e. similar to filopodia but contain slender filaments. Pseudopodia play important roles both in locomotion as well as in food uptake. Examples include *Amoeba proteus*, *Acanthamoeba*, *Entamoeba*.

Cilia and flagella   In contrast to pseudopodia, flagella and cilia are permanent microtubular organelles that are anchored within the plasma membrane of certain protista, and project from the cell surface. Flagella are long slender structures (50–200 μm) usually one to few on a cell, with a whip-like movements starting at the tip or the base of the cell. This results in forward, backward or spiral movements. Cilia are similar to flagella but smaller in size, i.e. 5–20 μm and usually more numerous. Cilia move with a back-and-forth stroke. Their thickness is approximately 0.2 μm. Both cilia and flagella exhibit bending motion, resulting in fluid propulsion and cellular movements. However, due to the extended length of the flagellum, bends are propagated along the flagellum pushing the surrounding water symmetrically on both sides of the cell. In contrast, with the shorter length and large numbers, cilia beat in coordination with one another.

Gliding movements Other forms of locomotion in protista are gliding. For example, several flagellates glide over surfaces. In such cases, flagella do not exhibit whip-like movement but make contact with the surface and slide over it with the aid of microtubules. Other examples include sporozoites stage during the life cycle of *Plasmodium* spp.

Locomotory proteins   The major proteins involved in locomotion are: i) microtubules that are cylindrical fibrils formed of tubulin molecules around 25 nm in diameter, and ii) microfilaments, also known as actin filaments, that are composed of actin molecules and are about 7 nm thick. The polymerization and depolymerization of the tubulin molecules in microtubules and their ability to interact with dynein adenosine triphosphatase proteins cause movement in some protista. In contrast, actin is polymerized and forms microfilaments. These microfilaments are pushed along by interacting with myosin ATPase molecules. Other proteins important in protista locomotion are myonemes that are centrin, which form filaments of about 10 μm thickness. Depending on calcium concentration, centrin exist in two states: filaments shorten in the presence of calcium (centrin binds to calcium) and extend when calcium is withdrawn. These contractions result in cellular movements, e.g. *Vorticella*, and *Stentor*.

Reproduction   To ensure survival, a species must reproduce in large numbers when conditions are favorable. Reproduction in protista can be sexual, i.e. formation and fusion of gametes producing new offsprings or asexual, i.e. mitotic division of a parent cell into two or more identical offsprings. Many protista only use asexual, while others use both during their life cycles.

Asexual   Asexual reproduction is the primary mode of reproduction in many protista.
reproduction   Asexual reproduction occurs by binary fission, multiple fission and budding. The process of cell division is divided into an S phase (DNA synthesis phase) and an M phase (division of nucleus that involves DNA condensation and organized distribution of chromosomes). Both phases are separated by gap phases, i.e. G1 and G2. In addition, cytoplasmic organelles are duplicated, followed by cytokinesis. The resulting daughter cells are identical to their parent cell and are produced in sufficient numbers for successful transmission. Asexual reproduction occurs in majority of amoebae, ciliates, and flagellates.

Binary fission

In this form, a single cell divides into two daughter cells and it is the most common type of asexual reproduction lasting from 6–24 hours. This results in a very large numbers of identical parasite populations within days. The DNA and cytoplasmic organelles are duplicated followed by nuclear division and cytokinesis, and finally a constriction ring bisects the cell producing two daughter cells. Examples include *Toxoplasma*, *Ichthyophthirius*, and *Acanthamoeba*.

Multiple fission

In this form of reproduction, mitotic nuclear division occurs several times. This results in several nuclei within the cytoplasm. Each nucleus together with a layer of cytoplasm gives rise to independent daughter cells that are released. In parasitic protista, it results in the rapid production of large protistan populations which overwhelm the host immune system. This occurs in various protista including *Pelomyxa palustris*, and *Volvox*, *Gonium*.

Budding

In this form of reproduction, nuclei divide and the daughter nuclei migrate into a cytoplasmic bud. This is followed by cytoplasmic fission and release of the cell, which develops into a mature reproductive organism. The resulting daughter cells may differ from the parent cells. Such reproductive schemes are limited to several protista such as *Trichophrya*, *Ephelota*.

Sexual   Many parasitic protista have complex life cycles and reproduce both sexually and
reproduction   asexually. Sexual reproduction involves formation of haploid gametes by meiosis that fuse to form a diploid zygote generating new organisms.

Gamete formation

Gametes with haploid chromosomes are formed in the vegetative stage, which fuse with one another to produce diploid zygotes. The zygote undergoes meiosis and numbers of chromosomes are halved followed by asexual reproduction to produce large populations of the organisms. Gametes formation directly from the original population followed by their fusion into a zygote is called 'hologamy'. Examples include *Chlamydomonas*, *Dunaliella*, and *Polytoma*. If gametes are identical (at least morphologically but there may be minor genetic or physiolog-ical differences), the term 'isogamy' is used. These include *Chlorogonium* as well as many foraminiferan sarcodinids produce isogametes and some sporozoa. When there are clear differences between male and female gametes, typically known as micro- and macrogametes (size differences, presence of flagella, physiological or biochemical properties), the gamete formation and their fusion into a zygote is called 'anisogamy'. This is a common form of gamete formation in protista. Examples include *Plasmodium* spp. The factors that induce gamete formation are not clearly known but may involve environmental conditions such as salinity, pH, temperatures, nutrients, etc.

If both gametes arise from the same clone, the species is called monoecious but if they arise from different clones, species is called diecious. However, if self-fertilization occurs, the species is called hermaphrodite. If the new population of an organism develops from unfused gamete without fertiliza-tion, it is known as parthenogenesis. Examples include genus *Volvox*, *Eucoccidium dinophili*.

Gametic nulei: conjugation

Some ciliated protista are unable to produce gametes. Instead they possess dual nuclei and during their sexual reproduction, the nuclei from two organisms (instead of gametes) fuse together yielding a zygotic nucleus. This is followed by asexual fissions producing large populations. Again, diverse factors are responsible for conjugation including temperatures, salinity, pH, etc. Examples include *Paramecium* and *Tetrahymena*.

Protista life cycles   The protista that are major human pathogens can be subdivided into three main categories, i.e. those which are bloodborne that are usually transmitted via insects (e.g. *Plasmodium*); those which are transmitted via contaminated food/water (e.g. *Giardia*, and *Acanthamoeba*) and those which are sexually transmitted (e.g. *Trichomonas*). The life cycles of bloodborne protista generally involve humans and insects. The life cycle in an insect may be necessary for parasite development or merely act as vectors to transmit to new host. In contrast, other protistan pathogens are acquired by new host through exposure to contaminated water/food or through sexual intercourse.

Plasmodium spp.

Although the precise life cycle of *Plasmodium* varies between species, it can be divided broadly into two hosts as follows:

Life cycle in vertebrate host

When an infected mosquito takes a blood meal, it injects saliva containing sporo-zoites (approximately 10–15 μm long and 1 μm in diameter). The sporozoites penetrate hepatocytes (liver cells) and undergo asexual reproduction, a process known as the pre-erythrocytic (PE) cycle or exoerythrocytic (EE) cycle. Within the hepatic cell, the parasite transforms into a trophozoite stage that feeds on the host cell cytoplasm. Within a few days, trophozoite matures and produces daughter nuclei and at this stage is called a schizont. A single schizont undergoes cytoki-nesis and produces many daughter cells called merozoites (2.5 μ EE m long and 1.5 μm in diameter). The merozoites are released and infect new hepatocytes or enter the erythrocytic cycle. Upon entry into an erythrocyte, a merozoite again transforms into a trophozoite and feeds on host cell cytoplasm, forming a large food vacuole giving the characteristic ring appearance. The trophozoite becomes a schizont again and produces many merozoites which infect new erythrocytes. After several generations, some merozoites infect erythrocytes and become macrogametocytes and microgametocytes. These are taken up by the mosquito, where the remaining life cycle continues (*Fig. 3*).

Life cycle in invertebrate host

The gametocytes are taken up by mosquito during their blood meal. If gametocytes are taken up by a susceptible mosquito (*Anophles* sp. in the case of human *Plasmodium*), gametocytes develop into gametes, i.e. micro- and macro-gametes. The nucleus of microgametocyte divides to produce 6 to 8 nuclei and exflagellates. The microgamete fuses with a macrogamete to form a diploid zygote. The zygote elongates to become a motile ookinete that is 10–15 µm. The ookinete penetrates the gut wall of mosquito and develops into an oocyst. The oocyst undergoes meiosis and produces sporozoites, which penetrate salivary glands and are injected into new a host at the next blood meal completing the cycle.

Acanthamoeba spp.

Acanthamoeba has two stages in its life cycle, i.e. an infective trophozoite stage and a dormant cyst stage (*Fig. 4*). The vegetative trophozoite stage is approximately 15–35 µm depending on the species. Trophozoites divide asexually by binary fission. *Acanthamoeba* remains in the trophozoite stage under favorable

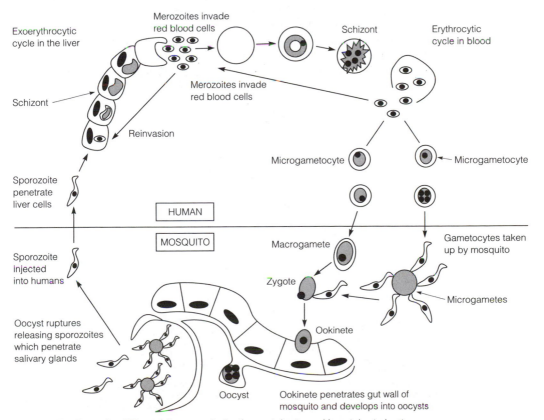

Fig. 3. The life cycle of Plasmodium *spp. indicating vertebrate and invertebrate hosts.*

conditions, e.g. plentiful nutrients and divide every 8–20 hours. The time of division may vary depending on the species. However, under harsh environmental conditions such as lack of nutrients, extreme temperatures, pH, and high osmolarity, *Acanthamoeba* transform into a resistant cyst form. Cysts are dormant, nonmotile and exhibit minimal metabolic activity. *Acanthamoeba* cysts are double walled, consisting of an outer ectocyst and an inner endocyst with an average diameter of 10–15 μm. The outer wall of Acanthamoeba cyst is largely made of cellulose and proteins. The outer walls of cysts in other protista may be formed of chitin, cellulose material, and occasionally tectin or gelatin. Cysts expel excess water and can also be airborne, suggesting their possible role in the dispersal of the species. *Acanthamoeba* cysts can remain viable for years while maintaining their pathogenicity. Cysts are resistant to various antimicrobial agents, thus presenting a problem in successful treatment of *Acanthamoeba* infections. In addition, biocide resistance may lead to recurrence of the disease. Under favorable conditions, Acanthamoeba emerge from the cyst to become the active trophozoite stage thus completing the life cycle (*Fig. 4*).

Strategies of immune evasion
Upon invasion into the human body, some pathogenic protista are recognized as foreign and have to cope with a highly professional host immune system. The involvement of the immune system in protozoan infections is further shown with the finding that the majority of parasitic infections are limited to babies/children or individuals with less developed (impaired) immune systems and these infections may be self-limiting in adults with a fully developed immune system. To this end, parasite-specific antibodies together with the presence of phagocytes provide a highly effective defense system. For example, the role of antibodies in malaria is demonstrated with the finding that passive transfer of antibodies from immune individuals to those suffering from acute malaria results in the reduction of parasitemia. In addition, B cell deficiency severely impairs the

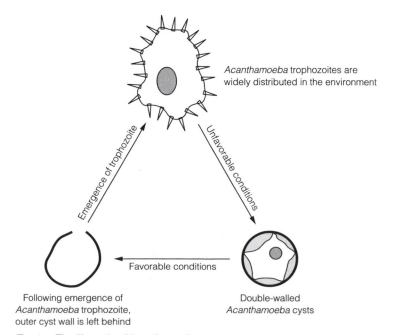

Fig. 4.   The life cycle of Acanthamoeba spp.

host's ability to clear malaria. Antibodies specifically bind to antigens on a microbial surface with their Fab region, and binds to phagocytes (neutrophils/macrophages) using their Fc region. Thus ideally phagocytes in the presence of specific antibodies effectively clear parasites from the human body.

Therefore parasites must develop mechanisms to evade the immune system. This is essential for some parasites to buy sufficient time to complete development (reproduction) and, in particular, for parasites that cause chronic infections, lasting months to years. There are several strategies used by pathogenic protozoa to overcome the host's immune response as described below.

Non-induction of immune system

Pathogens may target organs and/or tissues that have a limited immune response. For example, *Acanthamoeba* produces infection of the avascular cornea, or *Naegleria* produces infection of the brain tissues, both of which have limited immune responses.

Anatomical seclusion

Parasites can invade non-immune cells, replicate intracellularly and hide from immune cells, thus avoiding an overwhelming immune response. For example, some stages of *Plasmodium* spp. live inside liver cells. Other stages during its life cycle live inside erythrocytes, which do not have a nucleus, thus are not recognized by cytotoxic T-cells or natural killer cells. Another benefit of anatomical seclusion is to avoid being exposed to antimicrobials. Even if taken up by immune cells, some parasites have the ability to survive the onslaught. For example, *Leishmania* spp. are taken up by macrophages and produce antioxidases to counter products of the macrophage's oxidative burst.

Antigenic variation

Parasites can evade the immune response by changing the antigens that are expressed on their surface, every few generations. This is achieved by random activation or rearrangement of genes that code for surface antigens. Thus, antibodies produced against the former antigens become ineffective. This is best studied in the African trypanosomes. Shedding of surface antigens in *E. histolytica* provides immune evasion. This may be achieved even when antibody–antigen complex is formed and directs the immune response away from the parasite. Some parasites secrete protease that breakdown antigen–antibody complex or cleave an antibody before it binds to the parasite surface.

Host mimicry

Some parasites use molecular mimicry to escape host immunosurveillance. This can be achieved by expressing proteins/glycoproteins on the parasite surface that mimic host proteins/glycoproteins or by adsorbing host components on their surfaces. The ultimate purpose is to be recognized as host cells and thus avoid an overwhelming immune response.

Interference with host immune signaling molecules

Some protistan pathogens interfere with the host immune signaling molecules thus affecting host ability to build an effective immune response. For example, *Leishmania* spp. reduces the production of interferon-gamma by the host which is an important cytokine in the activation of macrophages, i.e. immune cells important in clearing protozoa from the host.

Strategies against protistan pathogens

To maximize the chances of controlling protozoan infections, a variety of approaches must be employed and continued research should identify novel strategies to prevent (prophylaxis), eradicate and/or control these infections. Some of these strategies are indicated below.

Table 1.   The spectrum of activity and mode(s) of action of the major anti-protista drugs

Drug	Mode of action	Spectrum of activity
Sulfonamides (a group of antibiotics such as sulfadiazine)	Inhibition of folic acid synthesis, which is crucial for DNA synthesis	*Plasmodium, Toxoplasma, Acanthamoeba*
Proguanil and Mefloquine	Inhibition of folic acid synthesis	*Plasmodium*
Suramin	Binds to enzymes in nucleus and inhibits DNA synthesis	*Trypanosoma*
Chloroquine	Inhibition of DNA synthesis	*Plasmodium*
Pentamidine	Inhibition of DNA replication and translation	*Trypanosoma, Leishmania*
Amphotericin B	Disrupts membrane function	*Leishmania, Acanthamoeba*
Eflornithine	Inhibition of protein synthesis	*Trypanosoma*
Tetracycline	Inhibition of protein synthesis	*Plasmodium, Entamoeba histolytica*
Albendazole	Inhibition of microtubule assembly	*Gardia intestinalis, Microsporidia*
Buparvaquone	Inhibition of energy production	*Plasmodium*
Megumine antimonate	Inhibition of energy production	*Leishmania*
Metronidazole	Inhibition of energy production	*Gardia intestinalis, Entamoeba histolytica*
Primaquine	Inhibition of energy production	*Plasmodium*
Difluoromethylornithine	Inhibition of polyamine synthesis, inhibiting cellular growth	*Trypanosoma*
Berenil	Inhibition of DNA synthesis	*Trypanosoma*
Nifurtimox and Benznidazole	Increases oxidant stress on intracellular parasites	*Trypanosoma*
Ketoconazole and Fluconazole and Itraconazole	Inhibits biosynthetic pathways	*Trypanosoma*
Pentostam	Inhibits macromolecular (DNA, RNA, Protein) synthesis	*Leishmania*
Glucantime	Inhibits protein synthesis	*Leishmania*
Diiodohydroxyquin	Similar to metronidazole but for longer periods	*Entamoeba histolytica*
Chlorhexidine and Polyhexamethylene biguanide	Disrupts membrane function	*Acanthamoeba*
Propamidine isethionate	Inhibits DNA synthesis	*Acanthamoeba*
Azithromycin	Inhibits protein synthesis	*Acanthamoeba*
Alkylphosphocholine	Induces apoptosis	*Acanthamoeba*
Qinghaosu (Artemisinin)	Produces oxidative effects	*Plasmodium, Naegleria*
Trimethoprim and Sulfamethoxazole (combination is called Co-trimoxazole)	Inhibits folic acid synthesis	*Pneumocystis*
Pyrimethamine	Inhibits DNA synthesis	*Toxoplasma*
Nitazoxanide	Inhibits energy metabolism	*Cryptosporidium*

Chemotherapy Chemotherapeutic approaches remain the most common form of treatment of parasitic diseases (*Table 1*). These provide the most direct and cheapest way of controlling these infections. Many of the currently available anti-parasitic drugs were identified by screening large number of compounds. A limited number of drugs against some parasitic diseases are available as chemoprophylaxis, which can be used before, during and after parasite exposure (especially for individuals who are not immune and are visiting endemic areas). A complete understanding of parasite metabolism should help identify novel targets for the rational development of drugs. A major problem with drug therapy is the ability of a sub-population of parasites to develop drug resistance, thus other control measures should be explored in conjunction with chemotherapeutic approaches.

Control measures Parasitic protozoa can be controled by a variety of measures applied at various stages of their life cycles.

Sanitary measures
The maintenance of good sanitary measures such as draining swamps, building sewage systems, and most importantly providing clean water supplies (in particular for waterborne protista), although expensive, are crucial for controlling parasitic infections.

Vector control
Human-vector contact can be reduced by use of repellent impregnated bed nets, vector repellents and protective clothing. Spraying houses with insecticides and installing screens are also helpful.

Reduction of vector capacity
Vector capacity can be reduced in a number of ways:

- Environmental modifications (vector habitat alterations) – measures that reduce the risks of flooding, which may provide breeding sites for vectors.
- Larvacides/insecticides during various stages of insect life cycle.
- Biological control – introducing genetically modified vector in the endemic area that can not act as host for parasites.

Reduction of the parasite reservoir
Early diagnosis of the disease (monitoring the populations in endemic areas) and aggressive treatment of parasitic infections should reduce the parasite reservoir.

L1 VIRUS STRUCTURE

Key Notes

Definitions

Viruses are obligate intracellular parasites and vary from 20–200 nm in size. They have varied shape and chemical composition, but contain only RNA or DNA. The intact particle is termed a 'virion' which consists of a capsid that may be enveloped further by a glycoprotein/lipid membrane. Viruses are resistant to antibiotics.

Methods of study

Virus morphology has been determined by electron microscopy (EM) (using negative staining), thin-section EM (using negative staining), immunoelectron microscopy (using negative staining), electron cryomicroscopy and X-ray crystallography.

Virus symmetry

Virus capsids have helical or icosahedral symmetry. In many cases the capsid is engulfed by a membrane structure (the virus envelope). Helical symmetry is seen as protein sub-units arranged around the virus nucleic acid in an ordered helical fashion. The icosahedron is a regular-shaped cuboid which consists of repetitions of many protein sub-units assembled so as to resemble a sphere.

Virus envelopes

Virus envelopes are acquired by the capsid as it buds through nuclear or plasma membranes of the infected cell. Envelopes may contain a few glycoproteins, for example, human immunodeficiency virus (HIV), or many glycoproteins, for example, herpes simplex virus (HSV). The virus envelope contains the receptor biological activity which allows the particle to attach to and infect the host cell.

Related topic

Bacteriophage (E7)

Definitions

Viruses are obligate intracellular parasites which can only be viewed with the aid of an electron microscope. They vary in size from approximately **20–200 nm**. In order to persist in the environment they must be capable of being passed from host to host and of infecting and replicating in susceptible host cells. A virus particle has thus been defined as a structure which has evolved to transfer nucleic acid from one cell to another. The nucleic acid found in the particle is either **DNA** or **RNA**, is single- or double-stranded and linear or segmented. In some cases the nucleic acid may be circular. The simplest of virus particles consists of a protein coat (sometimes made up of only one type of protein which is repeated hundreds of times) which surrounds a strand of nucleic acid. More complicated viruses have their nucleic acid surrounded by a protein coat which is further engulfed in a membrane structure, an envelope consisting of virally coded glycoproteins derived from one of several regions within the infected cell during the maturation of the virus particle. The genetic material of these complex viruses encodes for many dozens of virus specific proteins.

The complete fully assembled virus is termed the **virion**. It may have a glyco-protein envelope which has **peplomers** (projections) which form a 'fringe' around the particle. The protein coat surrounding the nucleic acid is referred to as the **capsid** (*Figs 1* and *2*). The capsid is composed of morphological units or capsomers. The type of capsomer depends on the overall shape of the capsid, but in the case of **icosahedral** capsids the capsomers are either **pentamers** or **hexamers**. Capsomers themselves consist of assembly units that comprise a set of structure units or **protomers**. **Structure units** are a collection of one or more non-identical protein subunits that together form the building block of a larger assembly complex (e.g. virus proteins VP1, VP2, VP3 and VP4 of picornaviruses). The combined nucleic acid–protein complex which comprises the genome is termed the **nucleocapsid**, which is often enclosed in a core within the virion.

Methods of study

While the **electron microscope** (EM) had been known for many years, the invention of the **negative-staining technique** in 1959 revolutionized studies on virus structure. In negative-contrast EM, virus particles are mixed with a heavy metal

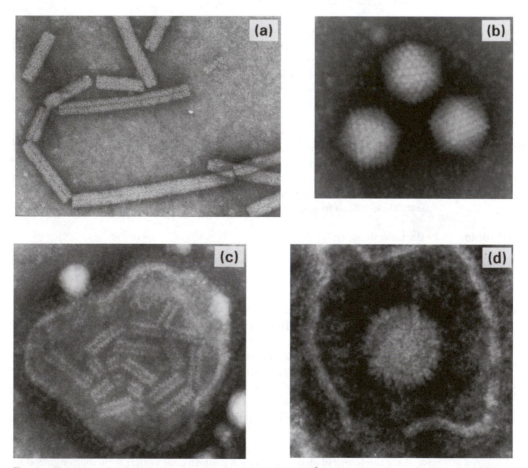

Fig. 1.   Examples of viruses from main groups according to 'standard' morphology: (a) unenveloped/helical (tobacco mosaic virus); (b) unenveloped/icosahedral (adenovirus); (c) enveloped/helical (paramyxovirus); (d) enveloped/icosahedral (herpesvirus). From Harper, D., Molecular Virology, 2nd edn, © BIOS Scientific Publishers Limited, 1998. Photographs courtesy of Dr Ian Chrystie, Department of Virology, St Thomas' Hospital, London, and Professor C. R. Madeley, Department of Virology, Royal Victoria Infirmary, Newcastle-upon-Tyne.

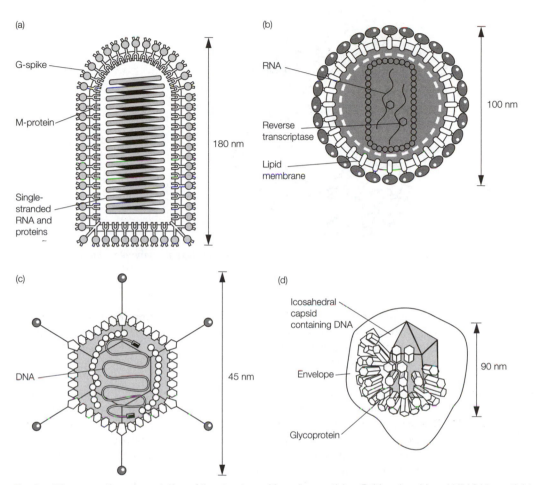

(a)

G-spike

M-protein

Single-stranded RNA and proteins

180 nm

(b)

RNA

Reverse transcriptase

Lipid membrane

100 nm

(c)

DNA

45 nm

(d)

Icosahedral capsid containing DNA

Envelope

Glycoprotein

90 nm

Fig. 2. Diagrammatic representation of the structure of four virus particles. Rabies virus (a) and HIV (b) have tight-fitting envelopes, herpes simplex (d) has a loose-fitting envelope, whereas adenovirus (c) is non-enveloped. The diagrams are based on electron microscopic observation and molecular configuration exercises. Redrawn from Phillips and Murray, Biology of Disease, 1995, with permission from Blackwell Science Ltd.

solution (e.g. sodium phosphotungstate) and dried onto a support film. The stain provides an electron-opaque background against which the virus can be visualized. Observation under the electron microscope has thus allowed the definition of **virus morphology** at the 50–77 Å resolution level. In addition, negative staining of thin sections of infected cells has allowed definition of structures which appear during virus maturation and their interactions with cellular proteins. **Immunoelectron microscopy** is used to study those viruses which may be present in low concentrations or grow poorly in tissue culture (e.g. Norwalk virus). These clumps of virus are more readily observed. **Electron cryomicroscopy** reduces the risk of seeing artifacts as may be unavoidable by negative staining. High concentrations of virus are rapidly frozen in liquid ethane while on carbon grids. Electron micrographs can be digitized and three-dimensional reconstruction performed. Resolutions of 9 Å have been achieved using this method. **X-ray diffraction** of virus crystals is the ultimate in determining the ultrastructure of virion morphology. At present, only simple viruses can be

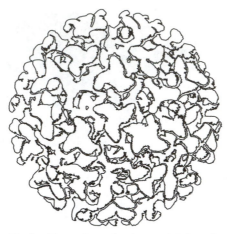

Fig. 3.   Three-dimensional reconstruction of an icosahedrally symmetric virus particle.

crystallized. More complex viruses are analyzed by attempting to form crystals of sub-particular molecules. The X-ray diffraction pattern of the virion particle allows mathematical processing, which can predict the molecular configuration of the virus particle (*Fig. 3*).

Virus symmetry   The capsids of virions tend to have one of two symmetries – **helical** or **cuboid.** Helical symmetry can be loosely described as having a 'spiral staircase' structure. The structure has an obvious axis down the center of the helix. The subunits are placed between the turns of the nucleic acid. A diagram of such a structure (tobacco mosaic virus) is shown in *Fig. 4*, and an electron micrograph in *Fig. 1a*. Animal viruses with a similar capsid structure include measles, rabies and influenza. Most animal viruses have spherical or cuboid symmetry. Obtaining a true sphere is not possible for such structures and hence subunits come together to produce a cuboid structure which is very close to being spherical. The 'closed shell' capsid is usually based on the structure referred to as an icosahedron. A regular icosahedron, formed from assembly of identical subunits, consists of **20 equilateral triangular faces, 30 edges and 12 vertices and exhibits 2-, 3- and 5-fold symmetry** (*Fig. 4*). The minimum number of capsomers required to construct an icosahedron is 12, each composed of five identical subunits. Many viruses have more than 12. A model of such a structure is shown in *Fig. 4*, although in adenoviruses projecting fibers are also present, which distinguishes this capsid from that of other viruses. The maturation and assembly of these structures is very complex; indeed, much of how it happens is unknown. Icosahedral structures are, however, usually formed via a complex but structured array of molecular-assembly procedures which eventually give rise to the mature capsid. These may be self-assembly processes or may involve virus non-structural proteins acting as **scaffolding** proteins which do not finish up in the mature capsid.

Virus envelopes   Many viruses in addition to having a capsid also contain a virus-encoded **envelope**. Most enveloped viruses bud from a cellular membrane (plasma membrane, e.g. influenza virus, or nuclear membrane, e.g. herpes simplex virus).

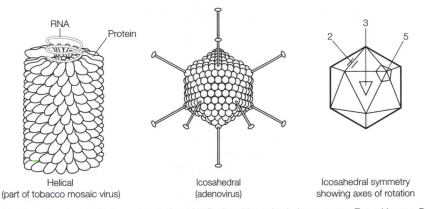

RNA   Protein

3
2   5

Helical
(part of tobacco mosaic virus)

Icosahedral
(adenovirus)

Icosahedral symmetry
showing axes of rotation

Fig. 4.   *Diagrammatical representation of helical and icosahedral symmetry. From Harper, D. Molecular Virology, 2nd edn, © BIOS Scientific Publishers Limited, 1998.*

Within this virus lipid/protein bilayer are a number of inserted virus-encoded **glycoproteins**. The envelopes can be amorphous (e.g. the herpes virion, *Figs 1* and *2*) or tightly bound to the capsid (e.g. HIV). Thus, the lipid of the envelope is derived from the cell, the glycoprotein being encoded by the virus. Quite how the process of budding occurs is not fully understood, but is known to involve specific domains (late domains) of a viral protein, which interact with cellular proteins, and also the cellular ubiquitin-proleasome system.

L2 VIRUS TAXONOMY

Key Notes

Virus taxonomy

The classification of viruses involves the use of a wide range of characteristics (morphology, genome, physicochemical and physical properties, proteins, antigenic and biological properties) to place viruses in orders, families, genera and species.

Virus orders

These are groupings of families of viruses that share common characteristics. There are three orders, Mononegavirales, Nidovirales and Caudovirales.

Virus families

These are groupings of genera of viruses that share common characteristics and are distinct from the members of other families. They are designated by the suffix -viridae.

Virus genera

These are groupings of species of viruses which share common characteristics. These are designated by the suffix -virus.

Virus species

These represent a polythetic class of viruses that constitutes a replicating lineage and occupies a particular ecological niche.

Related topics

Prokaryote taxonomy (D1)
Taxonomy (G1)

Virus taxonomy

Initial attempts to classify viruses were based on their pathogenic properties, and the only common feature to many of the viruses placed together in such groupings was that of **organ tropism** (e.g. viruses causing hepatitis or respiratory disease). There were more important aspects (e.g. **virus structure** and **composition**) which led virologists to believe that these initial attempts of classification were far from adequate. In the late 1950s and early 1960s, hundreds of new viruses were beginning to be isolated and the need for a different classification method became essential. In 1966 the **International Committee on Nomenclature of Viruses (ICNV)** was established at the International Congress of Microbiology held in Moscow. As the present classification scheme has evolved, an acceptance of the characteristics to be considered and their respective weighting has become universal.

The eighth report of the **ICTV (International Committee on Taxonomy of Viruses)** recorded a universal taxonomy scheme consisting of three orders, 73 families, 9 sub-families, 233 genera and 1938 species. The system still contains hundreds of unassigned viruses, largely because of lack of data. Also, new viruses are still being discovered.

As techniques in molecular biology, serology and EM have advanced, taxonomy has been based on more and more characteristics. Thus, factors now taken into consideration (*Table 1*) include virion morphology, physicochemical and physical properties, the genome, virus proteins, lipids, carbohydrates,

Table 1. Some properties of viruses used in taxonomy

Virion properties
    Morphology
        Virion size
        Virion shape
        Presence or absence and nature of peplomers
        Presence or absence of an envelope
        Capsid symmetry and structure
    Physicochemical and physical properties
        Virion molecular mass (M_r)
        Virion buoyant density (in CsCl, sucrose, etc.)
        Virion sedimentation coefficient
        pH stability
        Thermal stability
        Cation stability (Mg^{2+}, Mn^{2+})
        Solvent stability
        Detergent stability
        Irradiation stability
    Genome
        Type of nucleic acid (DNA or RNA)
        Size of genome in kb/kbp
        Strandedness: ss or ds
        Linear or circular
        Sense (positive-sense, negative-sense, ambisense)
        Number and size of segments
        Nucleotide sequence
        Presence of repetitive sequence elements
        Presence of isomerization
        G + C content ratio
        Presence or absence and type of 5′ terminal cap
        Presence or absence of 5′ terminal covalently linked protein
        Presence or absence of 3′ terminal poly (A) tract
    Proteins
        Number, size and functional activities of structural proteins
        Number, size and functional activities of nonstructural proteins
        Details of special functional activities of proteins, especially transcriptase, reverse transcriptase, hemagglutinin, neuraminidase and fusion activities
        Amino acid sequence or partial sequence
        Glycosylation, phosphorylation, myristylation property of proteins
        Epitope mapping
    Lipids
        Content, character, etc.
    Carbohydrates
        Content, character, etc.
    Genome organization and replication
        Genome organization
        Strategy of replication
        Number and position of open reading frames
        Transcriptional characteristics
        Translational characteristics
        Site of accumulation of virion proteins
        Site of virion assembly
        Site and nature of virion maturation and release

Antigenic properties
        Serologic relationships, especially as obtained in reference centers

Biologic properties
        Natural host range
        Mode of transmission in nature
        Vector relationships
        Geographic distribution
        Pathogenicity, association with disease
        Tissue tropisms, pathology, histopathology

Table 2. Taxonomic chart of selected virus families

Family	Characteristics	Typical members	Diseases caused
Poxviridae	dsDNA, 'brick'-shaped particles; largest virus	Vaccinia Variola	None known Laboratory virus Smallpox (now eradicated)
Herpesviridae	dsDNA, icosahedron capsid enclosed in an envelope, latency in host common	Herpes simplex Varicella-zoster Cytomegalovirus Epstein–Barr virus	'Cold' sores, genital infections Chicken pox, shingles Febrile illness or disseminated disease in immunosuppression Glandular fever. Virus is also associated with certain malignancies, e.g. Burkitt's lymphoma
Adenoviridae	dsDNA, icosahedron with fiber structures, non-enveloped	Adenoviruses (many types)	Respiratory and eye infections, tumors in experimental animals
Papillomaviridae	ds circular DNA, 72 capsomeres in capsid, non-enveloped	Human papilloma viruses	Warts, association with some cancers (e.g. cervical cancer)
Hepadnaviridae	One complete DNA minus strand with 5' terminal protein, DNA circularized by an incomplete plus strand, 42 nm enveloped particle	Hepatitis-B virus	Serum hepatitis, association with hepatocellular carcinoma
Paramyxoviridae	ssRNA, enveloped particles with 'spikes'	Parainfluenza virus Measles virus Respiratory syncytial virus	Respiratory tract infection ('croup') Measles Bronchiolitis
Orthomyxoviridae	Eight segments of ssRNA, enveloped particles with 'spikes', helical nucleocapsid	Influenza virus	Influenza
Reoviridae	10–12 segments of dsRNA, icosahedron, non-enveloped	Rotavirus	Infantile diarrhea
Picornaviridae	ssRNA, 22–30 nm particle of icosahedral symmetry, non-enveloped	Poliovirus Coxsackie virus Rhinovirus Heptatitis-A virus	Poliomyelitis Myocarditis Common cold Infectious hepatitis

Family	Description	Virus	Disease
Togaviridae	ssRNA, enveloped particles, icosahedron nucleocapsid	Rubella virus	German measles
Flaviviridae	ssRNA, enveloped particles, icosahedron	Yellow fever virus, Hepatitis C virus	Yellow fever, Hepatitis C
Rhabdoviridae	ssRNA, bullet-shaped, enveloped particle	Rabies virus	Rabies
Filoviridae	ssRNA, pleomorphic enveloped (mainly baciliform or filamentous)	Marburg	Hemorrhagic fever
Bunyaviridae	ssRNA, enveloped spherical or pleomorphic, helical capsids	Hantavirus	Hemorrhagic fever
Retroviridae	ssRNA, enveloped particles with icosahedral nucleocapsid, employ reverse transcriptase enzyme to make DNA copy of genome on infection	Human T lymphotropic virus-1; Human immunodeficiency virus (HIV)	Adult T cell leukemia and lymphoma; Acquired immune deficiency syndrome (AIDS)

and antigenic and biological properties. It has, however, been estimated that to define a virus appropriately some 500–700 characters must be determined. The rapidity of genome sequencing has made the classification of newly energing viruses so much easier, e.g. SARS virus was soon identified as being a Coronavirus. Over the next few years the ICTV plan is to create a readily accessible database which will allow cataloging of viruses down to strains. *Table 2* is a brief summary of some well known animal viruses and the diseases they cause.

Virus orders

These represent groupings of families of viruses that share common characteristics which make them distinct from other orders and families. Orders are designated by the suffix **-virales**. Three orders have been approved by the ICTV: **Mononegavirales,** which consists of the families Bornaviridae, Paramyxoviridae, Rhabdoviridae and Filoviridae (these are single-stranded negative-sense non-segmented enveloped RNA viruses): **Nidovirales**, which consist of the Coronaviridae and Arteriviridae (these are ss positive sense enveloped viruses): **Caudovirales** (which includes two phage families).

Virus families

These represent groupings of genera of viruses that share common characteristics and are distinct from the members of other families. Families are designated by the suffix **-viridae**. Families as groupings have proved to be excellent models for classification. Most of the families have **distinct virion morphology**, **genome structure** and **strategies of replication**. Examples include Picornaviridae, Togaviridae, Poxviridae, Herpesviridae and Paramyxoviridae. In some families (e.g. Herpesviridae) the complex relationships between individual members has led to the formation of sub-families, which are designated with the suffix **-virinae**. Thus, the Herpesviridae are further classified into the Alphaherpesvirinae (e.g. herpes simplex virus) the Betaherpesvirinae (e.g. cytomegalovirus) and the Gammaherpesvirinae (e.g. Epstein–Barr virus).

Virus genera

Virus genera are groupings of species of viruses which share common characteristics and are distinct from the members of other genera. They are designated by the suffix **-virus** (e.g. genus Simplexvirus and genus Varicellovirus of the Alphaherpesvirinae). The criteria for designating genera vary from family to family, but include genetic, structural and other differences.

Virus species

A virus species is defined as 'a polythetic class of viruses that constitutes a replicating lineage and occupies a particular ecological niche'. Members of a polythetic class are defined by more than one property. At present the ICTV is examining carefully the properties which can be included in determining species. The division between species and strains is a difficult one.

L3 VIRUS PROTEINS

Key Notes

Overview

Viral proteins, encoded by the viral genome, are either structural (capsid, envelope) or non-structural (e.g. enzymes, oncoproteins, inhibitors of cell macromolecular synthesis and interaction with MHC presentation). They may be essential or non-essential in tissue culture replication.

Methodology

Structural proteins are studied, following virus purification, using a range of techniques including sodium dodecyl sulfate (SDS)–polyacrylamide gel electrophoresis, Western blotting and immunoprecipitation. Non-structural proteins are examined by, for example, pulse–chase experiments, or the use of protease and glycosylation inhibitors.

Protein synthesis and complexity

Viral proteins are synthesized by the translation of viral mRNAs on cellular ribosomes. Proteins are often processed following synthesis (e.g. proteolytic cleavage, glycosylation, myristoylation, acylation phosphoralation and palmitoylation). In many viruses, protein synthesis is controlled at the levels of transcription and translation, which makes virus replication quite an efficient process. This control is usually directed by the virus genome.

Structural proteins

Structural proteins are either nucleocapsid, matrix or envelope proteins. They have a role in protecting the viral genome and in delivering the genome from one host to another via receptors on host cells. They also have a major role in the assembly of the virion. In most viruses, structural proteins are produced in abundance late in the replicative cycle.

Non-structural proteins

These may be carried in the virion (but are not part of the virion architecture) where they have enzymic activity which is necessary for initiating infection. Others are not destined for the virion but have roles in the infected cell. These roles include switching off host-cell nucleic acid and protein synthesis, polymerase, protease and kinase activities, DNA-binding activity and gene regulation.

Related topics

Virus replication (L7)	Virus vaccines (L10)
Viruses and the immune system (L9)	Antiviral chemotherapy (L11)

Overview

The **coding capacity** of viral genomes varies from <5 to >100 genes. This in turn is reflected in the different complexities of virus particles and their respective replicative cycles. Viral proteins are either **structural** (part of the virion architecture), **non-structural** but in the virion (usually enzymes) or **non-structural** and present only in infected cells, never the virion (these have a range of functions). The limited number of virus proteins synthesized means that many of them are multifunctional.

Virus proteins are often referred to as being **essential** or **non-essential**. The former are an absolute requirement for the virus in order for it to complete a replicative cycle and to produce infectious virions. The latter can be deleted from the virus genome without seriously affecting virus growth in tissue culture. However, this may not reflect the *in vivo* situation.

Methodology

The study of virus structural proteins as assembled virions has required the **purification of viruses** from infected cells or infected-cell supernatants, ensuring that they are free from surrounding contaminating proteins. This is usually achieved by a series of **differential centrifugation** steps, followed by **sucrose gradient-density centrifugation**. Centrifugation through a sucrose gradient usually results in a sharp band of virus at a specific location on the gradient. This is harvested for further studies. It is normal to study **radiolabeled virions** by a variety of techniques including **SDS–polyacylamide gel electrophoresis** (separation based on size), **Western blotting** (reaction with antibodies) and **immunoprecipitation** (precipitation of proteins with antibodies).

The location of virus proteins within cells can be determined by **differential staining, immunofluorescence and confocal microscopy** techniques, with monoclonal antibodies to specific virus protein epitopes being the key reagents. Use of, for example, **pulse–chase** experiments, **protease** or **glycosylation inhibitors** have allowed studies on protein processing.

Gene sequencing and **amino acid prediction** of proteins has allowed a number of predictions to be made concerning virus protein structure and function, including cleavage sites, membrane spanning ability, and glycosylation sites. Sequencing allows identification of conserved areas in genes.

Protein synthesis and complexity

As was outlined earlier, all proteins are synthesized from a mRNA template which is translated on the host-cell ribosomes. The mRNA may be (a) transcribed from the viral genome DNA (e.g. HSV), (b) complementary to the viral genome RNA (**negative-strand RNA viruses,** e.g. influenza) or (c) **viral** genomic RNA (**positive-strand RNA viruses,** e.g. polio).

During the growth cycle, in more complex viruses, virus-specific proteins are synthesized at various times post-infection, for example, structural proteins may be produced late on in infection. Simple viruses (e.g. polio) do not have such levels of **temporal control**.

The amino acid sequence of the protein contains a number of **motifs** which determine its **post-translational modification**, its location in the cell (by **signaling**) and its **secondary** and **tertiary** structure.

Proteins may be the result of **proteolytic cleavage** of a larger **precursor molecule** (e.g. polio polyprotein, hepatitis C polyprotein, HIV gag-polymerase complex). They may be **phosphorylated**, the degree of phosphorylation often determining the functional activity of the protein (e.g. N, NS proteins of rhabdoviruses). The attachment of sugar residues to proteins (**glycosylation**) is either N- or O-linked and for many virus proteins (usually envelope glycoproteins) the sugar residues may account for 75% of the protein weight. **Myristylation, acylation** and **palmitoylation** are other post-translational modifications.

Structural proteins

Whatever the complexity, size or shape of a virus particle it is the role of the structural protein to provide **protection** for the viral genome and to allow delivery of virus particles from one host to another. Structural proteins are in either the **nucleocapsid, matrix** or **envelope** of the particle (*Fig. 1*).

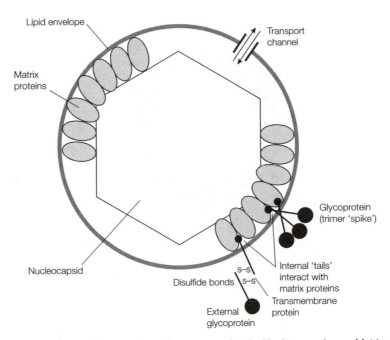

Fig. 1. Several classes of proteins are associated with virus envelopes. Matrix proteins link the envelope to the core of the particle. Virus-encoded glycoproteins inserted into the envelope serve several functions. External glycoproteins are responsible for receptor recognition and binding, while transmembrane proteins act as transport channels across the envelope. Host-cell derived proteins are also sometimes found to be associated with the envelope, usually in small amounts. Redrawn from Cann, A., Principles of Molecular Virology, 1993, with permission from Academic Press.

Nucleocapsid proteins either self-assemble (e.g. TMV, polio) or assemble with the help of scaffolding proteins (e.g. HSV) to form the helical and icosahedral structures described earlier (Topic J1). The polio capsid is relatively simple with four proteins, VP1, VP2, VP3 and VP4. These assemble via a precursor structure (the procapsid) which contains VP0 (a precursor protein) and VP1 and VP3. VP0 cleaves to give VP2 and VP4 as the capsid acquires its nucleic acid. The reovirus capsid is much more complex and is composed of a double-shelled icosahedron structure (*Fig. 2*). Capsid proteins, in addition to protecting the genome and 'assembling' to form structural virions, must also interact with the nucleic acid genome during the assembly process. **Packaging** of the large piece of nucleic acid into a confined area requires complex folding of the genome and intimate association through chemical bonding with selected capsid proteins. Examples are rhabdovirus N protein and influenza virus NP protein whose positively charged amino acids interact with the negatively charged nucleic acid to encourage packaging. Nucleic acids also contain **packaging sequences** which result not only from nucleotide sequence but also from the **secondary** and **tertiary structures** of the genome. Packaging becomes even more complex when one considers a multigenome virus (e.g. influenza) where eight different segments of RNA have to be packaged into the virion nucleocapsid.

The **envelope** which surrounds many virus particles is acquired by nuclear membrane, plasma membrane or endoplasmic reticulum budding. Within the envelope are glycoproteins.

Matrix proteins are internal proteins whose function, when present, is to link the internal nucleocapsid proteins to the envelope. They are usually not glycosylated

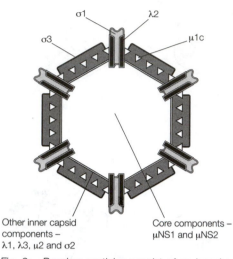

σ1   λ2

σ3   μ1c

Other inner capsid
components –
λ1, λ3, μ2 and σ2

Core components –
μNS1 and μNS2

Fig. 2.   Reovirus particles consist of an icosahedral, double-shell arrangement of proteins surrounding the core. Redrawn from Cann, A., Principles of Molecular Virology, *1993, with permission from Academic Press.*

and may contain **transmembrane anchor domains** or are associated with the membrane by hydrophopic patches on their surface or by protein–protein interactions with envelope glycoproteins. In HSV the space between the envelope and the capsid is referred to as the **tegument**.

External glycoproteins are anchored in the envelope by **transmembrane domains**. Most of the structure of the protein is on the outside of the membrane, with a relatively short internal tail. Many of these glycoproteins are **monomers** which group together to form the tiny spikes visible in the electron microscope. They are the major **antigens** of enveloped viruses and include the G spikes of rabies virus, the gp120 spike of HIV and the HA spike of influenza.

The detailed structure of many of these molecules has been determined by **X-ray crystallography** and **cryoelectron microscopy**. One of the first molecules to be analyzed in this detail was the **hemagglutinin** of influenza. This excellent work resulted in the detailed molecule shown in *Fig. 3*.

Certain structural proteins in either naked capsids (e.g. polio) or enveloped viruses (e.g. HSV) initiate viral infection by attaching to **receptor sites** on host cells. Immune responses to these virus proteins often protect against viral infection and our understanding of this interaction forms the basis of modern vaccine design (see Topic L10). *Table 1* lists examples of specific receptor sites available on host cells for attachment by viruses.

Non-structural proteins

Non-structural proteins may be carried within the virus particle (but not as part of the 'architecture') or appear only in infected cells. The number and function of these proteins varies greatly from one virus family to another, depending on the complexity of the viral genome and the replicative cycle. Many have enzymic activity (e.g. the **reverse transcriptase, protease** and **integrase** of retroviruses, the **thymidine kinase** and **DNA polymerase** of HSV). These enzymes are key targets for antiviral drugs (see Topic L11). Others have a role in virion assembly, acting as scaffolding proteins upon which capsids are assembled. Many (e.g. HIV **Tat** HSV **tegument proteins**) have regulatory roles in transcription and others form part of nucleic acid synthesis complexes (e.g. **DNA helicase, DNA binding**

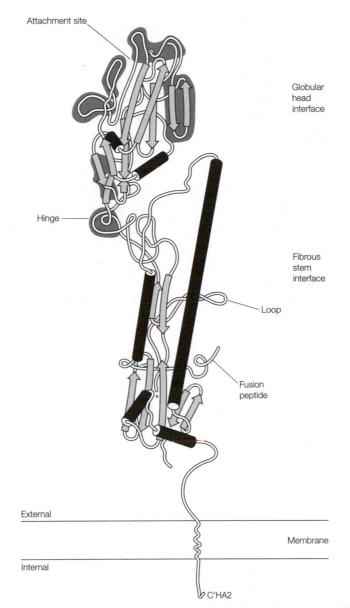

Fig. 3.   The structure of the hemagglutinin of influenza. Redrawn from Wiley et al., Nature, vol. 289, pp. 373–378, 1981, with permission from Macmillan Magazines Limited.

proteins). The shut-off of host cell macromolecular synthesis is usually the function of viral non-structural proteins. Viruses which induce tumors in their host by **transforming** normal cells do so by producing **viral oncogenes** (non-structural virus proteins) or by activating cellular oncogenes. Other proteins associated with cell transformation include, for example, the large **T antigen** of SV40 and the **EBNA** protein of Epstein–Barr virus. More recently, non-structural proteins have been associated with **anti-apoptosis**, **anti-cytokine** activity and interference with MHC antigen presentation (i.e. immune evasion – see Topic L9): *Table 2* lists the functions of HIV proteins as an example of selected activities.

Table 1.   Examples of receptors for viruses that infect humans

Family	Virus	Cellular Receptor
Adenoviridae	Adenovirus type 2	Integrins $\alpha_v\beta_3$ and $\alpha_v\beta_5$
Coronaviridae	Human coronavirus 229E	Aminopeptidase N
Coronaviridae	Human coronavirus OC43	N-Acetyl-9-O-acetylneuraminic acid
Hepadnaviridae	Hepatitis B virus	IgA receptor
Herpesviridae	Herpes simplex virus	Heparan sulfate proteoglycan plus mannose-6-phosphate receptor
Herpesviridae	Varicella zoster virus	Heparan sulfate proteoglycan?
Herpesviridae	Cytomegalovirus	Heparan sulfate proteoglycan plus second receptor
Herpesviridae	Epstein–Barr virus	CD21 (CR2) complement receptor
Herpesviridae	Human herpesvirus 7	CD4 (T4) T-cell marker glycoprotein
Orthomyxoviridae	Influenza A virus	Neu-5-Ac (neuraminic acid) on glycosyl group
Orthomyxoviridae	Influenza B virus	Neu-5-Ac (neuraminic acid) on glycosyl group
Orthomyxoviridae	Influenza C virus	N-Acetyl-9-O-acetylneuraminic acid
Paramyxoviridae	Measles virus	CD46 (MCP) complement regulator
Picornaviridae	Echovirus 1	Integrin VLA-2 ($\alpha_2\beta_1$)
Picornaviridae	Poliovirus	IgG superfamily protein
Picornaviridae	Rhinoviruses	ICAM-1 adhesion molecule
Poxviridae	Vaccinia	Epidermal growth factor receptor
Reoviridae	Reovirus serotype 3	β-Adrenergic receptor
Retroviridae	Human immunodeficiency virus	CD4 (T4) T-cell marker glycoprotein and Chemokine co-receptor
Rhabdoviridae	Rabies	Acetylcholine receptor

Table 2.   Proteins of the human immunodeficiency virus (HIV) type 1

Structural proteins

gag	Polyprotein, cleaved by the viral aspartyl proteinase, produces the core proteins
pol	Polymerase/reverse transcriptase, aspartyl proteinase and integrase
env	Polyprotein (gp160), cleaved by cellular enzymes to produce gp41 and gp120

Auxiliary genes: non-structural proteins

vif	Required for production of infectious virus
vpr	Function unknown
tat	Up-regulates viral mRNA synthesis by binding to the TAR (*trans*-activator response element) in the transcripts from the long terminal repeat. Made from a spliced mRNA
rev	Required by mRNA transport. Made from a spliced mRNA
vpu	Virion release, receptor degradation
nef	Down-regulates CD4 & MHC, essential for virus pathogenicity

L4 VIRUS NUCLEIC ACIDS

Key Notes

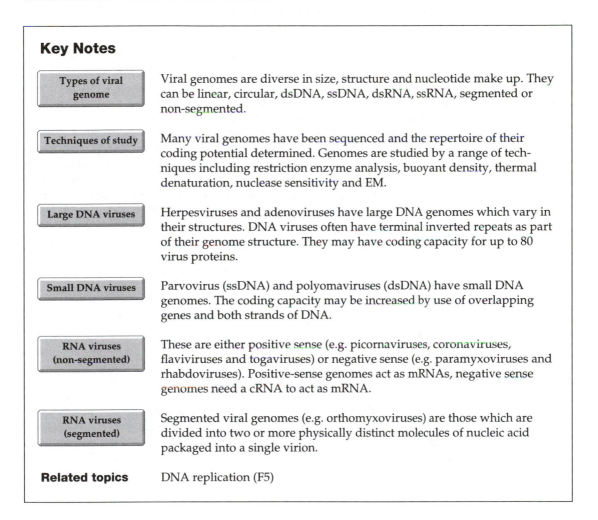

Types of viral genome

Viral genomes are diverse in size, structure and nucleotide make up. They can be linear, circular, dsDNA, ssDNA, dsRNA, ssRNA, segmented or non-segmented.

Techniques of study

Many viral genomes have been sequenced and the repertoire of their coding potential determined. Genomes are studied by a range of techniques including restriction enzyme analysis, buoyant density, thermal denaturation, nuclease sensitivity and EM.

Large DNA viruses

Herpesviruses and adenoviruses have large DNA genomes which vary in their structures. DNA viruses often have terminal inverted repeats as part of their genome structure. They may have coding capacity for up to 80 virus proteins.

Small DNA viruses

Parvovirus (ssDNA) and polyomaviruses (dsDNA) have small DNA genomes. The coding capacity may be increased by use of overlapping genes and both strands of DNA.

RNA viruses (non-segmented)

These are either positive sense (e.g. picornaviruses, coronaviruses, flaviviruses and togaviruses) or negative sense (e.g. paramyxoviruses and rhabdoviruses). Positive-sense genomes act as mRNAs, negative sense genomes need a cRNA to act as mRNA.

RNA viruses (segmented)

Segmented viral genomes (e.g. orthomyxoviruses) are those which are divided into two or more physically distinct molecules of nucleic acid packaged into a single virion.

Related topics

DNA replication (F5)

Types of viral genome

The viral genome carries the nucleic acid sequences which are responsible for the **genetic code** of the virus. In infected cells the genome is **transcribed** and **translated** into those amino acid sequences which make up the viral proteins, be they structural or non-structural products. The genome, or in some cases a transcript of it, forms the basis of the template upon which new genomes are synthesized before assembly of progeny virions. Viral genomes are either **linear** or **circular** and composed of **RNA** or **DNA**. They vary in size from 3500 nucleotides (e.g. small phage) to 560 000 nucleotides (e.g. some herpesviruses). These sequences must be capable of being decoded by the host cell and thus the control signals must be recognized by host factors usually in association with viral proteins. Because of their small size, viral genomes have evolved to make **maximal** use of their nucleotide coding potential. Thus, **overlapping** genes and **spliced mRNAs** are common.

Characterization of the viral genome is based on a number of parameters: (1) **composition of the nucleic acid (i.e. DNA or RNA);** (2) **size and number of strands**; (3) **terminal structures**; (4) **nucleotide sequence**; (5) **coding capacity**; (6) **regulatory signal elements, transcriptional enhancers, promoters and terminators.**

Some points of note are:

- dsDNA genomes (e.g. Poxviridae, Herpesviridae and Adenoviridae) are usually the largest genomes, ssDNA genomes (e.g. Parvoviridae) being smaller.
- dsRNA genomes (e.g. Reoviridae) are all segmented.
- ssRNA genomes are classified as being positive (+) sense or negative (–) sense. Viral genomes that are + sense can act as mRNAs and are usually infectious without the nucleocapsid proteins (e.g. Picornaviridae, Caliciviridae, Coronaviridae, Flaviviridae and Togaviridae). Negative-sense RNA genomes are usually not infectious unless accompanied by nucleocapsid proteins which have enzymic activity (transcriptases). These enzymes transcribe the negative-sense RNA into a complementary strand (cRNA) which acts as mRNA (e.g. Orthomyxoviridae, Paramyxoviridae, Rhabdoviridae and Filoviridae).
- Most ssRNA genomes are single molecules with the exception of, for example, Orthomyxoviridae (influenza) which have segmented genomes. Retroviridae have a diploid ssRNA viral genome which replicates via a dsDNA intermediate, this being facilitated by a reverse transcriptase enzyme carried within the virion.
- The DNA genome of Hepadnaviridae is synthesized within hepatocytes via an RNA intermediate. These different modes of viral replication are summarized in *Fig. 1.*

Techniques of study

The advances in molecular-biological techniques over the last decade have made analysis of viral nucleic acids simpler, more efficient and quick. Viral genomes from most of the families have been totally **sequenced** and the open reading frames, and in many cases the gene products, characterized. This allows comparison with known sequences of genes stored on computer databases and has revealed fascinating similarities between viral and various eukaryotic genes which have been conserved through the process of evolution. Genes can be cloned into a variety of vectors and analyzed by a number of techniques (e.g. site-directed mutagenesis) to study the role of individual amino acids in determining the structural and functional integrity of the coded protein (see *Instant Notes in Biochemistry* in this series for an excellent review of these techniques).

DNA viruses are often shown diagramatically as linear molecules with **restriction enzyme** sites scattered throughout the genome. There are dozens of restriction enzymes, which are used to digest DNA into small segments at very specific nucleotide sequences. These segments are separated by size on agar gels following electrophoresis. Each DNA genome has a restriction enzyme **map** which characterizes it. This was not possible for RNA genomes until a reverse transcriptase enzyme was used to make a copy of a DNA molecule from the RNA template, the cDNA being 'cut' with restriction enzymes. This technique is reffered to as 'reverse genetics'.

Viral nucleic acids can be characterized by their melting temperature (T_m), their buoyant density in cesium chloride gradients, their 'S' value in sucrose gradients, their infectivity (or lack of), their nuclease sensitivity and their electron microscopic appearance.

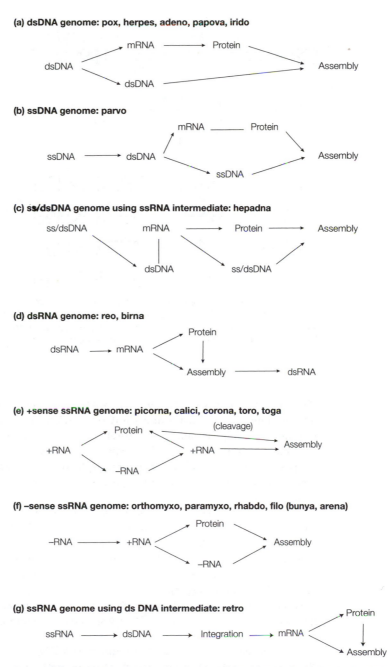

(a) dsDNA genome: pox, herpes, adeno, papova, irido

(b) ssDNA genome: parvo

(c) ss/dsDNA genome using ssRNA intermediate: hepadna

(d) dsRNA genome: reo, birna

(e) +sense ssRNA genome: picorna, calici, corona, toro, toga

(f) –sense ssRNA genome: orthomyxo, paramyxo, rhabdo, filo (bunya, arena)

(g) ssRNA genome using ds DNA intermediate: retro

Fig. 1. General methods of viral replication. From Harper, D., Molecular Virology, 2nd edn, © BIOS Scientific Publishers Limited, 1998.

Large DNA viruses

Herpesviridae vary in size from herpes simplex virus and varicella zoster virus (120–180 kbp) to cytomegalovirus and HHV-6 (180–230 kbp). The DNA codes for >30 virion proteins and >40 non-structural proteins (found only in infected cells). The structure of the DNA, while showing minor variations between members of the group, is unique in that several **isomers** of the same molecule can exist.

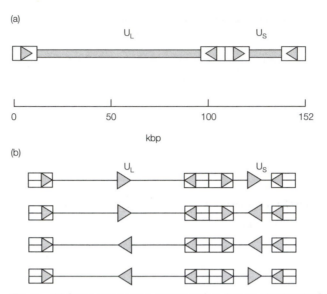

Fig. 2. (a) Some herpes virus genomes (e.g. herpes simplex virus) consist of two covalently joined sections, U_L and U_S, each bounded by inverted repeats. (b) This organization permits the formation of four different forms of the genome. Reproduced from Cann, A., Principles of Molecular Virology, 1993, with permission from Academic Press.

The genome of herpes simplex virus consists of two covalently-joined sections, a **unique long** (U_L) and a **unique short** (U_S) region, each bounded by **inverted repeats**. These repeats allow structural rearrangements of the unique regions, thus creating four isomers all of which are functionally equivalent (*Fig. 2*). Herpesvirus genomes also contain **multiple repeated sequences** which vary greatly between viruses, thus giving them a larger or smaller molecular weight than average.

The genome of adenoviruses is smaller, with a size of 30–38 kbp. Each virus contains 30–40 genes. The terminal sequence of each strand is a 100–140 bp **inverted repeat** and thus the denatured single strands can combine to form a **pan-handle** structure, which has an important role in replication. The genome has a 55 kb terminal protein at the 5′ end which acts as a primer for the synthesis of new DNA strands (*Fig. 3*).

Whereas in herpesviruses each gene has its own promoter the adenovirus genome has **clusters** of genes which are served by a common promoter.

Small DNA viruses

Parvovirus genomes are linear non-segmented ssDNA of about 5 kb. Most of the strands that appear in virions are of negative (–) sense. These very small genomes contain only two genes, **rep**, which encodes proteins involved in transcription, and **cap,** which encodes the coat proteins. The ends of the genomes have **palindromic sequences** of about 115 nucleotides which form hairpin structures essential for the initiation of genome replication.

Polyoma viruses have dsDNA, 5 kbp in size and circular. The DNA in the virion is **super-coiled** and is associated with four cellular histones (H2A, H2B, H3 and H4). The virus has six genes, this having been achieved by using **both strands** of the DNA for coding and making use of **overlapping genes** (*Fig. 4*).

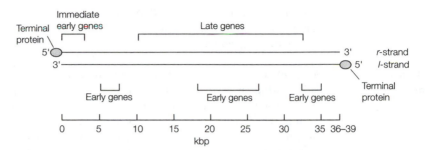

Fig. 3.   Organization of the adenovirus genome. Reproduced from Cann, A., Principles of Molecular Virology, 1993, with permission from Academic Press.

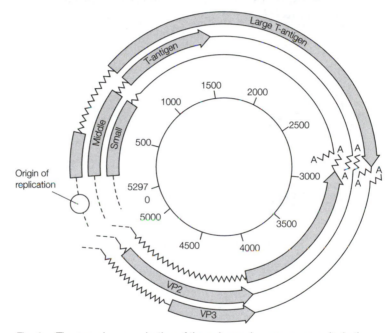

Fig. 4.   The complex organization of the polyomavirus genome results in the compression of much genetic information into a relatively short sequence. Reproduced from Cann, A., Principles of Molecular Virology, 1993, with permission from Academic Press.

RNA viruses (non-segmented)

RNA viruses tend to have smaller genomes than most DNA viruses. They vary in size from 30 kb (for example, coronaviruses) to 3.5 kb (small phage, for example, Qβ). They are either **positive sense** or **negative sense** as described earlier. Most (not all) positive-sense ssRNA genomes have a **cap** that protects the 5' terminus of the RNA from attack by phosphatases and other nucleases and promotes mRNA function at the level of translation. In the 5' cap the terminal 7-methylguanine and the penultimate nucleotide are joined by their 5'-hydroxyl groups through a triphosphate bridge. This 5'–5' linkage is inverted relative to the normal 3'–5' phosphodiester bonds in the remainder of the polynucleotide chain. In most picornaviruses this **cap** is replaced by a small protein, VPg. Some viruses, e.g. picornaviruses, flaviviruses, have a complex hairpin-like RNA structure – an **internal ribosome entry site** (IRES) where translation is initiated. The 3' end of most positive-sense viral genomes, like most eukaryotic mRNAs, is **polyadenylated.**

Viruses with negative-sense RNA genomes are usually larger and encode more genetic information than positive genomes. The genomic organization of selected RNA viruses is shown in *Figure 5*.

RNA viruses (segmented)

Segmented virus genomes are those which are divided into two or more **physically distinct molecules** of nucleic acid, all of which are packaged into a single virion. They have the advantage of carrying genetic information in smaller strands, which makes them less likely to break due to shearing. This breakage is a hazard for large RNA molecules. On the downside, of course, is the fact that these viruses must have an elaborate packaging mechanism which allows a representative of each strand to be packaged into the virion. Influenza virus, a negative ssRNA virus, has eight segments each coding for one or two viral proteins (*Fig. 6* and *Table 1*).

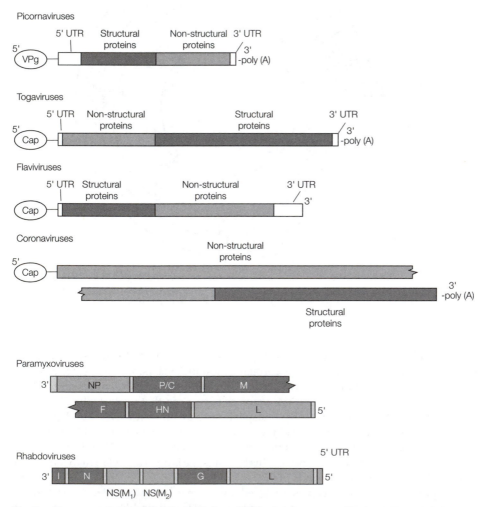

Fig. 5.   Diagrammatic representation of selected RNA virus genomes showing structural and non-structural genes. 5′ UTR are regions of the genes that are not translated into proteins. Reproduced from Cann, A., Principles of Molecular Virology, *1993, with permission from Academic Press.*

Orthomyxoviruses

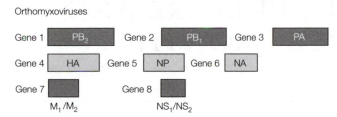

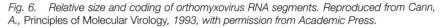

Fig. 6.   Relative size and coding of orthomyxovirus RNA segments. Reproduced from Cann, A., Principles of Molecular Virology, *1993, with permission from Academic Press.*

Table 1.   Segments of the influenza virus genome

Segment	Size (nt)	Polypeptide(s)	Function
1	2341	PB$_2$	Transcriptase: cap binding
2	2341	PB$_1$	Transcriptase: elongation
3	2233	PA	Transcriptase: (?)
4	1778	HA	Hemagglutinin
5	1565	NP	Nucleoprotein: RNA binding; part of transcriptase complex
6	1413	NA	Neuraminidase
7	1027	M$_1$	Matrix protein: major component of virion
		M$_2$	Integral membrane protein – ion channel
8	890	NS$_1$	Non-structural (nucleus): effects cellular RNA transport; disables innate immunity
		NS$_2$	Non-structural (nucleus + cytoplasm): mediates export of vision RNA from cell nucleus

L5 CELL CULTURE AND VIRUS GROWTH

Key Notes

Historical perspective	The major thrust in virology research was only possible after the derivation of defined synthetic growth medium and the *in vitro* culture of cells. This allowed viruses to be grown, purified and studied outside of the host, leading to the development of the first virus vaccines.
Method of cell culture	Single-cell suspensions are seeded into tissue culture quality glass or plastic vessels where they attach to the surface and divide to form a monolayer of cells. Cells can be grown in a full range of vessels but must be carefully nurtured and repassaged, ensuring they are free from contamination. Most cells grow at 37°C.
Media and buffer	Defined growth media contain a balanced salts solution supplemented with amino acids, vitamins, glucose, serum, antibiotics, a buffer and pH indicator. Such media are used routinely to support the growth of tissue-culture cells and can be either manufactured from the individual constituents or bought ready-made commercially.
Cell culture types	Cells are either primary (very limited cell passage), diploid cell strains (up to 50–60 cell passages) or continuous cell lines (unlimited cell passage).
Virus growth in culture	Small aliquots of virus are added to cell monolayers at low or high multiplicity of infection and, following replication, progeny virus is harvested and assayed. The virus is either cell-associated or released into the culture medium. Following titration, the virus is stored at either –20°C or –70°C until required.
Use of embryonated eggs	For some viruses, growth in embryonated hens eggs is the preferred culture method. Influenza virus is grown in the allantoic cavity of the embryonated egg.
Related topics	Virus assay (L6) Virus replication (L7)

Historical perspective

In addition to helping solve the mysteries of virus diseases, knowledge of virus structure, replication and host interactions, virus research has extended our understanding of the fundamentals of eukaryotic biology (e.g. protein:protein interactions, nucleic acid transcription and translation, evolution). Few of these studies would have been possible without the means of growing viruses outside their normal hosts. For most viruses this means growing them in **cell culture**. Cell culture is the art of growing cells *in vitro*. In addition to deriving cells for culture, this technique relied heavily on the development of **media** and **buffers** that supported cell growth *in vitro*. Most of these techniques did not become available until the early 1950s, which consequently saw an upsurge in the study of viruses.

There are now over 3200 characterized cell lines derived from over 75 species, which are kept in, for example, the American Type Culture Collection and the European Collection of Animal Cell Cultures.

Method of cell culture

Cells are propagated on **glass** or **plastic flasks** of various sizes (as required) or in vast vessels or vats. The technique relies on good **aseptic technique**, thus keeping the cultures free of fungal and bacterial contamination.

Single-cell suspensions of cells of known concentration are **seeded** into a sterile flask along with appropriate **growth medium**. The flask, usually a plastic or glass bottle, is incubated at the appropriate temperature (usually 37°C) in the flattened position. Cells adhere to the surface and begin to replicate, forming a **monolayer** of cells which adhere to each other, in addition to being anchored to the surface of the flask. After a few days the metabolic activity of the cells means that the growth medium will be 'spent' and the cells, unless **reseeded**, will deteriorate and die. Thus, the cell monolayer is treated with trypsin and/or versene solution to create single cells again. These cells are used for **seeding** into fresh flasks. The continued seeding of cells is referred to as cell passage. Cell monolayers are used for virus growth and assay and to examine many aspects of virus–host interactions (*Fig. 1*).

In addition to growing as a monolayer, some cell types may also be able to grow in **suspension**, where they do not anchor themselves to the surface of the flask or adhere to each other (e.g. hybridoma cells which secrete monoclonal antibodies).

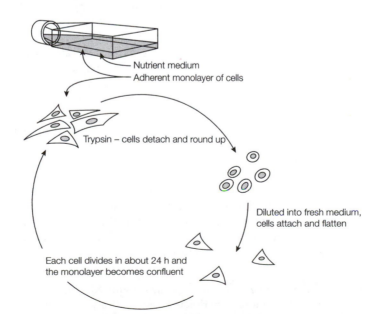

Nutrient medium
Adherent monolayer of cells

Trypsin – cells detach and round up

Diluted into fresh medium, cells attach and flatten

Each cell divides in about 24 h and the monolayer becomes confluent

Fig. 1. Cell culture. Redrawn from Dimmock and Primrose, Introduction to Modern Virology, *4th edn, 1994, with permission from Blackwell Science Ltd.*

Media and buffer   Most growth media in use today are chemically defined, but are usually supplemented with 5–20% serum (which contains the stimulants necessary for cell division). Serum-free medium with added stimulants is used for some purposes. Media contain an isotonic **balanced salts solution**, supplemented with amino acids, vitamins and glucose, for example, Eagle's minimal essential medium (MEM), formulated by Eagle in the 1950s. In addition to serum, MEM is also supplemented with antibiotics, usually penicillin and streptomycin, to help prevent bacterial contamination. Generally, cells grow well at pH 7.0–7.4 and hence phenol red is added to act as a pH indicator. It is red at pH 7.4, orange at pH 7.0 and yellow at pH 6.5. In the alkaline direction it becomes bluish at pH 7.6 and purple at pH 7.8.

Culture media also require buffering under two sets of conditions: (a) use of open flasks, in which exposure to oxygen causes the pH to rise; and (b) high cell concentrations when CO_2 and lactic acid are produced, causing the pH to fall. A buffer is thus incorporated into the medium, and in 'open flask' conditions the incubator is purged with an exogenous source of CO_2. The buffer used by most laboratories relies on the bicarbonate–CO_2 system, and hence bicarbonate solution is added to the growth medium.

Reagents for use in media preparation and cell culture must be sterilized. This is achieved by autoclaving (moist heat), hot-air ovens (dry heat), membrane filtration or, in the case of plastic ware, by irradiation.

Cell culture   **Primary cells** are freshly isolated cells that are derived directly from the tissue of
types   origin. The tissue source for the majority of primary cell cultures is either laboratory animals (e.g. monkey kidney cells) or human pathology specimens (e.g. human amnion cells). Tissue samples are incubated with a proteolytic enzyme (usually trypsin) overnight with a number of washes, etc., in order to produce a single-cell suspension. The harvested cells are then seeded into appropriate flasks with growth medium. Cell cultures are usually either epithelial (cuboid-shaped) or fibroblastic (spindle-shaped), but it is usual for primary cultures to contain both morphological types. Primary cultures are sensitive to a wide range of viruses and are routinely used in diagnostic laboratories for growth of fresh virus isolates (from patients). Unfortunately, primary cells usually die after only a few passages. Some cells, when grown *in vitro* change so as to lose their ability to support virus growth. Hepatocytes are an example, where *in vivo* they do, but *in vitro* do not, support hepatitis C replication.

Many **cell lines** will continue to grow for more than four to five passages but subsequently die after 50–60 passages. These cell lines, usually referred to as **diploid cell strains**, are often derived from fetal lung tissue. They have the normal chromosome number and are usually fibroblastic (e.g. MRC-5 cells).

Continuous cell lines are capable of continued passage in tissue culture. They may be epithelial (e.g. Hela cells) or fibroblastic (e.g. baby hamster kidney cells, BHKs) and are derived from tumorous tissue (e.g. Hela cells) or by the sudden **transformation** of a primary cell (e.g. BHK cells). They are **heteroploid** (i.e. have aberrant chromosome numbers).

Virus growth in   Most experiments in virology have required virus growth in culture although,
culture   nowadays, more experiments rely entirely on cloned genes and expressed proteins outside of cell cultures. Historically, however, it has been those viruses

which grow well in cell culture that have been studied in most detail. Lack of *in vitro* growth has seriously curtailed progress in research, vaccine production and development of antiviral drugs, for example hepatitis B and C viruses.

Viruses are grown in cultures to create **virus stocks**. The passaged virus is stored at –70°C and referred to as a **master-stock, sub-master stock**, etc., depending on its **passage number**. It is important in virology to record the passage history in tissue culture of the virus being used.

Virus stocks are grown by infecting cells at a low **multiplicity of infection (m.o.i.)** that is, approximately 0.1–0.01 infectious units per cell. Virus attaches to cells and goes through several replicative cycles in the cell culture. After a few days, virus is harvested from the extracellular medium surrounding the cultured cells or from the cells themselves, which are lysed by freezing and thawing or by using an ultrasonic bath. The virus so derived is quantitated by an infectivity assay.

When large numbers of virus particles are required (e.g., for virus purification), cell cultures are infected at a high m.o.i. (e.g. 10 infectious units per cell). This ensures that all cells are synchronously infected and that only one round of replicative cycle will ensue. Virus is harvested as above at the end of the cycle. Infected cells yield various numbers of new (progeny) virus particles ranging from 10–10 000 particles per cell.

Use of embryonated eggs

For some viruses (e.g. influenza virus) cell culture is not the chosen procedure for virus growth and instead a fertilized chick embryo is used. The fertilized embryo has a complex array of membranes and cavities which will support the growth of viruses (*Fig. 2*). Small aliquots of influenza virus are inoculated into the allantoic cavity of the egg. The virus then attaches to and replicates in the epithelial cells lining the cavity. Virus is released into the allantoic fluid and harvested after two days growth at 37°C. Influenza vaccines are propagated in this way.

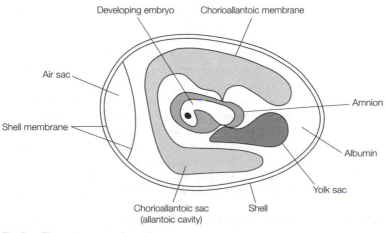

Fig. 2.   The embryonated egg.

L6 VIRUS ASSAY

Key Notes

Virus infectivity	This is defined as the ability of a virus particle to attach, penetrate and undergo an infectious cycle in a susceptible host cell, usually resulting in cell damage.
Virus dilution	Virus is diluted using factors of two, five or 10 in an appropriate solution (e.g. buffer or growth medium) prior to an infectivity assay.
Plaque assay	This is a focal assay used to detect zones of cytopathic effect (CPE), also known as plaques, in a monolayer of healthy cells. One infectious virus produces one plaque.
$TCID_{50}$	The tissue culture infective dose$_{50}$ ($TCID_{50}$) is defined as that dilution of virus which will cause CPE in 50% of a given batch of cell cultures.
Particle counting	The total number of virus particles (infectious and non-infectious) is determined by counting with the aid of an electron microscope. This allows determination of the particle/infectivity ratio.
Hemagglutination	The agglutination of red blood cells (RBCs) by some viruses is referred to as hemagglutination (HA). This agglutination forms the basis of an assay which measures the number of HA units per unit volume in a given suspension.
Related topics	Cell culture and virus growth (L5) Virus replication (L7)

Virus infectivity

In order to 'reproduce', viruses need to be capable of replicating in a susceptible host cell. This **replicative cycle** is accompanied by a number of biochemical and morphological changes within the cell which usually results in the death of the cell. The accompanying morphological changes (e.g. cell rounding or fusion) are referred to as the cytopathic effect (CPE). A particular type of CPE is often a characteristic of specific virus growth and can be used when attempting to identify an unknown virus. The appearance and detection of CPE regularly forms the basis of **infectivity assays**, designed to determine the number of **infectious units of virus per unit volume**, and is the infectivity **titer** (e.g. **plaque forming units** (pfus) per milliliter). An infectious unit is thought of as being the smallest amount of virus that will produce a detectable biological effect in the assay (e.g. a pfu). Infectivity assays are either **quantal,** an 'all or none' approach (e.g. tissue culture infective dose 50 ($TCID_{50}$)) or **focal,** detection of a focus of infection (e.g. a plaque assay).

Virus dilution

Virus titers are determined by making accurate serial dilutions of virus suspensions. Such dilutions are usually done using factors of two, five or 10. For routine use, **10-fold dilutions** are usually carried out. It is important to use a new sterile pipette for the transfer of volumes between each dilution and to mix thoroughly the dilution before further transfer. Once diluted, virus should be assayed as soon as possible as most viruses rapidly lose infectivity at room temperature.

Plaque assay

The plaque assay quantifies the number of **infectious units** in a given suspension of virus. **Plaques** are localized discrete foci of infection denoted by zones of cell lysis or cytopathic effect (CPE) within a monolayer of otherwise healthy tissue culture cells. Each plaque originates from a **single infectious virion,** thus allowing a very precise calculation of the virus titer. The most common plaque assay is the **monolayer** assay. Here, a small volume of virus diluent (0.1 ml) is added to a previously seeded confluent tissue culture cell monolayer. Following adsorption of virus to the cells, an **overlay medium** is added to prevent the formation of **secondary** plaques. Following incubation, the cell sheets are 'fixed' in **formol saline** and stained, and the plaques counted. For statistical reasons, 20–100 plaques per monolayer are ideal to count, although the actual number that can be easily counted is often dependent on the size of the plaque and the size of the vessel used for the assay. Typical plaques are shown in *Fig. 1,* and the assay procedure is summarized in *Fig 2*. The **infectivity titer** is expressed as the number of plaque forming units per ml (**pfu ml^{-1}**) and is obtained in the following way:

$$\text{pfu ml}^{-1} = \frac{\text{plaque number}}{\text{dilution} \times \text{volume (ml)}}$$

For example, if there is a mean number of 100 plaques from monolayers infected with 0.1 ml of a 10^{-6} dilution then the calculation is:

$$\frac{100}{10^{-6} \times 0.1} = 1 \times 10^{9} \text{ pfu ml}^{-1}$$

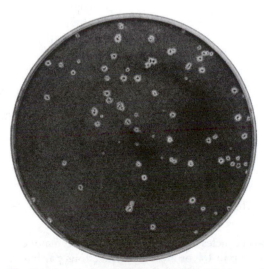

Fig. 1.   Herpes virus plaques on a tissue culture monolayer.

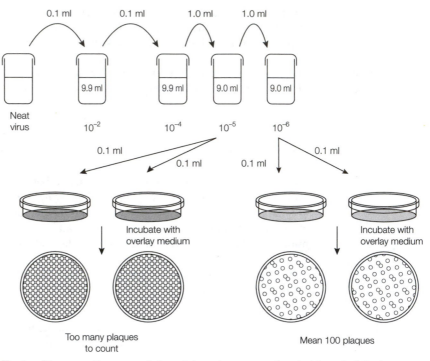

Fig. 2.   Diagrammatic representation of virus plaque assay (see text for calculation).

TCID$_{50}$

The TCID$_{50}$ is defined as **that dilution of virus required to infect 50% of a given batch of inoculated cell cultures**. The assay relies on the presence and detection of CPE. Host cells are grown in confluent healthy monolayers, usually in tubes, to which aliquots of virus dilutions are made. It is usual to use either five or 10 repetitions per dilution. During incubation the virus replicates and releases progeny virus particles into the supernatant which in turn infect other healthy cells in the monolayer. The CPE is allowed to develop over a period of days, at which time the cell monolayers are observed microscopically. Tubes (the 'test units') are scored for the presence or absence of CPE. In this quantal assay the data are used to calculate the TCID$_{50}$, i.e. the dilution of virus which will give CPE in 50% of the cells inoculated. *Table 1* shows some typical data.

By using the data in *Table 1* the following calculation can be made:

$$\text{TCID}_{50} = \log_{10} \text{ of highest dilution giving 100\% CPE}$$

$$+ \frac{\frac{1}{2} - \text{total number of test units showing CPE}}{\text{number of test units per dilution}}$$

$$= -6 + \frac{1}{2} - \frac{4}{5} = -7.3 \text{ TCID}_{50}$$

or $10^{-7.3}$ TCID$_{50}$ unit vol.$^{-1}$

The titer is therefore $10^{7.3}$ TCID$_{50}$ per unit vol.$^{-1}$

Particle counting

Not all virus particles are infectious. Indeed, in many cases, for every one infectious particle up to 100 or more **non-infectious particles** may be produced from an infected cell. The total number of particles can only be determined by counting

Table 1.   Data used to calculate $TCID_{50}$ (see text for calculation)

Log$_{10}$ of virus dilution	Infected test units (e.g. infected tubes)
−6	⅘
−7	⅗
−8	⅕
−9	0/5

them with the aid of an **electron microscope**. The counting procedure relies on the use of **reference particles** which are usually latex beads of uniform diameter. The principle is that if viruses can be mixed with reference particles of known concentration (i.e. a number per unit volume), a simple determination of the ratio of virus to reference particles will yield the virus count. Latex and virus particles are distinguished after **negative staining** with phosphotungstate. The ratio of total particles to infectious particles is termed the **particle/infectivity ratio,** which is important to know when, for example, monitoring virus purification, or determining the state or age of a virus suspension.

Hemagglutination   Many viruses have the ability to agglutinate RBCs, this being referred to as **hemagglutination**. In order for the reaction to occur, the virus should be in sufficient concentration to form cross-bridges between RBCs, causing their agglutination. Non-agglutinated RBCs will form a **pellet** in a hemispherical well, whereas agglutinated RBCs form a **lattice-work** structure which coats the sides of the well. This phenomenon forms the basis of an assay which determines the number of **hemagglutinating particles** in a given suspension of virus. It is not a measure of infectivity, but is one of the most commonly used **indirect methods** for the determination of virus titer. The assay is done by **end-point titration**. Serial two-fold dilutions of virus are mixed with an equal volume of RBCs and the wells are observed for agglutination. The end point of the titration is the **last dilution showing complete agglutination,** which by definition is said to contain one **HA unit**. The HA titer of a virus suspension is therefore defined as being the reciprocal of the highest dilution which causes complete agglutination and is expressed as the number of HA units per unit volume. An example upon which a calculation of the HA titer can be made is shown in *Fig. 3*. The end point in this figure is ¹⁄₅₁₂. If 0.2 ml virus dilution was added per well the HA titer would be 512 HA units per 0.2 ml or 2560 HA units ml^{-1}.

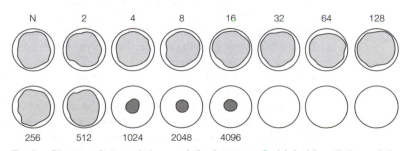

Fig. 3.   Diagram of a sample hemagglutination assay. Serial doubling dilutions of virus shows agglutination end-point at 1:512.

L7 VIRUS REPLICATION

Key Notes

Replicative cycle

As obligate intracellular parasites, viruses must enter and replicate in living cells in order to 'reproduce' themselves. This 'growth cycle' involves specific attachment of virus, penetration and uncoating, nucleic acid transcription, protein synthesis, maturation and assembly of the virions and their subsequent release from the cell by budding or lysis.

Attachment, penetration and uncoating

Attachment is a very specific interaction between the virus capsid or envelope and a receptor on the plasma membrane of the cell. Virions are either engulfed into vacuoles by 'endocytosis' or the virus envelope fuses with the plasma membrane to facilitate entry. Uncoating is usually achieved by cellular proteases 'opening up' the capsid.

Transcription and translation

Using cellular and virus-encoded enzymes and 'helper' proteins, nucleic acid is usually transcribed in a controlled fashion. Control is also exercised at the level of mRNA concentration (apart from some simple viruses, e.g. polio). Nucleic acid is synthesized by virus-encoded enzymes. Translated proteins may undergo post-transitional modification (e.g. cleavage, glycosylation, phosphorylation).

Maturation, assembly and release

Subunits of capsids assemble via 'sub-assembly' structures, with or without the help of scaffolding proteins. Envelopes, when present, are acquired by capsids budding through the nuclear or plasma membrane.

Related topics

Cell culture and virus growth (L5)   Antiviral chemotherapy (L11)
Virus assay (L6)

Replicative cycle

The complexity and range of virus types is echoed in the various strategies they adopt in their replicative cycles. Viruses, as **obligate intracellular parasites,** must **attach** to or **enter** host cells in order to undergo a 'reproductive' cycle. This cycle is highly dependent on the metabolic machinery of the cell, which in most cases the virus takes over and orchestrates towards its own replication, usually **inhibiting host-cell protein and nucleic acid synthesis**. The outcome is the production of hundreds of progeny virions which leave the infected cell (by **lysis** or **budding**), killing the cell and spreading to infect more host cells and tissues. This replicative or **growth cycle** can be analyzed in tissue culture cells and is often referred to as the **one-step** growth cycle. The cycle has a number of stages – attachment and penetration, nucleic acid synthesis and transcription, protein synthesis, maturation, assembly and release. A typical pattern for a growth curve is shown in *Fig. 1*. Following attachment and penetration by virus, cells are lysed and titrated for infectious virus particles (pfus) at various times post-infection. Plotting $\log_{10}$ pfu versus time gives the characteristic curve which has an **eclipse period** (where no new virions have been formed) followed by a **logarithmic expansion phase** until **peak** virus titers are reached when the cell usually dies

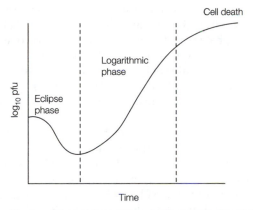

Fig. 1.   A typical 'one-step' growth or replicative cycle of a virus.

and virions are released. The shape of this curve varies greatly between viruses – for most bacteriophage it takes less than 60 min and for many animal viruses it can exceed 24 h before maximum titers are reached. The replication cycles of three animal viruses have been selected below for further study and to compare their strategies – herpes simplex (a DNA virus), poliovirus (RNA virus) and HIV (a single-stranded RNA virus which replicates via a DNA intermediate). Their strategies are typical of many viruses, although divergent strategies also exist!

Attachment, penetration and uncoating

Attachment is mediated by a **specific interaction** between the virus and a **receptor** on the plasma membrane of the cell (*Fig. 2*). Indeed it is the presence of such a receptor that determines the **cell tropism** and **species tropism** of the virus. These receptors have cellular functions other than providing a binding

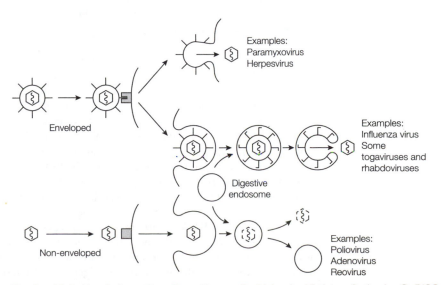

Fig. 2.   Methods of virus entry. From Harper, D., Molecular Virology, 2nd edn, © BIOS Scientific Publishers Limited, 1998.

site for viruses, but have been paramount in affecting virus evolution. Thus, the herpes simplex virus (HSV) binds to a heparin sulfate proteoglycan plus mannose-6-phosphate receptor via at least two virus-coded envelope glycoproteins. One of the four proteins in the poliovirus capsid attaches to a receptor of the **Ig protein superfamily** (only on primate cells). HIV, via its major envelope glycoprotein (**gp120**), attaches to the **CD4 receptor** found predominantly on human T4 lymphocytes. Two further ligand receptors were shown to have a role in HIV attachment.

For HSV and HIV, penetration of the virus across the plasma membrane is achieved by **fusion** of the viral envelope with the membrane, releasing the nucleocapsid into the cytoplasm of the cell (*Fig. 2*). The naked capsid of poliovirus, however, is taken up by the process of **endocytosis**, the membrane invaginating to engulf the capsid, resulting in the formation of a **vacuole** which transports the capsid into the cytoplasm. The virion is later released from this vacuole (*Fig. 2*).

Transcription and translation

(i) **HSV**. HSV, a large DNA virus, replicates mainly in the **nucleus** of the cell, although of course protein synthesis and post-translational modification take place in the cytoplasm. The genome encodes for dozens of virus-specific proteins, many with **enzymic** activity (e.g. **thymidine kinase, DNA polymerase**) – such proteins are usually **non-structural** (i.e. will not finish up in the virion). Others are structural proteins and will form the capsid, envelope and tegument (the structure between the capsid and the envelope).

Viruses need proteins at different times and in different concentrations throughout their growth cycles and hence the virus shows transcriptional control. Depending on the timing of their expression in the virus replication cycle, HSV genes are classified as either **immediate early**, **early** or **late** (α, β or γ). The mRNA produced encodes proteins that have control functions, switching on subsequent genes. All proteins, of course, are formed on cytoplasmic host-cell ribosomes and remain in the cytosol or are directed to the endoplasmic reticulum where they undergo the **post-translational** events (e.g. **glycosylation, phosphorylation**) that give them their final identity. Eventually, these proteins find their way (specifically directed, i.e. **chaperoned**) back to the nucleus for assembly.

The double-stranded viral DNA is synthesized by the viral **DNA polymerase** in association with a number of **DNA-binding proteins**. This enzyme has formed the target for a number of **antiviral drugs** as it is significantly different to the host-cell DNA polymerase (*Fig. 3)*.

(ii) **Poliovirus**. Poliovirus is much simpler in its replicative procedures and control, indeed it lacks fine control! The poliovirus genome RNA strand also acts as mRNA (termed a **positive**-sense RNA virus) and is immediately translated into one long **polyprotein** which is subsequently cleaved into a number of structural and non-structural poliovirus proteins. Included are the structural proteins VP1 and VP3 and the **precursor** protein VP0. Non-structural proteins include a **protease** and **RNA polymerase**. Replication takes place in the **cytoplasm**, indeed enucleate cells will support poliovirus replication. The ss RNA, under the direction of the viral **RNA polymerase** and cellular factors, replicates via a series of ds **replicative intermediate** molecules which act as template for the synthesis of new positive strands. These are then destined to act as mRNA for further rounds of protein synthesis, or become genomes in newly formed progeny virions (*Fig. 4*).

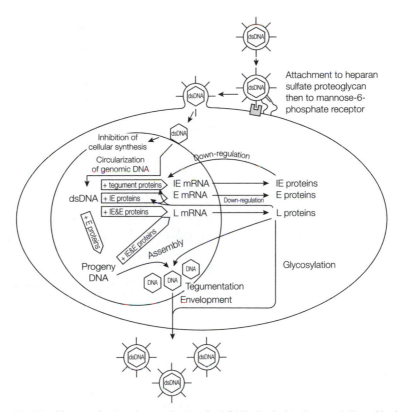

Fig. 3. Herpes simplex virus replication (a dsDNA virus) showing regulation of by immediate early (IE), early (E) and late (L) proteins. From Harper, D., Molecular Virology, 2nd edn, © BIOS Scientific Publishers Limited, 1998.

(iii) **HIV**. HIV is a member of the unique family of viruses (**Retroviridae**) that carry a **reverse transcriptase enzyme** in their particles, the enzyme catalyzing the formation of DNA from a RNA template. Thus, HIV goes through a series of events (*Fig. 5*), which terminates in the formation of a ds DNA circularized molecule, formed from an initial input of diploid HIV ss RNA. This molecule, under the direction of a virally encoded integrase, is inserted into the host DNA as a **provirus**. DNA synthesis is now under the control of the cell – when a daughter cell is produced the provirus is reproduced at the same time. The transcription of viral mRNA is under viral control (from the long terminal repeat (LTR) region of its genome) and a series of **mRNA molecules of various sizes** are transcribed. Translated proteins are in some cases **proteolytically cleaved** by virus **protease** into smaller functional proteins. HIV envelope proteins (gp120 and gp41) are further processed in the endoplasmic reticulum before being laid down in the plasma membrane of the cell. **Full-length copy RNA** molecules are also transcribed from the provirus DNA, these forming **progeny RNA strands** destined to be encapsidated.

Maturation, assembly and release

As proteins and nucleic acid are synthesized in the infected cell they are channeled to various locations for virion assembly. Capsids assemble in the **nucleus** or the **cytoplasm**. The steps of capsid assembly vary, depending on the complexity

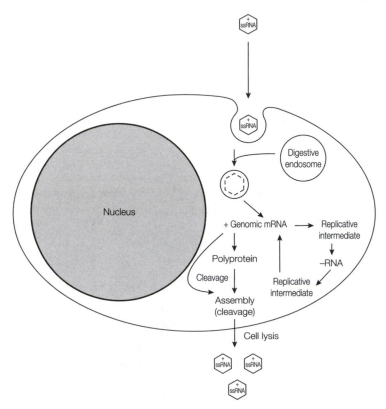

Fig. 4. Poliovirus replication. From Harper, D., Molecular Virology, © *BIOS Scientific Publishers Limited, 1994.*

of the mature capsid, and for enveloped viruses the site of envelope acquisition is either the **nuclear membrane**, **plasma membrane** and rarely the **endoplasmic reticulum**. Prior to capsids 'budding' through these membranes, virus-specific proteins would have been laid down in the membrane.

(i) **HSV**. Mature HSV capsids are assembled in the nucleus via a number of **precursor forms**. Assembly is assisted by **scaffolding proteins,** which do not finish up in the mature capsid but facilitate protein–protein and protein–nucleic acid bonding as the subunits of the capsid come together. The mature capsid (containing the DNA genome) buds through the nuclear membrane to acquire its envelope. Virions are channeled through the ER to the plasma membrane where they are released. Many herpesviruses invade adjacent cells by the process of **cell–cell fusion**. Thus, the plasma membrane of an infected cell fuses with an adjacent normal cell, facilitating the entry of progeny virions which undergo a further replication cycle. In tissue culture this phenomenon can be seen as large areas of **multinucleate fused cells (syncytia)**.

(ii) **Poliovirus**. Poliovirus is a relatively simple icosahedral capsid with four proteins making up the capsid (VP1, VP2, VP3 and VP4). A further virus-coded protein, VPg, is attached to the ssRNA, serving as a recognition protein. This capsid self-assembles without the need for scaffolding proteins, but does produce an 'immature' capsid form prior to the RNA being inserted into the virion. Thus, the subunits of the capsid assemble via 'pentamers' to form a nucleic

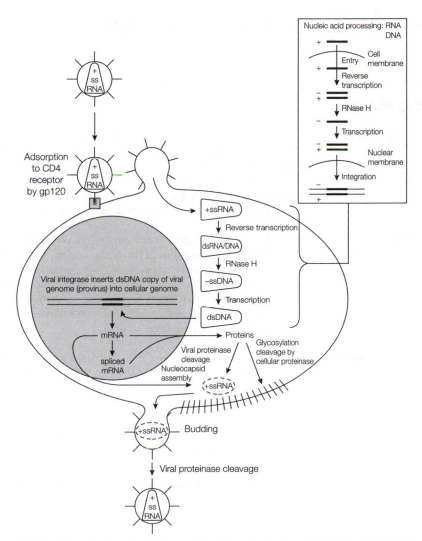

Fig. 5. HIV replication. From Harper, D., Molecular Virology, 2nd edn, © BIOS Scientific Publishers Limited, 1998.

acid-free capsid shell containing VP0, VP1 and VP3. As the RNA is sequestered into the capsid, VP0 is cleaved into VP2 and VP4 and the mature virion is formed. Poliovirus exits the cell by lysis, releasing several hundreds of progeny virions.

(iii) **HIV**. The capsid of HIV assembles in the cytoplasm. Capsid assembly of the icosahedron follows the normal pattern, but the control mechanisms are obviously important here as the mature virion contains within its central core many components: proteins, two identical strands of RNA and the reverse transcriptase, integrase and protease complex of polymers. Budding of HIV is via the plasma membrane.

Many viruses (e.g. influenza, reovirus) have multisegmented genomes. Each segment, during infection, codes for at least one virus protein essential in replication. In order to be infectious the virion must, therefore, contain a copy of each segment of RNA. Quite how this is achieved remains a mystery!

L8 VIRUS INFECTION

Key Notes

Virus spread	Viruses can gain access to the host through the skin and mucous membranes, via the respiratory or gastrointestinal tracts or through sexual contact. Without spread from host to host, viruses will cease to replicate and become extinct. A knowledge of how viruses spread and their routes of entry and exit from the host has allowed a study of the epidemiology of infections, which in turn has helped control the spread.
Clinical results of infection	The outcome of virus infection is dependent on a number of factors (e.g. age and immune status of the host). Infections are either localized at or near the site of entry, or systemic, where the virus spreads from its point of entry to involve one or more target organs. The outcome of a viral infection follows one of several patterns: (a) inapparent infection; (b) disease syndrome, virus eradication and recovery; (c) latency; (d) carrier status; (e) neoplasia; (f) death.
Related topics	Viruses and the immune system (L9) Virus vaccines (L10) Antiviral chemotherapy (L11)

Virus spread

In order to persist and evolve in nature, viruses need a large population of susceptible hosts and an efficient means of spread between these hosts. The normal route is termed **horizontal** spread. The most common route of entry and exit of viruses is the **respiratory route**. Following their inhalation, viruses usually infect and replicate in the epithelial cells of the upper and lower respiratory tract (e.g. rhinoviruses, coronaviruses, influenza, para-influenza and respiratory syncytial virus). The viruses produced in these airways exit their host via sneezing and coughing. Many viruses which enter and exit via this route are not 'respiratory viruses' (e.g. chicken pox, measles, German measles). In these cases the virus leaves the respiratory tract to set up infection in other target organs. The **oral–gastrointestinal** route is used mainly by those viruses responsible for gut infections (rotavirus, Norwalk virus and the enteroviruses, including polio and coxsackie viruses). Vast numbers of virus particles can be excreted in fecal material (e.g. in the order of 10^{12} particles g^{-1}), facilitating the easy spread of these viruses in conditions of poor sanitation. Thus, the drinking of fecally contaminated water and consumption of contaminated shellfish or other food prepared by unhygienic food handlers are ways in which these viruses are spread.

Whilst the skin normally provides an impenetrable barrier to virus invasion, infectious viruses can enter following **trauma to the skin**. This may be from the bite of an animal vector (e.g. rabies via an infected canine, yellow fever and dengue via an infected mosquito). HIV, hepatitis B and hepatitis C may be transmitted by the injection of blood or blood products either in the form of a blood transfusion, a needle-stick injury, or by intravenous drug abuse.

Sexual transmission of viruses is an important route for the spread of HSV, the papilloma viruses, hepatitis B and HIV.

Viruses may also be transmitted **vertically** – that is, from mother to offspring via the placenta, during childbirth, or in breast milk. Examples are rubella virus (German measles) and cytomegalovirus (CMV), acquired by the mother during pregnancy and transmitted to the developing embryo, often leading to severe congenital abnormalities and/or spontaneous abortion. Some viral infections, for example, HSV infections, if acquired *in utero* or during birth, can present as an acute disease syndrome in the neonate. In the case of HIV and hepatitis B transmission, the neonate may be born with an asymptomatic infection, the virus persisting in a **carrier state** and developing into disease much later.

Clinical results of infection

The outcome of a viral infection is dependent on a number of factors including **age**, **immune status** and **physiological well-being** of the host. Thus, HSV infection is usually fatal in the neonate but not in the older child. Epstein–Barr virus (EBV) causes a very mild febrile illness in young children but infectious mononucleosis (glandular fever) in teenagers. CMV in a healthy individual may cause a mild febrile illness, but in immunosuppressed individuals can lead to fatal pneumonia. Measles rarely causes severe complications in healthy well nourished children, but kills around 900 000 children per year in 'developing' countries where malnutrition is a problem. Upon infection viruses either remain **localized** or become systemic (*Fig. 1*). The outcome of a virus infection is considered below.

Inapparent (asymptomatic) infection
Many virus infections are **sub-clinical**, there being no apparent outward symptoms of disease. This is virtually always true in the immune host where recovery from a previous infection or vaccination protects the host from virus growth following reinfection by the wild-type virus. However, several viruses (e.g. respiratory and enteroviruses) may not produce clinical symptoms in some non-immune individuals. Thus, polio virus, in 80% of infected individuals, replicates in the epithelial cells of the gastrointestinal tract, is excreted in the feces, but causes no symptoms.

Disease syndrome, virus eradication and recovery
This is the pattern following most viral infections in otherwise healthy individuals – clinical symptoms of various severity (i.e. a **disease syndrome**) followed by virus eradication by the immune system, recovery and often life-long immunity. This is true of most childhood infections, for example, measles, mumps and German measles, and most respiratory diseases, of which there are a high number of viruses responsible. A vast spectrum of other viruses also follows this pattern, including those of hepatitis A virus (infectious hepatitis), rotavirus (gut infections) and coxsackie virus (myocarditis, pericarditis, conjunctivitis).

Fig. 2 shows the possible routes of infection by respiratory viruses. One important respiratory pathogen of infants (**respiratory syncytial virus**, RSV) causes severe necrosis of the bronchiolar epithelium, which sloughs off, blocking the small airways. This leads to obstruction of air flows and respiratory disease. Children recovering from acute RSV bronchitis are often left with a weakened and vulnerable respiratory system, predisposing them to a lifetime of chronic lung disease. Following respiratory infection with measles, the virus replicates in local lymph nodes that drain from the infected tissue. The virus thus enters the blood (**a primary viremia**) where it grows on epithelial surfaces before

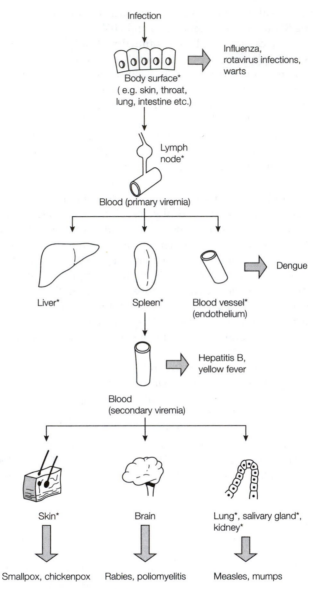

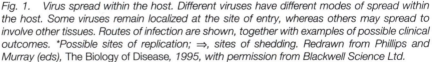

*Fig. 1. Virus spread within the host. Different viruses have different modes of spread within the host. Some viruses remain localized at the site of entry, whereas others may spread to involve other tissues. Routes of infection are shown, together with examples of possible clinical outcomes. *Possible sites of replication; ⇒, sites of shedding. Redrawn from Phillips and Murray (eds),* The Biology of Disease, *1995, with permission from Blackwell Science Ltd.*

entering the blood again (**a secondary viremia**). At this point the patient is highly infectious but does not have the distinctive measles rash, which appears about 14 days post-infection (*Fig. 3* for course of events).

Latency
A restricted range of viruses, most being in the **herpesvirus family** (HSV, varicella zoster, Epstein–Barr virus and CMV) are not eradicated from the body

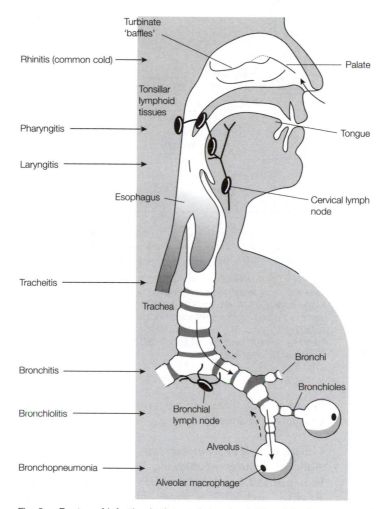

Fig. 2.   Routes of infection in the respiratory tract. Virus infections can produce a variety of respiratory disorders, depending on the area of respiratory tract infected. Dotted arrow, mucociliary flow; solid arrow, air flow. Redrawn from Phillips and Murray (eds), The Biology of Disease, 1995, with permission from Blackwell Science Ltd.

following recovery, but instead become **latent** within the host. Virus replication may be initiated some time later (**reactivation**) and cause clinical symptoms which are either similar or different to those observed in primary infection.

HSV can reactivate many times during the life of an individual and produce the typical painful cold sore lesions on the mouth or genitals. The virus lays dormant in the trigeminal or sacral ganglia between periods of reactivation. It is not understood what happens to reactivate the virus at the cellular level, although a number of stimuli including menstruation, exposure to UV light and stress are responsible for initiating it.

A very different clinical syndrome may result from the reactivation of varicella zoster virus. Thus, the primary infection produces chicken pox, whereas reactivation is associated with the development of **shingles**. Shingles is characterized by a localized area of extremely painful vesicles which clear after 1–2 weeks.

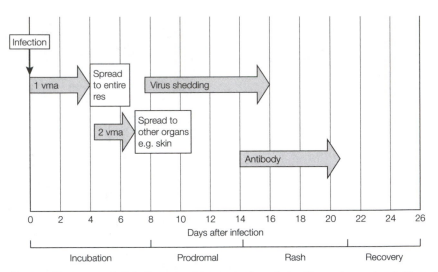

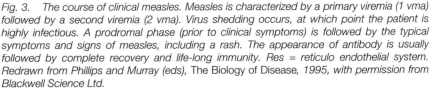

Fig. 3.   *The course of clinical measles. Measles is characterized by a primary viremia (1 vma) followed by a second viremia (2 vma). Virus shedding occurs, at which point the patient is highly infectious. A prodromal phase (prior to clinical symptoms) is followed by the typical symptoms and signs of measles, including a rash. The appearance of antibody is usually followed by complete recovery and life-long immunity. Res = reticulo endothelial system. Redrawn from Phillips and Murray (eds)*, The Biology of Disease, *1995, with permission from Blackwell Science Ltd.*

However, the patient may suffer from very severe **post-herpetic neuralgia** which can persist for months or years.

Carrier, chronic or persistent state
Although a rare event following viral infection, the virus carrier state is induced by **hepatitis B**, **hepatitis C** and **HIV**; 5–10% of individuals infected with hepatitis B virus, and 80% of those infected with hepatitis C virus will carry the infective particles in their blood for months or years. Estimates suggest that >300 million individuals worldwide carry these viruses in their blood and body fluids. These viruses have adapted to avoid the clearance mechanisms of the host immune system. Hepatitis B is transmitted in body fluids but is also passed vertically from mother to offspring. The exact details on hepatitis C routes of transmission, other than in blood and body fluids, are not known.

Unfortunatley, chronic carriers of these viruses have a greatly increased risk of developing **cirrhosis of the liver** and **hepatocellular carcinoma**. At present, in the UK, hepatitis C infection is the predominant condition that leads to the need for liver transplantation.

HIV, following a primary infection, replicates at a low level in **T4 lymphocytes** and other cells. The infected individual becomes **HIV antibody-positive** and excretes virus in a range of body fluids. The virus, in most individuals if untreated, will eventually progress to more rapid replication and cause the clinical syndrome known as **AIDS**. Approximately 10% of individuals infected with the virus to date, and after >20 years untreated viral infection have not progressed to AIDS. These individuals are important in our studies on HIV and its interaction with the host, **i.e.** why are they not progressing?

Hepatitis B and C and HIV are often referred to as persistent viruses.

Neoplastic growth

Introduction of genetic material (**viral oncogenes**) and the rearrangement or switching on of cellular genes (**cell oncogenes**) are events which can be mediated by viruses. In some situations this can contribute to the development of **neoplasia**. **Feline leukemia** virus causes a range of lymphoblastic leukemias in cats, indeed is the second highest cause of death in cats (the highest is road accidents!). Many viruses (e.g. **hepatitis B**, **hepatitis C**, **EBV**, **HSV** and **papilloma viruses)** have been implicated as being co-factors in the development of a range of malignancies.

Death

Whilst many viruses are fatal in distinct circumstances and others in a percentage of victims, some are always fatal. **Rabies**, **HIV** infection leading to **AIDS**, and a range of **neurological conditions** resulting from viral infections (e.g. subacute sclerosing panencephalitis) are examples.

While the incubation period of rabies is 30–90 days, after initial symptoms of the disease are manifest, patients die within 7–12 days. Following a bite, the virus may enter the peripheral nerves and then moves to the spinal cord and brain where it replicates. The virus leaves the cells and spreads to virtually all the tissues of the body including the salivary glands where it is shed in the saliva. The patient develops a variety of abnormalities including **hydrophobia** (aversion to water) **rigidity**, **photophobia** (aversion to light), focal or generalized convulsions and a variety of **autonomic disturbances**. Development of a flaccid paralysis and onset of coma precede death.

L9 VIRUSES AND THE IMMUNE SYSTEM

Key Notes

The immune system	The whole armory of the immune system is harnessed to combat viral infections. Viruses thus react specifically with B cells and T cells. In addition, non-specific responses (e.g. interferon and natural killer cells), have a role in combating virus infection.
Virus antigens	During their replication, viruses code for a range of viral proteins (structural and non-structural) which are foreign to the host. Many induce protective antibody and cytotoxic T lymphocyte (CTL) responses. Others, while not protective, are important in diagnostic tests. The smallest part of a molecule with antigenicity is referred to as an epitope.
Antigen recognition	Antibodies secreted from B lymphocytes (assisted by T4 helper cells) specifically react with antigenic sites on virus particles and may neutralize virus infectivity. CTLs recognize antigens on and lyse infected cells. Antigen–antibody recognition is extremely specific and is determined by the detailed molecular structure of the antigen (epitope) inducing a unique response from a lymphocyte.
Primary response to viruses	Interferon and natural killer cells respond first in combating virus-infections. Specific CTL responses take 7–10 days and antibodies slightly longer. The secondary response is almost immediate. The primary antibody response is usually an IgM response with smaller concentrations of IgG. The secondary response consists mainly of the more efficient IgG molecules.
Viruses and the immunocompromised	Individuals who are immunosuppressed suffer worse than most from viral infections where the clinical syndrome may be more severe than in an immunocompetent host. HIV causes immunosuppression. In addition, many individuals are born with genetic defects which leave them immunosuppressed; others are suppressed by medication.
Virus-induced immunopathology	The immune response mounted against an invading virus may itself be responsible for the damage and disease state which follows infection.
Evasion of the immune system	Many viruses have adapted to counteract the various host defense mechanisms.
Related topics	Virus proteins (L3) Virus vaccines (L10) Virus infection (L8)

The immune system

Virus infections are countered by the host **immune** system. The immune system consists of a wide range of morphologically and antigenically distinct white cells with different functional responses to viral invasion. Simplistically, the major type of white cells are **lymphocytes**, these being **T cells** or **B cells**. T cells (**cytotoxic (CD8)** or **helper (CD4)** cells) pass through the thymus during their maturation. B cells give rise to antibody-producing plasma cells. T4 helper cells help B cells react to highly specific antigenic stimuli (e.g. viral proteins). In addition to these specific responses the host mounts a series of non-specific responses, for example, production of **cytokines** (e.g. **interferon**), **natural killer cells**, **mucociliary responses** and **macrophages**. It is the outcome of the competition between the immune system and the progress of viral replication that determines the outcome of the infection. In **self-limiting** infections viruses are 'cleared' by the immune system. Good humoral immunity requires the induction of high levels of IgG antibodies.

Virus antigens

During the virus-replication cycle, the virus genome is expressed to give a number of proteins which are recognized by the host as being foreign. Many of these proteins induce a **protective response** in the host (important in **vaccine design**). In many viruses these proteins are laid down in the plasma membrane of the infected cells where they are recognized by both T cells and B cells (antibodies). It is usually the viral **structural proteins** (e.g. envelope and capsid proteins) that induce this protective antibody response.

Antigenic recognition

Antibodies, secreted by B cells, recognize whole viral protein antigens, usually in the fluid phase. The specific interaction between the antibody and the antigen has been likened to a 'hand-in-glove' reaction. Antibodies **neutralize** the infectivity of virus particles by agglutination, conformational change, etc. This, of course, reduces dramatically the number of host cells infected by virus. Cytotoxic T cells recognize virus antigens via a **cell-surface heterodimer** (the **T-cell receptor**) in association with the **major histocompatability complex** (MHC-class 1) which is present in almost all nucleated cells with the exception of neurons. Thus, cytotoxic T cells interact with the antigens present on the plasma membrane of infected cells, killing the cell before high concentrations of virus can be produced.

In order for an antibody to be secreted by B cells, the specific antigen must be presented to T4 helper cells, this being achieved by macrophages and dendritic cells (**antigen-presenting cells**). In this instance the specific recognition of the antigenic site is made in association with the **MHC-class II** protein (present on selected cells in the body, e.g. macrophages and dendritic cells). Primary antigen recognition usually results in the formation of **memory B and T cells** which persist, making antigenic recognition much faster during re-infection.

Primary response to viruses

During a primary infection, the mass of antigen available to stimulate the immune response increases as the organism replicates. Initial responses are made by the host via **interferon** and **natural killer** (NK) cells. Interferons are host cell proteins with antiviral activity. It is α and β interferon that are recognized as having an antiviral effect. Once induced they have an effect against a wide range of viruses. Interferon acts in two stages: **induction** which results in the derepression of the interferon gene and the release of interferon which produces an **antiviral state** in other cells. NK cells exist as large granular lymphocytes in peripheral blood and mediate cell lysis. They are not antigen-specific and do not

Table 1.   Mechanisms to combat virus infection

Stage of infection	Immune response	Mechanism
Early infection (first line of defense)	Interferon, natural killer cells Soluble mucosal surface IgA antibody	Inhibits virus replication Kills virus
Viremia (virus in the blood)	Antibodies, complement Macrophages	Kills virus (neutralizes infectivity) Reduces spread Digests antibody complexes
Target organs	Antibody, complement Cytotoxic T cells, complement	Lysis of infected cells Kills virus-infected cells, reducing virus replication
	Interferon	Inhibits virus replication

exhibit immunological memory. They appear to have an important role in the control of some virus infections.

The specific T-cell responses peak early (7–10 days) and decline within 3 weeks post-infection. The initial (primary) antibody response usually peaks later than the rise in cytotoxic T lymphocytes (CTL). Antibodies are often barely detectable during the acute stage of infection but increase dramatically 2–3 weeks post-infection. High levels of antibody may linger for several months. Indeed, where virus cannot be isolated from the host, a dramatic increase in **specific antibody** titer from the **acute** to the **convalescent** stage of infection is the main indicator of the cause of the infection. Upon reinfection the **secondary** response is virtually immediate in terms of CTL and antibody response. The mechanisms that combat viral infections are outlined in *Table 1*.

Viruses and the immuno-compromised

Individuals with natural or artificially induced immunosuppression are at risk from a range of viral infections but particularly those which are responsible for latent infections (e.g. HSV, CMV, chicken pox virus) where the virus reappears to cause disease (**recrudescence**). Thus infants with **severe combined immuno-deficiency** (**SCID**) develop recurrent infections early in life (e.g. rotavirus in the gut, which induces prolonged diarrhea). **Immune suppression** which follows transplant surgery may lead to, for example, generalized shingles (reoccurrence of chicken pox), CMV pneumonia or genital warts. Individuals with little or no antibody production (**hypogammaglobulinemia**) often excrete viruses for many years (e.g. poliovaccines from their gut).

Some viruses induce immunosuppression. Measles is known to slightly suppress T cell responses but the significant culprit is the **human immuno-deficiency virus** (HIV). This virus infects T4 helper cells, predominantly by attachment to the **CD4 receptor** where it persists for several months or years before, usually, progressing to high concentrations of virus which seriously deplete the T4 cell population, thus resulting in serious immunosuppression and susceptibility to a wide range of pathogens (AIDS).

Virus-induced immunopathology

The immune response to viruses can result in damage to the host, either by the formation of immune complexes or by direct damage to infected cells. Complexes can form in body fluids or on cell surfaces. **Chronic immune complex glomerulonephritis** can occur in mice infected neonatally with lymphocytic choriomeningitis virus. In adult mice, direct damage to infected and non-infected brain cells by a T-cell-dependent mechanism is responsible for most of the fatal tissue damage. This is also thought to be true for liver damage in chronic active hepatitis in man.

Potentially fatal hemorrhagic fever can result when an individual previously infected with Dengue virus is later infected with a different strain of Dengue, the pathology resulting from the immune response to the second strain. Viruses may also evoke autoimmunity, probably via **molecular mimicry** (production of an antigen which shares conserved sequences with a host cell protein). As a result, antibodies or T cells are produced which also react with host proteins.

Evasion of the immune system

During coevolution with their hosts, viruses have adopted several measures to counteract the various host defense mechanisms. These include inhibition of peptide processing, resistance to serum inhibitors, resistance to or poor inducers of interferon, inhibition of phagocytes, suppression of immune responses, low capacity to evoke an immune response, alteration of lymphocyte traffic, effects on cellular modulators, depression of complement activity, resistance to the immune response and antigenic variation. Thus, the high mutation rate of many RNA viruses, e.g. HIV, creates mutants that are no longer recognized by antibody molecules. Influenza virus undergoes constant change in the environment (so-called antigenic 'drift' and 'shift' of its hemagglutinin molecule) leading to the appearance of different strains of virus, capable of replication in previously immune hosts. Most of the herpesviruses escape the immune system whilst latent in their host and many during replication (e.g. Epstein-Barr virus, EBV) may specifically inhibit the intracellular transport mechanism by which viral peptide fragments are presented to the cell-mediated immune system. Adenovirus, poxvirus and cytomegalovirus interfere with peptide presentation by downregulating the expression of the MHC class I protein. Vaccinia virus, reovirus, adenoviruses and EBV are examples of viruses capable of inhibiting the action of interferon, usually by interfering with the action of a cellular protein kinase enzyme. Both herpes simplex virus (HSV) and EBV structural proteins interfere with the complement cascade mechanism. HSV, in addition, has a structural protein, present on infected cell surfaces, which has Fc-receptor activity, thus preventing complement fixation or opsonization by phagocytes.

The ability of viruses to evolve more rapidly than their host means that virus replication and its subsequent disease will continue to threaten the health of the world's population. The evolutionary process has also led to the emergence of new viruses capable of replicating in different species, including humans.

L10 VIRUS VACCINES

Key Notes

Vaccination	Vaccination is the use of a vaccine to stimulate the immune response to protect against challenge by wild-type virus. A good vaccine is designed so as to mimic the host response to infection seen with the wild-type infection and to protect against disease but to have minimal side effects. Vaccines are either live (attenuated) virus or inactivated (killed) virus. Sub-virion protein or DNA vaccines are being developed at present, although most are at an experimental stage.
Live (attenuated) vaccines	Produced by continued passage in tissue culture or by genetic manipulation (e.g. gene deletion). These vaccines are effective in stimulating the full range of immune responses. They are the preferred type of vaccine. Examples are measles, mumps and German measles vaccines.
Inactivated (dead) vaccines	Produced by chemical inactivation of wild-type virulent virus, they are generally less effective than live vaccines but do give significant protection against virus challenge. Examples are influenza and hepatitis A vaccines.
'Sub-virion' vaccines	These are mainly experimental (apart from hepatitis B surface antigen vaccine) and are composed of 'subunits' of the virion. They are used when other vaccines are ineffective or technologically not possible to manufacture.
DNA vaccines	These are totally experimental vaccines and consist of injecting a host with a plasmid containing DNA which encodes antigenic portions of the virus.
Related topics	Virus proteins (L3) Viruses and the immune system (L9) Virus infection (L8)

Vaccination

The early experiments of **Jenner** with cowpox were the basis upon which many of today's vaccination (*vacca* is Latin for cow) programs were derived. Jenner introduced the mild **cowpox** virus into some small boys whom he later challenged with **smallpox** virus. The boys were protected against the disease. Cowpox shares a number of antigens with smallpox but does not cause severe disease in humans. Hence, cowpox replication had induced an immune response which, upon subsequent challenge with smallpox, was sufficient to prevent the smallpox virus undergoing any significant infection. A hybrid cowpox virus, of unknown origin, vaccinia virus, was used in a vaccination program which in 1977 succeeded in eradicating smallpox from the world. Vaccines are basically of two types, live (**attenuated**) or inactivated (**dead**) vaccines. More recently, vaccines which contain selected virus proteins (e.g. hepatitis B vaccine) have been developed.

Many, however, are only at the experimental stage of development. Even more recently have been experiments using DNA as a source of vaccination (see later). Vaccines are administered by a number of routes – **intramuscular**, **intradermal**, **subcutaneously**, **intranasal** or **oral**.

Live (attenuated) vaccines

It is generally accepted that live vaccines are the preferred vaccines and examples are shown in *Table 1*. Initially, they were derived by continued passage of virulent wild-type virus in tissue culture, this process selecting for **avirulent viruses** which will replicate *in vivo* but not cause disease. Such viruses could, however, revert back to wild-type phenotype and indeed in some cases during the early years of their use did. This rather empirical approach has now been replaced by the genetic 'manufacture' of viruses with deleted or mutagenized genes. The risk of reversion with these vaccines is minimal. However, many (e.g. Sabin polio vaccine) are the original mutants created by tissue culture passage over 40 years ago.

Live vaccines have distinct advantages over killed. They require small amounts of input virus (which, however, must replicate) they can induce local immunity (i.e. mucosal IgA), they may be given by the natural route of infection (e.g. Sabin polio vaccine which is given orally) and they are usually cheaper. The drawbacks to their use include reversion to virulence, ineffectiveness if not kept live (i.e. refrigerated or freeze-dried), ineffectiveness in individuals with, for example, infections of the gut, where the polio virus vaccine will not grow, contamination with adventitious agents and limited use in immune-suppressed patients. Recent experiments have shown that, for poliovirus type 3, reversion to a virulent form involves only two amino acid changes.

The World Health Organization has an ambitious vaccination program which seeks to get vaccines to most children of the world and to eradicate many viruses in the next decade. Polio is now endemic in only a few countries, and by using National Immunization Days many millions of people are being made immune to polio and other viruses. It is hoped to eradicate polio globally by 2008.

Table 1. Currently available live attenuated viral vaccines

Vaccine	Vaccine type	Uses
Oral polio	Attenuated trivalent	Routine childhood immunization; mass campaigns
Measles	Attenuated (Schwarz, Moraten, others)	Routine childhood immunization; mass campaigns
Rubella	Attenuated (RA 27/3)	Routine childhood immunization; adolescent girls; susceptible women of childbearing age
Mumps	Attenuated (Urabe or Jeryl Lynn)	Routine childhood immunization
Measles, mumps, rubella (MMR)	Attenuated	Routine childhood immunization (1 or 2 doses)
Varicella	Attenuated (Oka)	Routine childhood immunization (USA); vaccination of susceptible people
Yellow fever	Attenuated (17D)	Routine immunization or mass vaccination in endemic areas; vaccination of travellers to endemic areas

Inactivated (dead) vaccines

The early polio vaccine (**Salk vaccine**) still used in many countries typifies the approach to dead vaccination; other examples are shown in *Table 2*. High titers of wild-type virus are grown in tissue culture and inactivated chemically by the use of, for example, β-propiolactone or formaldehyde. The killed virus is administered parenterally (subcutaneously or intradermally) in order to stimulate an immune response. With no subsequent virus replication (as seen with live vaccines) the immune response follows that of a 'primary response' to an inert antigen. The necessary high levels of IgG are therefore stimulated only by multiple injections.

Killed vaccines have the advantage of non-reversion to virulence, are not 'inactivated' in the tropics, can be administered to immune-suppressed patients and are usually not affected by 'interference' from other pathogens (important in many developed countries). They are, however, expensive, do not stimulate local immunity, are considered dangerous to manufacture (prior to inactivation) and require multiple doses. The recently derived **hepatitis A vaccine** is a dead vaccine, primarily because it was technically impossible to develop an attenuated virus capable of liver growth but no disease.

It is considered that the use of dead vaccines alone, which often allow some wild-type virus replication at local sites (e.g. polio) will not be sufficient to totally eradicate viruses from the world.

'Sub-virion' vaccines

Attempts to produce these vaccines, often referred to as **subunit** vaccines have been prompted by a number of factors. Many viruses do not grow in tissue culture (e.g. hepatitis B and C); some are considered by some scientists to be too dangerous for use as live or killed vaccines (e.g. HIV); the vaccine may become latent in the body and become reactivated (e.g. HSV); the vaccine may be poorly effective and have side effects (e.g. influenza virus vaccines).

Table 2. Currently available inactivated viral vaccines

Vaccine	Vaccine type	Uses
Inactivated polio	Killed whole virus	Immunization of immunocompromised people; universal childhood immunization (some developed countries)
Hepatitis B	Purified Hepatitis B surface antigen produced by recombinant yeast, mammalian cells or from plasma	Routine childhood or adolescent immunization: immunization of high-risk adults; post-exposure immunization
Influenza	Killed whole or split virus	Vaccination of high-risk individuals or the elderly
Rabies	Killed whole virus	Post-exposure vaccination; pre-exposure; veterinarians/travelers
Hepatitis A	Killed whole virus	Pre-exposure vaccination: high-risk people and travelers
Japanese B encephalitis	Killed whole virus	Pre-exposure vaccination of travelers to endemic areas; routine or mass vaccination in endemic areas
Tick-borne encephalitis	Killed whole virus	Pre-exposure vaccination of travelers to endemic areas; (?)routine or mass vaccination in endemic areas

The basis of the approach is to develop a vaccine product which, when injected into the host, will induce protective immunity. The virus protein of choice for such a vaccine is usually a capsid or envelope protein, often the protein which binds to the cell receptor (e.g. gp120 of HIV, hemagglutinin of influenza). These proteins can be extracted from the virion by chemical treatment (e.g. influenza hemagglutinin), concentrated from the plasma of infected patients and inactivated (e.g. hepatitis B surface antigen), engineered by recombinant DNA technology (e.g. hepatitis B surface antigen) or synthesized as a peptide (still experimental, but examples include foot and mouth disease virus).

These vaccines may suffer from some of the drawbacks of dead whole virion vaccines in as much as they do not replicate in the host. They are not good inducers of immunity and presentation of virus antigens in such a way as to optimally stimulate the immune system is a very important area of research. Antigens may be injected with adjuvants (immune stimulating chemicals, e.g. alum) or as chimeric proteins linked to e.g. host interleukins. Another approach under study at present is to introduce the selected viral gene into a virus vector (e.g. vaccinia virus). The vaccination virus is then used to vaccinate the host, the virus replicating and expressing a number of antigens (including the cloned gene product) which in turn stimulate immunity.

The only subunit vaccine in general use at present is the hepatitis B surface antigen which has been genetically engineered into and expressed by yeast cells. The vaccine is used routinely to induce protection against hepatitis B in members of the medical profession.

DNA vaccines   These appear to be an exciting innovative approach to vaccination. DNA-encoding antigenic portions of viruses (e.g. protective antigens) are inserted into a plasmid which can be injected into the host as 'naked' DNA free of protein or nucleoprotein complexes. Host cells take up the DNA and may express the virally encoded proteins. Such vaccines stimulate excellent cytotoxic T-cell responses as the plasmid DNA provides internal adjuvant action through its 'immunostimulating sequences', i.e. unmethylated CpG. Experimental influenza vaccines in animals have proved effective in challenge experiments and clinical trials are underway with e.g. HIV.

L11 ANTIVIRAL CHEMOTHERAPY

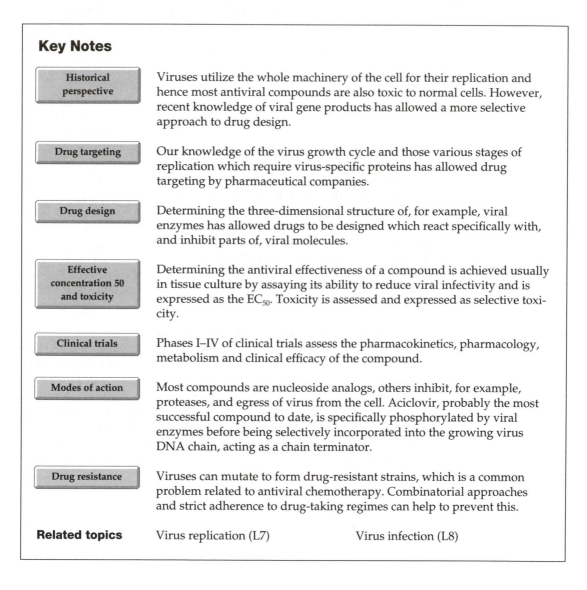

Key Notes

Historical perspective	Viruses utilize the whole machinery of the cell for their replication and hence most antiviral compounds are also toxic to normal cells. However, recent knowledge of viral gene products has allowed a more selective approach to drug design.
Drug targeting	Our knowledge of the virus growth cycle and those various stages of replication which require virus-specific proteins has allowed drug targeting by pharmaceutical companies.
Drug design	Determining the three-dimensional structure of, for example, viral enzymes has allowed drugs to be designed which react specifically with, and inhibit parts of, viral molecules.
Effective concentration 50 and toxicity	Determining the antiviral effectiveness of a compound is achieved usually in tissue culture by assaying its ability to reduce viral infectivity and is expressed as the EC_{50}. Toxicity is assessed and expressed as selective toxicity.
Clinical trials	Phases I–IV of clinical trials assess the pharmacokinetics, pharmacology, metabolism and clinical efficacy of the compound.
Modes of action	Most compounds are nucleoside analogs, others inhibit, for example, proteases, and egress of virus from the cell. Aciclovir, probably the most successful compound to date, is specifically phosphorylated by viral enzymes before being selectively incorporated into the growing virus DNA chain, acting as a chain terminator.
Drug resistance	Viruses can mutate to form drug-resistant strains, which is a common problem related to antiviral chemotherapy. Combinatorial approaches and strict adherence to drug-taking regimes can help to prevent this.
Related topics	Virus replication (L7) Virus infection (L8)

Historical perspective

As obligate intracellular parasites with very restricted genetic-coding capacity viruses rely heavily on utilizing the metabolic machinery the cell for their replication. It was therefore considered by many that the concept of **selective toxicity** was an unattainable goal and that interference with viral replication would always bring unacceptable damage to the host. Viruses by definition are resistant to the action of antibiotics.

In the last two decades a series of discoveries based on our increasing knowledge of the viral replication cycle and the specific gene products encoded by viruses have resulted in a number of successful antiviral agents being prescribed.

However, some still have a level of toxicity which, while acceptable for some diseases (e.g. AIDS), would not be tolerated for less severe diseases.

Chemotherapeutic agents fall into three broad groups. **Virucides** directly inactivate viruses (e.g. detergents, solvents), **antivirals** strive to inhibit viral multiplication but not host-cell metabolism and **immunomodulating agents** attempt to enhance the immune response against viruses (e.g. administration of interleukins). In this chapter **antivirals** will be discussed.

Drug targeting

Viral replication requires the virus to pass through a number of stages which are common to all viruses: **attachment and penetration, uncoating of the nucleic acid, transcription and translation, replication of nucleic acids** and **release of mature progeny**. This is demonstrated in diagrammatic fashion for the growth cycle of HIV in *Fig. 1*.

To date, most antivirals are directed at inhibiting one of the steps shown in *Fig. 1*, although the most common target is interference with nucleic acid metabolism by using many **nucleoside analogs**. Several compounds that interfere with influenza effect the disassembly of particles by blocking a virus protein complex that acts as an ion channel. Many viruses produce a virus-specific protease to process their proteins and this has been targeted in the search to develop an inhibitor of HIV. A series of inhibitors of picornaviruses act by directly binding to the virion capsid, blocking the interaction between the virion and the receptor on the cell surface that facilitates its entry and disassembly. A recent anti-influenza drug targets the virus neuraminidase and inhibits egress of virus from the cell.

Drug design

In the early days, most new antiviral compounds were discovered in an **empirical** fashion. Chemists would produce a wide range of, for example, nucleoside analogs, originally in many cases as anti-cancer drugs, which were tested in tissue culture to determine their antiviral activities. However, with our knowledge of the molecular basis of virus replication, the availability of the entire **nucleotide sequences** of virus genomes and the **three-dimensional protein structure** derived from **X-ray diffraction analysis**, compounds can be designed to interact with specific targets involved in the replicative cycle of viruses. In practice the most useful targets are **virus-induced enzymes** which often have properties different to those of the counterpart enzymes induced by the host cell (e.g. **thymidine kinase, DNA polymerase, reverse transcriptase** and **protease**). When the enzyme can be crystalized and its three-dimensional structure determined the synthesis of appropriate molecules to interact with particular sites on that enzyme can be determined. However, future antiviral compounds should lack toxicity and should be able to be produced cheaply from available precursors. The chance of a wide range of new products appearing is therefore remote.

Effective concentration 50 and toxicity

Antiviral compounds are usually assessed at an early stage for their ability to interfere with viral growth in tissue culture. The virus is assayed by, for example, $TCID_{50}$ or plaque production with and without the drug. The percentage of reduction in infectivity is plotted against log_{10} of the drug concentration. The concentration of compound that reduces virus titer by 50% is measured and expressed as the effective dose 50 concentration (ED_{50}) (*Fig. 2*). Tissue culture can also be used to assess the toxicity of compounds, although this is often tested in animals and humans. Animals play a vital role in the study of toxicity and a number of statutory tests are carried out to determine the risks and side

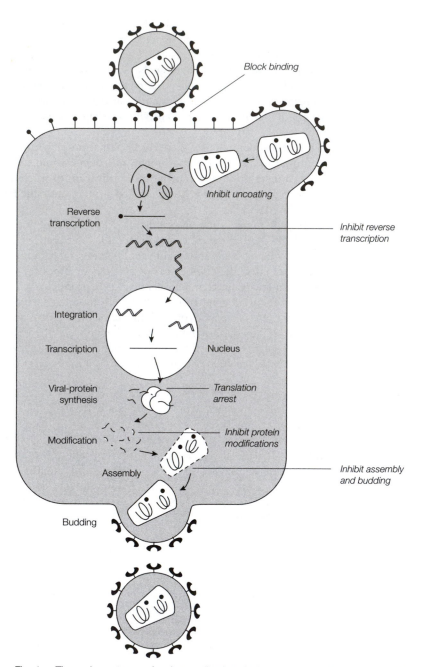

Fig. 1.   The various stages of a virus replication. Antiviral drugs can be used to inhibit any of these steps. Redrawn from Yarchoan, R. et al., Scientific American, vol. 259, no. 4, p. 90, 1988.

effects before new compounds enter clinical trials. The ratio of the 50% toxic concentration to the 50% inhibitor concentration is termed the **selective index**. If this is close to unity the compound is toxic. A high selective toxicity suggests a useful compound. A **benefit/risk** ratio is also considered when using a compound, that is, side effects will be tolerated more if the risk from the disease is high (e.g. AIDS).

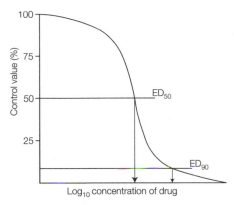

Fig. 2.   Determination of effective dose (ED$_{50}$, ED$_{90}$) for an antiviral drug in tissue culture.

Clinical trials

Phase I involves administering the drug to healthy human volunteers where studies on the pharmacokinetics, pharmacology and metabolism of the compound are monitored.

In **Phase II** the compound is administered to diseased patients, and data similar to those above are obtained as the metabolism may be different in the diseased patient. Usually <100 patients are used for these trials.

In **Phase III** the drug is usually tested for its clinical efficacy by comparing with **placebos** or existing drugs. The main aim is to determine the **benefit/risk ratio** for the therapeutic course, this requiring 100–1000 patients.

Phase IV studies are usually conducted following marketing approval and increased experience of treating patients, providing more information on safety and efficacy.

Unfortunately, most potential antiviral compounds, although good inhibitors of viral growth in tissue culture, fail to pass successfully through all of the phases of clinical studies.

Modes of action

Table 1 highlights the modes of action of selected antiviral compounds. Perhaps the most successful antiviral compound is the nucleoside analog **acyclovir** (ACV – now named **aciclovir**) with more than 40 million patients treated. The drug inhibits the replication of HSV, and has been administered prophylactically for over 10 years to individuals, with no ill effects, to suppress recurrences of genital herpes. The compound is related to the natural nucleoside guanosine. To become active the compound must be converted into the triphosphate form by three phosphorylation steps (i.e. ACV mono-, di- and then triphosphate). These steps are not carried out in normal uninfected cells and hence ACV is a poor substrate for cellular enzymes. In contrast, the **HSV thymidine kinase** can convert ACV to ACV-monophosphate (ACV-MP). Cellular enzymes convert ACV-MP to the triphosphate (ACV-TP) where it enters the nucleoside pool and competes with guanosine triphosphate as a substrate for **HSV DNA polymerase**. Cellular DNA polymerases are much less sensitive to inhibition. As the ACV residue is linked to the growing chain of viral DNA it forms a **chain terminator** as there is no 3'-OH group on the ACV sugar moiety to link the next residue to the growing chain of virus DNA.

Table 1. Mechanism of action of antiviral drugs

Drug	Mechanism(s) of action	Virus
Aciclovir	Nucleoside analog, inhibits nucleic acid synthesis: active form produced by viral thymidine kinase	HSV, VZV
Ribavirin	Nucleoside analog, inhibits nucleic acid synthesis, possibly by inhibiting viral RNA polymerase	RSV, Hep C
Amantadine Rimantadine	Inhibit virus coating, maturation and egress from cell	Influenza
Lamivudine	Reverse transcriptase inhibitor – inhibits replication of virus	Hep B
Tamiflu	Inhibits neuraminidase and blocks release of virus	Influenza
Ganciclovir	Nucleoside analog, blocks nucleic acid synthesis by inhibiting viral thymidine kinase and other enzymes	CMV
Azidothymidine Dideoxyinosine Dideoxycytidine	Nucleoside analogs, inhibit nucleic acid synthesis by inhibition of reverse transcriptase	HIV
Indinavir	Anti-protease inhibitor, prevents cleavage of virus proteins	HIV
Foscarnet	Blocks protein synthesis by inhibition of RNA and DNA	CMV, HSV
Idoxuridine	Nucleoside analog, inhibits DNA synthesis	HSV
Vidarabine	Nucleoside analog, blocks DNA synthesis by inhibiting DNA polymerases	HSV
Interferon	Renders normal cells 'immune' to infection by interfering with virus transcription	Hep C

HSV, herpes simplex; VZV, Varicella–Zoster; RSV, respiratory syncytial; CMV, cytomegalovirus; HIV, human immunodeficiency virus.

Interferon is a natural human product (a cytokine) which acts on the surface of normal cells to render them immune to virus replication. The compound used to treat e.g. hepatitis C infections, is, however, toxic, the side effects mimicking those of an influenza infection. A new version, polyethylene glycol interferon, is more effective and less toxic and is administered subcutaneously once weekly for one year. Effective anti-HIV treatments are those involving combinations of drugs, e.g. two nucleoside analogs plus an anti-protease inhibitor. Such regimes have been highly effective in reducing virus loads and raising CD4 counts in HIV-infected individuals.

Drug resistance   Viruses adapt to become resistant to anti-viral drugs with examples in all areas of chemotherapy. Mutants resistant to Azidothymidine occur with a very high frequency, whereas combinatorial approaches appear to reduce the risk of viral resistance. Most important are measures which ensure that patients adhere to the regime of drug-taking, which in some cases is highly complex and involves a vast number of tablets.

L12 PLANT VIRUSES

Key Notes

Historical aspects	Plant viruses were discovered over a century ago and have featured greatly in contributing to our knowledge of virus structure (e.g. tobacco mosaic virus, turnip yellow mosaic virus and tomato bushy stunt virus).
Plant viruses	Plant viruses are diverse in size, shape and biochemistry and are present in many virus families which include animal viruses (e.g. rhabdoviridae).
Disease and pathology	Viruses are singly responsible for grave economic loss estimated at being over $70 billion worldwide. They cause necrosis, wilting, mosaic formation and other damage, which reduces yields and value of crops, etc.
Transmission, infection and systemic spread	Plant viruses are transmitted mainly by invertebrate animals (e.g. aphids, leaf hoppers) or through infected seeds or 'manually' by contaminated implements. They gain entry by penetrating cuticles of plant cells and need to spread systemically to cause disease (via plasmodesmata).
Control of plant virus disease	Infected plants are virtually impossible to 'cure'. Control is by use of naturally resistant plant varieties or more recently genetically manufactured resistant varieties and by eradication of the transmission vector.
Viroids	These are 'virus-like' infectious agents composed solely of RNA with a complex tertiary structure (e.g. potato spindle tuber viroid).

Related topics	Virus structure (L1)	Virus nucleic acids (L4)
	Virus taxonomy (L2)	Virus replication (L7)
	Virus proteins (L3)	

Historical aspects

Tobacco mosaic virus (TMV) has figured predominantly in early studies on virus structure and replication. The virus, which causes mosaicing of the leaves of the tobacco plant, was first described as being a 'contaguim virus fluidum' in 1898. In 1935 the virus was crystalized and shown to be a 'globular protein'. TMV was the first virus to be observed under the electron microscope, and the first to demonstrate the intrinsic infectivity of extracted RNA and to be assembled *in vitro* from purified preparations of viral RNA and coat protein molecules.

It was studies on turnip yellow mosaic virus and tomato bushy stunt virus that revealed the morphological details of icosahedral viruses.

Plant viruses

There are over 1000 plant viruses which have been classified by the ICTV. The virus usually receives its name by a combination of the host plant and the type of disease produced (e.g. **tobacco mosaic**, **turnip yellow mosaic**, **turnip bushy stunt**, **cauliflower mosaic**, **tomato spotted wilt**).

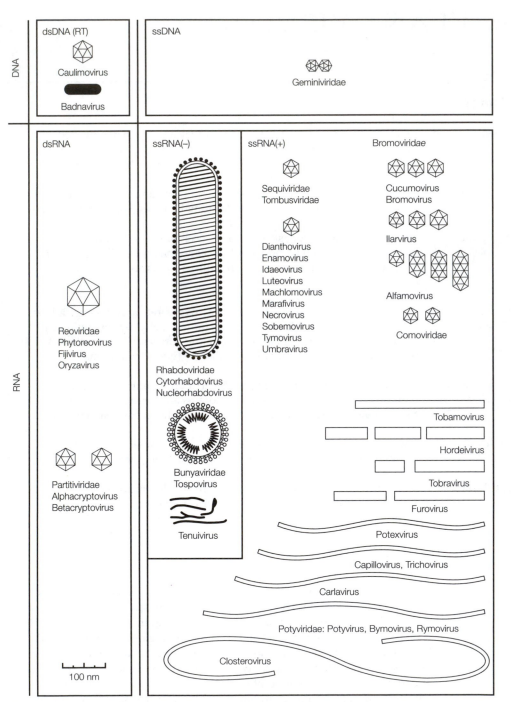

Fig. 1.   *Families and genera of plant viruses. Reproduced from* Sixth Report of the International Committee on Taxonomy of Viruses, *Springer-Verlag.*

Plant viruses are diverse in their morphology, nucleic acid composition and replication patterns. It is beyond the scope of this book to detail all viruses but *Fig. 1* highlights the various morphologies and groupings of plant viruses. They may be **dsDNA** (Badnavirus, e.g. rice tungrobacilliform virus), **ssDNA** (geminivirus, e.g. maize streak virus), **dsRNA** (reoviruses, e.g. clover wound tumor virus), **RNA viruses with ambi sense RNA** (rhabdoviridae, e.g. potato yellow dwarf virus), **positive-sense ssRNA viruses** with monopartite genomes (e.g. Tobamoviridae tobacco mosaic virus), **bipartite ssRNA** viruses (e.g. Comoviridae cowpamosaic virus) and **tripartite ssRNA** viruses (e.g. Cucumoviridae cucumber mosaic virus).

Disease and pathology

The impact of viruses globally on the world's plants is enormous and estimates of $70 billion have been quoted as being the monetary loss per year as a result of virus infections. It appears that crops of all types can be affected. One disease, **tristenza**, has wiped out or made non-productive more than 50 million citrus trees worldwide. Swollen shoot disease has destroyed more than 200 million cocoa trees in parts of western Africa. Rice plants, barley, vegetable and field crops all succumb to infections. Many perennial plants are chronically infected with virus. The morphological outcome of a plant infection can be seen as mosaic formation, yellowing, molting or other color disfiguration, stunting, wilting and necrosis. These conditions result from cell and tissue damage which will significantly reduce the yields and commercial value of crops. Thus, photosynthesis, respiration, nutrient availability and hormonal regulation of growth can all be affected. How plant viruses cause such changes remains for the main part a mystery. Plant viruses code for few proteins, but as is the case in many bacteriophage proteins, they are multifunctional and many of their interactions with host proteins, although not part of the virus-replication process, result in pathology.

Transmission, infection and systemic spread

Most plant viruses depend on some kind of 'agent' for their dissemination in the natural setting. Invertebrate animals are the most important of these agents, examples being **aphids**, **leaf hoppers**, **mealybugs**, **whiteflies**, **thrips**, **mites** and **soil nemotodes**. Some viruses are passed directly to progeny plants through infected pollen or seeds and others are present in tubers, bulbs and cuttings. A few viruses can be transmitted as a result of contact with **contaminated implements** (e.g. hoes) or by direct contact between neighboring plants. The stem and leaf surfaces of plants can in many ways be compared to the human skin. As such they do not have receptors for virus attachment and have a protective function. For plant viruses to be infective they must penetrate this barrier and gain access to the metabolic machinery of the plant cell. It is thus through wounds that viruses enter plant tissue, either via a vector or mechanically by manual damage. In the laboratory, infection is achieved by rubbing the leaf with a cloth soaked in a virus suspension, using a fine mesh abrasive method to create the necessary wounds. In effect this procedure mimics the plaque assay described for animal viruses, but is not as accurate! (*Fig. 2*). On entry the virus particle is uncoated and goes through a replication cycle similar to that of animal viruses, this cycle varying from virus to virus. Many plant viruses code for a **movement protein** in addition to enzymes and structural proteins. *Fig. 3* represents a diagram of the cycle of TMV. Plant viruses rarely cause significant damage and disease unless they become **systematically** distributed throughout the plant. Failure for this to

Fig. 2. *Focal assay of tobacco mosaic virus on the leaf of a plant. Reproduced from Dimmock and Primrose,* Introduction to Modern Virology, *4th edn, 1994, with permission from Blackwell Science Ltd.*

happen explains why some plants are resistant to particular virus infections. Movement is facilitated by a virus protein which is involved in the transport of virus or virus nucleic acid through the fine pores (**plasmodesmata**) in the cell walls that interconnect plant cells. Movement also occurs through the companion and sieve cells of the phloem, this being facilitated by the viral coat protein.

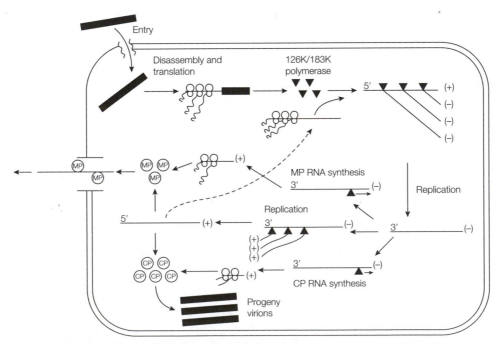

Fig. 3. *Diagram of stages of TMV infection. All the events shown are presumed to occur in the cytoplasm of infected cells. MP, movement protein; CP, coat protein. Redrawn from Fields* et al. *(eds),* Fundamental Virology, *3rd edn, 1996, with permission from Lippincott-Raven Publishers.*

Control of plant virus disease   Once infected, it is almost impossible to 'clear up' a virus infection in a plant by use of antiviral agents. Likewise plants do not mount immune responses. Until recently it was a reliance on **horticultural practice** (use of virus-free seeds, eradicating vectors, choosing the time of planting, etc.) that reduced the extent of virus infection. More recently **genetic engineering** has allowed the construction of plants which show **natural resistance** to virus infections. Other scientific measures made use of the fact that previous infections with a non-pathogenic virus appeared to protect the plant from an infection by a more pathogenic strain (but not because of immunity!). This phenomenon has been extended by transforming the plant with the coat protein gene of TMV and then challenging the plant with infectious TMV. The resulting **transgenic** plant was protected from challenge by the wild-type TMV. The phenomenon is referred to as **pathogen-derived resistance** to virus diseases. How and why this phenomenon works is not clearly understood, although in practice the method is receiving a great deal of attention from plant breeders and molecular biologists.

Viroids   Viroids are small, **unencapsidated ssRNA molecules** – the smallest known pathogens of plants. There are some 25 viroids which vary in nucleotide sequence. The first to be examined in detail was **potato spindle tuber viroid** (PSTVd) responsible for significant loss to the potato industry.

The RNA is a covalently closed circle and ranges in size from 246 to 357 nucleotides in length. The RNA has a complex secondary and tertiary structure which gives it a rod-like shape and **resistance to nucleases**. The RNA does not have a characteristic open reading frame and so does not act as mRNA. How this piece of RNA causes disease is largely unknown. It replicates in plant cells with the aid of host-cell enzymes (e.g. RNA polymerase II).

Viroids are spread by plant propagation (e.g. cuttings and tubers) through seeds and by manual mishandling with contaminated implements.

L13 PRIONS AND TRANSMISSIBLE SPONGIFORM ENCEPHALOPATHIES

Key Notes

The agent

Prions (proteinaceous infectious particles) are not conventional infectious agents. They have no nucleic acid but consist entirely of protein. The infectious form (PrP^{sc}) arises from a modified cellular protein (PrP^c). Prions induce fatal transmissible spongiform encephalopathies (TSEs).

Pathogenesis of TSEs

TSEs are fatal chronic degenerative diseases with a very specific underlying pathology denoted by the laying down of protein deposits as plaques or fibrils (known as amyloids) in the kidneys, spleen, liver and significantly the brain. Post-mortem examination of brain tissue reveals a spongy appearance, which reflects the formation of holes in the tissue due to the cytotoxic effects of such deposits. TSEs have a long incubation period and are invariably fatal.

Molecular nature of prions

Evidence that prions are composed entirely of protein comes from experiments designed to selectively inactivate proteins or nucleic acids. Prions are resistant to heat (135°C for 18 minutes), ultraviolet light, ionizing radiation, DNAse and RNAse and Zn^{2+} catalyzed hydrolysis, treatments that selectively destroy DNA and/or RNA. However, they are sensitive to urea, SDS, phenol and other protein-denaturing chemicals. Evidence suggests they are proteins of approx. 254 amino acids in length. They 'replicate' by converting the cellular PrP (PrP^c) to the infectious (PrP^{sc}) form.

Animal TSEs

Scrapie, the most extensively studied, causes a neurological disease in sheep. Probably transmitted orally (e.g. by placentas) sheep have different genetic susceptibilities. Bovine spongiform encephalopathy (BSE) appeared in 1986 as a result of feeding cattle with scrapie-contaminated food-stuff. BSE ('mad cow' disease) appears to have infected humans.

Human TSEs

The neurological condition of Kuru resulted from the cannibalistic ritual of the Fore people of New Guinea. The disease, orally transmitted, has an incubation period of up to 30 years. Creutzfeldt-Jakob disease is sporadic, iatrogenic and familial and affects 1 in 10^6 individuals worldwide. New variant CJD (vCJD) appears to be different and is probably caused by the BSE agent.

The agent

Prions are not viruses and appear to be a new class of infectious agents that lead to chronic progressive infections of the nervous system, inducing common pathological effects, the results of which, after a long incubation period of perhaps several years, are invariably fatal. The clinical syndromes for which they are responsible are known collectively as **transmissible spongiform encephalopathies** (TSEs). Examples, in animals, include **scrapie** in sheep, **transmissible mink encephalopathies** (TME**), feline spongiform encephalopathy** (FSE) and **bovine spongiform encephalopathy** (BSE or 'mad cow' disease). There are four forms of human TSE – **Creutzfeldt-Jakob disease** (CJD), **fatal familial insomnia** (FFI – an inherited disease), **Gerstmann-Strausster-Scheinker disease** (GSS) and **Kuru**.

There has been much speculation as to the molecular nature of the infectious agents responsible for these disease conditions, but in 1972 Stanley Prusiner penned the term **prion (proteinaceous infectious particle)**. His hypothesis was that these agents are totally free of nucleic acid and consist solely of protein, a hypothesis for which much supportive evidence has accumulated. In 1997 Prusiner was awarded the Nobel Prize for his studies.

Pathogenesis of TSEs

Whilst the various conditions induced by prions have subtle differences there are many features in common. All have long incubation periods, with the agent replicating in a number of tissues, e.g. spleen, liver, before appearing in high concentrations in the brain and CNS towards the terminal stages of the disease. The agents are responsible for severe degeneration of the brain and spinal cord and, once clinical signs appear, the condition is invariably fatal. Pathologically the disease is characterized by the appearance of abnormal protein deposits **(amyloids)**, in various tissues, e.g. kidney, spleen, liver and brain. Amyloids arise from the accumulation of various proteins, which take the form of **plaques** or **fibrils**. Amyloidosis is also a characteristic feature of e.g. Alzheimer's disease, although this condition is not transmissible. The protein deposits are cytotoxic and are responsible for the resultant pathology, namely the sponge-like appearance of the brain where small holes can be visualized by microscopy in thin sections of brain tissue taken at post-mortem. It is this appearance which gave the spongiform encephalopathies their name. There is no conventional immune response to the agent, although the immune system plays an important part in the development of the disease before the agent gets into the CNS.

A diagnosis of TSE can only be made by demonstrating the presence of prion proteins in tissue taken at post-mortem. This is usually by immunohistochemical staining of the prion protein or by transfer of the agent to a permissive experimental animal.

Molecular nature of prions

Prions differ from all other forms of infectious agent in several respects. Primarily they lack any nucleic acid and, therefore, have no genomes to code for their progeny and yet they do 'reproduce' themselves.

Evidence for the proteinaceous nucleic acid-free nature of the agent has come from a number of experiments in which the chemical and physical nature of the infectious agents have been examined. They are resistant to heat inactivation, indeed high temperature autoclaving at 135°C for 18 minutes does not eliminate infectivity. Further evidence suggests that heating at 600°C (dry heat) does not totally kill a suspension of the agent. They are resistant to both ultraviolet light and ionizing radiation, treatments which normally damage microbial genomes (nucleic acid). Resistance is also shown to DNAse and RNAse treatment and to Zn^{2+} catalyzed hydrolysis. They are, however, sensitive to urea, SDS, phenol and other protein-denaturing chemicals. These characteristics are indicative of an agent which is of a proteinaceous character, lacking nucleic acid, rather than a virus, i.e. a novel agent or a so-called **prion (PrP)**.

The only known and demonstrated agent is a modified protein coded for by a cellular gene. The cellular isoform, **PrPᶜ**, undergoes conformational change to the so-called infectious or scrapie form **PrPˢᶜ**. **Priˢᶜ** is the term used for the infectious form of any TSE, although some authors use specific terms e.g. PrPᶜʲᵈ. The two proteins have different confirmations. PrPˢᶜ can propagate itself by inducing normal prion protein molecules to adopt the abnormal conformation. The PrPᶜ structure is that of a **30% helix** with no β sheet whereas on converson the PrPˢᶜ adopts the **30% helix, 45% β sheet** configuration (*Fig. 1*) PrPᶜ exists in all mammals and birds examined to date and the protein is found anchored to external surfaces of cells by a glyolipid moiety. Its exact function is unknown.

However, transgenic 'knock-out' mice which lack the PrPᶜ gene appear to be normal. Infection of these mice with PrPˢᶜ does not lead to disease, which is good evidence that PrPᶜ is necessary in the pathogenesis of TSEs. It also suggests that PrPˢᶜ is itself not capable of replication.

As with conventional microbes there are different strains of prion. In humans there are two common versions of the prion protein that differ in a single amino acid (**valine or methionine at coding position 129**) but there are also rare mutant forms, many of which are associated with inherited susceptibility to prion disease. Sheep too have several different forms linked to susceptibility, whereas cattle have two forms which do not appear to be associated with susceptibility to BSE.

The infectious agents causing FSE and disease in exotic ruminants in zoos were found to be indistinguishable from the BSE strain, and although classical CJD is

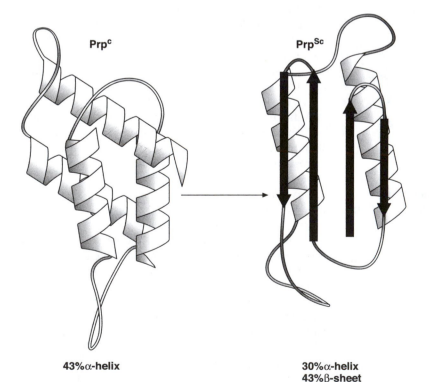

43%α-helix   **30%α-helix**
                                                       43%β-sheet

Fig. 1. Conformational changes in PrP. Redrawn from Principles of Molecular Virology, *A. Cann, Academic Press, London, 2001.*

distinct, the transmissible agent that causes **new variant CJD** (vCJD) is the same as that which causes BSE.

There are a number of questions to be answered as to the cross-species barriers that may or may not exist in the infectivity of these agents. It appears that BSE arose by feeding cattle with scrapie-contaminated food-stuffs i.e. cross-species infectivity. Some strains of BSE can be propagated by several animal species, each of which have their own but different normal prion protein. The propagated prion (PrP^{sc}) is identical in conformation regardless of the host species, demonstrating that its final structure is independent of the host.

Certainly the BSE prion appears to have infected humans and resulted in vCJD. To what extent this has happened nationwide we still do not know.

Animal TSEs

Scrapie has been recognized as a distinct infection in sheep for over 250 years. A major investigation into its etiology followed the vaccination of sheep for louping-ill virus with formalin-treated extracts of ovine lymphoid tissue, unknowingly contaminated with scrapie prions. Two years later, more than 1500 sheep developed scrapie from this vaccine. The scrapie agent has been extensively studied and experimentally transmitted to a range of laboratory animals, e.g. **mice and hamsters**. Infected sheep show severe and progressive neurological symptoms, such as abnormal gait. The name owes itself to the fact that sheep with the disorder repeatedly scrape themselves against fences and posts. The natural mode of transmission between sheep is unclear although it is readily communicable in the flocks. The placenta has been implicated as a source of prions, which could account for horizontal spread within flocks. In Iceland scrapie-infected flocks of sheep were destroyed and the pastures left vacant for several years. However, reintroduction of sheep from flocks known to be free of scrapie for many years eventually resulted in scrapie! Sheep have also been infected by feed-stuff contaminated with BSE, to which they are susceptible.

Bovine spongiform encephalopathy (BSE, 'mad cow' disease) appeared in Great Britain in 1986 as a previously unknown disease. Affected cattle showed altered behavior and a staggering gait. Post-mortem revealed protease-resistant PrP in the brains of the cattle and the typical spongiform pathology. To date >190 000 cattle have been infected with this agent which, it is thought, initiated from the common source of **contaminated meat and bone meal** (MBM) given to cattle as a nutritional supplement. MBM was initially prepared by rendering the offal of sheep and cattle using a process that involved steam treatment and hydrocarbon solvent extraction. In the late 1970s, however, the solvent procedure was eliminated from the process, resulting in high concentrations of fat in the MBM; it is postulated that this high fat content protected scrapie prions in the sheep offal from being completely inactivated by the steam. Thus the initial MBM contained only sheep prions, and the similarity between bovine and sheep PrP was probably an important factor in initiating the BSE epidemic. Bovine PrP differs from sheep PrP at 7 or 8 residues. As the BSE epidemic expanded, infected bovine offal began to be rendered into MBM that contained bovine prions, this being fed to cattle!

Since **1988** the practice of using dietary protein supplements for domestic animals derived from rendered sheep or cattle offal has been forbidden in the UK. Statistics argue that this food ban has been effective in getting the epidemic under control (*Fig. 2*). British beef is now reported as being free of BSE (June 2006).

Brain extracts from BSE cattle have transmitted disease to mice, cattle, sheep and pigs after intracerebral inoculation. Disease has also followed in mink, domestic cats, pumas and cheetahs after oral consumption of BSE prions in food-stuffs.

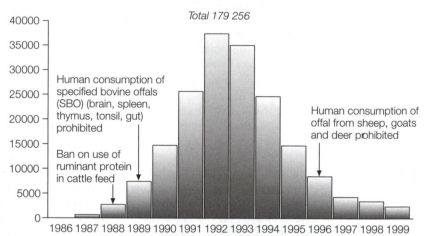

Incidence of BSE in the UK

Total 179 256

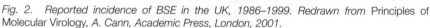

Fig. 2.   Reported incidence of BSE in the UK, 1986–1999. Redrawn from Principles of Molecular Virology, *A. Cann, Academic Press, London, 2001.*

Evidence is accumulating that the BSE prion has been transferred to humans and is responsible for vCJD. The oral route is suspected. The source of the infected material is unknown, although BSE-contaminated cattle products are the probable source. In **1989** the human consumption of specified bovine offals (brain, spleen, thymus tonsil, gut) was prohibited in the UK. In **1996** this ban was extended to sheep. Furthermore, the brain and CNS are now removed from all cattle at abattoirs prior to the distribution of meat.

Human TSEs

Kuru was the first human TSE to undergo extensive investigation. The first cases were recorded in the 1950s and occurred in male and female adolescents and adult women members of the **Fore people** in the highlands of New Guinea. The Fore people practised **ritual cannibalism** as a rite of mourning for their dead, the women and children but not the adult men taking part in the ceremony. Those with the disease demonstrated progressive loss of voluntary neuronal control, followed by death less than one year after the onset of symptoms. The incubation period for Kuru can be up to 30 years although normally it is shorter. At one point Kuru was the leading cause of death of women in the tribes. As the cannibalistic ritual ceased so did transmission of the disease. However, Kuru demonstrates that a human form of TSE can be acquired via the oral route.

Patients with **Creutzfeldt-Jakob disease** present with a progressive sub-acute or chronic decline in cognitive or motor function. CJD is rare, with an annual worldwide incidence of approximately **one per million population**. The disease is acquired **sporadically** (transmission route largely unknown), is **iatrogenic** (accidental transmission) or **familial** (10% of cases). Thus CJD has been transmitted by corneal transplantation, contaminated EEG electrode implantation, surgical operations using contaminated instruments, human pituitary growth hormone, human pituitary gonadotrophin hormone and dura mater grafts.

Recently accumulated evidence suggests that caucasian patients who are homozygous for **Met** or **Val** at **codon 129** of the PrP gene are more susceptible to sporadic CJD.

In April 1996 a new variant of CJD (**vCJD**) was described in the UK. The disease has features which distinguish it from other forms of CJD. These include an **early age of onset** (average 27 years as opposed to 65 for CJD), **prolonged period of illness** (average 13 months as opposed to 3 months for CJD) **a psychiatric presentation as opposed to neurological symptoms and the absence of the typical EEG appearances of CJD**. In terms of pathology vCJD bears a close resemblance to that of Kuru, particularly with regard to the type of amyloid plaque formation.

Up to January 2006, 159 cases of vCJD have been reported and there is strong epidemiological evidence that the epidemic has peaked. Although it is considered highly unlikely, the possibility of human-to-human transmission remains.

The vCJD scenario has raised a number of political, sociological and scientific questions which will continue to be debated into the future.

FURTHER READING

General reading:

Madigan, M.T. and Martinko, J.M. (2005) *Brock Biology of Microorganisms*, 11th Edn. Prentice Hall Inc., Upper Saddle River, NJ.

Prescott, L., Harley, J.P. and Klein, D.A. (2004) *Microbiology*, 4th Edn. McGraw-Hill Higher Education, Columbus, OH.

Singleton, P. and Sainsbury, D. (2002) *Dictionary of Microbiology and Molecular Biology*, 3rd Edn. John Wiley & Sons, New York.

Tortora, G.J., Funke, B.R. and Case, C.L. (2003) *Microbiology: An Introduction*, 6th Edn. The Benjamin-Cummings Publishing Co., Redwood City, CA.

Advanced Reading:

Section B Systematics

Feselstein, J. (2003) *Inferring Phylogenies*. Sinauer Associates., New York.

Hall, B.G. (2004) *Phylogenetic Trees Made Easy: A How-To Manual*, 2nd Edn. Sinauer Associates Inc., New York.

Page, R.D. (1998) *Molecular Evolution: A Phylogenic Approach*. Blackwell, Oxford.

Stackebrandt, E. (2006) *Molecular Identification, Systematics And Population Structure*. Springer Verlag, Berlin.

Section C Microbiology

Alberts, B., Johnson, A., Lewis, J., Raff, M., Roberts, K. and Walter, P. (2002) *Molecular Biology of the Cell*, 4th Edn. Garland

Atlas, R.M. and Bartha, R. (1997) *Microbial Ecology*, 4th Edn. Prentice Hall Inc., The Benjamin-Cummings Publishing Co., Redwood City, CA.

Cappucino, T.G. and Sherman, N. (1996) *Microbiology: A Laboratory Manual*, 4th Edn. The Benjamin-Cummings Publishing Co., Redwood City, CA.

Isaac, S. and Jennings, D. (1995) *Microbial Culture*. BIOS Scientific Publishers, Oxford.

Maier, R.M. (2000) *Environmental Microbiology*. Academic Press.

Moat, A.G., Foster, J.W. and Spector, M.P. (2002) *Microbial Physiology*, 4th Edn. Wiley-Liss, New York.

Section D Microbial Growth

Eriksson, L.A. (2001) *Theoretical Biochemistry*. Elsevier.

Smith, H.L. and Waltman, P. (1995) *The theory of the Chemostat*. Cambridge University Press, Cambridge.

Section E Microbial Metabolism

Haimes, B.D. (2000) *Instant Notes in Biochemistry*, 2nd Edn. BIOS Scientific Publishers, Oxford.

Nelson, D.L. and Cox, M.M. (2004) *Lehninger Principles of Biochemistry*, 4th Edn. W.H. Freeman

Nicholls, D.G. and Ferguson, S.J. (2002) *Bioenergetics 3*, 3rd Edn. Academic Press

Elliot, W.H. and Elliot D.C. (2005) *Biochemistry and Molecular Biology*, 3rd Edn. Oxford University Press, Oxford.

Section F Prokaryotic DNA and RNA metabolism

Abedon, S.T. and Lane-Calender, R. (2005) *The Bacteriophages*. Oxford University Press, Oxford.

Brown, T. A (2001) *Gene Cloning and DNA Analysis: An Introduction*. Blackwell Publishing, Oxford.

Brown, T. (2005) *Genomes 3*, 3rd Edn. BIOS Scientific Publishers, Oxford.

Howe, C. (1995) *Gene Cloning and Manipulation*. Cambridge University Press, Cambridge.

Latchman, D. (2006) *Gene Regulation*. BIOS Scientific Publishers, Oxford.

Lindahl, T.R. and West S.C. (1995) *DNA Repair and Recombination*. The Royal Society, London.

Russell, P.J. (2005) *iGenetics: A Molecular Approach*. The Benjamin-Cummings Publishing Co., Redwood City, CA.

Turner, P.C., McLennan, A.G., Bates, A.D. and White, M.R.H. (2005) *Instant Notes in Molecular Biology*, 3rd Edn. BIOS Scientific Publishers, Oxford.

Section G Industrial Microbiology

Waites, M.J., *Morgan, N.L., Rockey, J.S. and Higton, G. (2001) Industrial Microbiology: An Introduction*. Blackweel Scientific Publishing, Oxford.

Section H Bacterial Infections

Cedric, M., Nash, A. and Stephen, J. (2001) *Mim's Pathogenesis of Infectious Diseases*. Academic Press.

Eduardo A. Groisman. 2001.*Principles of Bacterial Pathogenesis*. Academic Press.

Stephen Gillespie; Kathleen Bamford. 2003. *Medical Microbiology and Infection at a Glance* Blackwell Publishing.

Donald R. Demuth; Richard Lamont (2006) *Bacterial Cell-to-Cell Communication: Role in Virulence and Pathogenesis (Advances in Molecular & Cellular Microbiology S.)* Cambridge University Press.

Section I Eukaryotic microbes: an overview

http://phylogeny.arizona.edu/tree/eukaryotes/eukaryotes.html

Sadava, D. (1993) *Cell Biology: Organelle Structure and Function*. Jones and Barlett, London.

Smith, C.A. and Wood, E.J. (1996) *Cell Biology*, 2nd Edn. Kluwer Academic Publishers

Section J The Fungi and Related Phyla

http://phylogeny.arizona.edu/tree/eukaryotes/fungi/fungi.html

Alexopoulos, C.J., Mims, C.W. and Blackwell, M. (1996) *Introductory Mycology*, 4th Edn. John Wiley & Sons.

Charlie, M.J. and Watkinson, S.C. (2001) *The Fungi*, 2nd Edn. Academic Press, London.

Deacon, J.W. (1997) *Introduction to Modern Mycology*, 3rd Edn. Blackwell Science, Oxford.

Jennings, D.H. and Lysek, G. (1996) *Fungal Biology*, 2nd Edn. Springer Verlag, Berlin.

Kavanagh, K. (2005) *Fungi: Biology and Applications*. John Wiley.

Manners, J.G. (1993) *Principles of Plant Pathology*. Cambridge University Press, Cambridge.

Section K Chlorophyta and Protista

http://phylogeny.arizona.edu/tree/eukaryotes/green-plants.html

Bold, H.C. and Wynne, M.J. (1997) *Introduction to Algae*, 2nd Edn. Prentice Hall, New Jersey.

Heelan, J.S. and Ingersoll, F.S. (2001) *Essentials of Human Parasitology*. Delmar.

Bogitsh, B.J., Carter, C.E. and Oeltmann, T.N. (2005) *Human Parasitology*. 3rd Edn. Elsevier Academic Press.

South, G. and Whittick, A. (1987) *Introduction to Phycology*. Blackwell Scientific Publications, Oxford.

Stephen A Berger; John S Marr, J.S. (2006) *Human Parasitic Diseases Sourcebook*. Jones and Bartlett.

Julius P. Kreier; John R. Baker. (1987) *Parasitic Protozoa*. Kluwer Academic Publishers.

van den Hoek, C., Mann, D.G. and Jahns, H.M. (1996) *Algae: An Introduction to Phycology*. Cambridge University Press, Cambridge.

Section L The Viruses

Cann, A.J. (2005) *Principles of Molecular Virology*. Academic Press, London.

Collier, L., and Oxford, J. (2006) Human Virology, 3rd Edn. Oxford University Press, Oxford.

Digard, P., Nash, A. and Kandall, R. (2005) *Molecular Pathogenesis of Viral Infections, 64th Symposium of the Society for General Microbiology*. Cambridge University Press, Cambridge.

Dimmock, N., Easton, A. and Leppard, K. (2001) *Introduction to Modern Virology*. Blackwell Scientific Publishing, Oxford.

Strauss, J., and Strauss, E. (2002) *Viruses and Human Disease*. Academic Press, London.

Wagner, E. and Hewlett, M.J. (2004) *Basic Virology*, 2nd Edn. Blackwell Scientific Publishing, Oxford.

INDEX